Material and Energy Recovery from Solid Waste for a Circular Economy

Material and Energy Recovery from Solid Waste for a Circular Economy describes solid waste to material and energy recovery to bridge the gap between theoretical possibilities and on-field criticalities. It deals with various resource recovery possibilities from numerous waste streams such as municipal solid, hazardous waste, human faecal sludge, construction and demolition waste, and electronic waste. The practical issues of resource recovery and possible remedies derived through onsite practice and experience are incorporated. It includes real-life feasibility analysis and implementation of waste-to-energy systems supported by case studies.

Features:

- Provides comprehensive discussion on both energy and material recovery
- Addresses the missing linkage between the techno-commercial feasibility of existing systems and environmental impact
- Discusses techno-commercial feasibility and environmental impacts
- Offers balance between theoretical knowledge sharing and practical execution-related issues
- Includes case study, LCA, and technical feasibility chapters

This book is aimed at graduate students and researchers in environmental, civil, and chemical engineering.

Material and Energy Recovery from Solid Waste for a Circular Economy

Edited by
Atun Roy Choudhury and Sankar Ganesh Palani

CRC Press is an imprint of the
Taylor & Francis Group, an informa business
A CHAPMAN & HALL BOOK

Designed cover image: www.shutterstock.com

First edition published 2025
by CRC Press
2385 NW Executive Center Drive, Suite 320, Boca Raton FL 33431

and by CRC Press
4 Park Square, Milton Park, Abingdon, Oxon, OX14 4RN

CRC Press is an imprint of Taylor & Francis Group, LLC

ISBN: 9781032399768 (hbk)
ISBN: 9781032428215 (pbk)
ISBN: 9781003364467 (ebk)

DOI: 10.1201/9781003364467

Typeset in Times
by Newgen Publishing UK

Contents

Foreword (by V. Ramgopal Rao)

Efficiently managing the millions of tons of solid waste generated daily is crucial as we strive for a sustainable planet. While proper treatment is a necessity, it is equally imperative to focus on recovering materials and energy from solid waste to foster a circular economy.

For numerous decades, BITS Pilani has been at the forefront of advocating sustainable waste management practices across its campuses. The on-campus sewage treatment plants play a pivotal role in utilizing treated sewage to supplement irrigation and flush tank water. Notably, the organic portion of solid waste, including food waste, is harnessed for biomethane production, bolstering energy resources for cooking purposes. The resulting slurry is thoughtfully composted and repurposed as an organic fertilizer for horticultural needs. This book seamlessly aligns with our mission of creating a greener and more resource-efficient future. It not only benefits our campuses but also imparts vital knowledge to students and readers, empowering them to replicate these initiatives wherever they may be.

Precisely, this book serves as a guiding compass for readers to achieve sustainability through effective waste management. In a world grappling with escalating waste generation and environmental degradation, this book becomes an indispensable resource for industry professionals, researchers, and policymakers in search of innovative circular economy solutions. The collaborative efforts of over 50 authors from 23 institutions across seven countries and five continents have culminated in this comprehensive volume.

The book delves into a diverse array of circular approaches, including food waste composting, bacterial concrete, repurposing construction and demolition waste, recycling packaging materials, insightful techno-commercial analyses of biomedical waste, varied solid waste treatment methodologies, the recovery and reuse of landfill gas, as well as biomethane and biohydrogen production from organic waste and fecal sludge, all while adhering to regulated processes. It also covers critical topics like life cycle assessments of waste treatment facilities, establishing waste recycling parks, and appraising the market value of waste-derived products. Additionally, the book delves into policy interventions aimed at promoting the circular economy.

Its content presents an inspiring vision of a society that values waste as a resource, and the insights shared within undoubtedly pave the way for transformative changes in waste management practices. These efforts drive us toward a more sustainable and circular future.

I extend my heartfelt appreciation to the editors for their exceptional work in curating this invaluable resource. Their dedication to advancing sustainable waste management is truly commendable, and this book stands as a testament to their unwavering passion for environmental stewardship.

With genuine optimism, I anticipate that this book will spark collaboration among stakeholders, nurture innovation, and accelerate the widespread adoption of sustainable waste management practices.

Thanking you
Prof V. Ramgopal Rao, PhD,
Fellow of IEEE, TWAS, INAE, INSA, IASc, NASI
Former Director (2016–2021), IIT Delhi
J. C. Bose National Fellow
Vice-Chancellor & Senior Professor
BITS Pilani

Foreword (by Rajarshi Shah)

As the District Collector responsible for overseeing all waste management projects within the jurisdiction of our district, it is with great enthusiasm and pride that I introduce this book.

Waste management is a critical issue facing communities worldwide, and our district is no exception. As we strive to create a sustainable and eco-friendly environment for our residents, this book serves as a beacon of knowledge, offering innovative solutions and insights into harnessing the potential of solid waste for a circular economy.

The journey toward a circular economy demands collaborative efforts and a forward-thinking approach. I commend the authors and editors for their dedication and expertise in shedding light on the transformative power of material and energy recovery from solid waste. Their work not only aligns with our district's vision for a cleaner and greener future but also serves as a guiding force for other regions facing similar challenges.

Readers will find a comprehensive exploration of cutting-edge technologies, sustainable practices, and successful case studies that demonstrate the tangible benefits of waste-to-resource initiatives. From waste-to-energy projects to the valorization of recyclables, the editors present a roadmap for us to move beyond traditional waste disposal and embrace a circular economy that values resource recovery.

As we seek to bridge the gap between environmental responsibility and economic growth, this book offers an indispensable resource for policymakers, researchers, waste management practitioners, and concerned citizens. The knowledge shared within these pages empowers us to make informed decisions, foster public engagement, and drive positive change toward a more sustainable future.

I extend my heartfelt appreciation to the editors for their dedicated efforts in crafting this valuable contribution to the field of waste management. I hope this book will inspire and catalyze meaningful action as we work together toward a waste-free and circular district, setting an example for the rest of the nation.

May the wisdom and insights shared within this book ignite a wave of transformative initiatives that will leave a lasting positive impact on our environment, society, and economy.

With warm regards,
Rajarshi Shah
District Collector & Magistrate, Adilabad

Preface

Welcome to *Material and Energy Recovery from Solid Waste for a Circular Economy*. This book delves into the critical aspects of waste management, focusing specifically on the recovery of valuable materials and energy from solid waste. In an era where sustainability and resource conservation have become imperative, the concept of a circular economy has emerged as a powerful framework for rethinking our approach to waste.

Solid waste, generated from various sources such as households, industries, and commercial establishments, poses significant environmental challenges. However, within this waste lies an untapped potential for recovering valuable resources and minimizing our reliance on virgin materials. The concept of a circular economy aims to close the loop, transforming waste into a valuable resource by adopting innovative strategies and technologies.

In this book, we embark on a comprehensive exploration of the processes, technologies, and best practices that facilitate material and energy recovery from solid waste. We bring together a diverse range of perspectives, insights, and case studies from experts in the field, offering a holistic understanding of the subject matter.

Throughout the chapters, we examine key topics such as waste characterization, waste-to-energy conversion technologies, recycling and upcycling methodologies, and the integration of waste management systems into the broader framework of a circular economy. We delve into the technical aspects, environmental considerations, economic viability, and policy implications that shape the implementation of sustainable waste management practices.

Our aim is to equip readers with the knowledge and tools necessary to navigate the complexities of waste management and foster a circular economy mindset. Whether you are an academic researcher, waste management professional, policymaker, or simply an individual seeking to contribute to a more sustainable future, this book offers valuable insights and practical guidance.

We would like to express our gratitude to the contributing authors who have shared their expertise, as well as the readers whose curiosity and dedication drive the pursuit of sustainable waste management solutions. It is our hope that this book serves as a source of inspiration, guiding us toward a future where solid waste becomes a valuable resource and our actions align with the principles of a circular economy.

Together, let us embark on this journey of material and energy recovery from solid waste, working toward a sustainable, circular future.

Editor Biographies

Atun Roy Choudhury is an accomplished professional currently serving as Assistant General Manager at Cube Bio Energy Private Limited. With a distinguished career spanning esteemed organizations, he is widely recognized as a leading researcher specializing in waste management and water sanitation. With numerous publications and extensive experience guiding dissertations, Atun enjoys a stellar reputation within the research community. He is a sought-after author, speaker, editor, and reviewer for international scientific platforms. His exceptional contributions have earned him prestigious titles and awards, including a gold medal from the University Grant Commission. Atun's expertise extends to environmental auditing and quality management. His groundbreaking work on microbial waste digestion has garnered global acclaim, being listed in reputable records and endorsed by the United Nations.

Sankar Ganesh Palani is Professor of Environmental Biotechnology at BITS Pilani, Hyderabad Campus. Prof. Ganesh received his PhD in Environmental Science and Engineering from Pondicherry (Central) University, India and had his post-doctoral training at the National Institute of Applied Sciences, Toulouse, France. Prof. Ganesh teaches ecology, environmental science, environmental biotechnology, waste management, sanitation technology, and gene toxicology. His research focuses on waste and wastewater treatment, biomethanation, and biohydrogen production. He has been involved in several projects and has published several books, book chapters and research papers in reputed international and national journals and conference proceedings. Prof. Ganesh is a "Faculty-preneur" being the Co-Director of a startup company that has spun off from his research on high-rate thermophilic anaerobic digestion of organic waste.

Contributors

Azam Akhbari
Department of Thematic Studies-Environmental Change and Biogas Research
Center
Linköping University
Sweden

Vishal Akula
Department of Civil Engineering
Gandhi Institute of Technology & Management
Hyderabad, Telangana, India

Shriya Bhatia
Ashank Desai Centre for Policy Studies (ADCPS)
Indian Institute of Technology, Bombay (IITB)
Powai, Mumbai, India
Balwant Sheth School of Architecture (BSSA)
Mumbai, Maharashtra, India

Kasturi Bidkar
Department of Biotechnology
National Institute of Technology Warangal
Hanamkonda, Telangana, India

Soheli Biswas
Department of Biotechnology
Brainware University
Barasat, Kolkata, West Bengal, India

Gabriela Campos
Polymer Science and Engineering Group
Institute of Materials Science and Technology (INTEMA),University of Mar Del Plata
Mar Del Plata, Argentina

Surajit Chakraborty
Department of Environment Management
Indian Institute of Social Welfare and Business Management, Management House College Square (West)
Kolkata, West Bengal, India

Adam Connell
Townsville, Queensland
Queensland, Australia

Alejandra Costantino
Polymer Science and Engineering Group
Institute of Materials Science and Technology (INTEMA), University of Mar Del Plata
Mar Del Plata, Argentina

Bimal Das
Department of Chemical Engineering
National Institute of Technology Durgapur
Durgapur, India

Deepshikha Datta
Department of Chemistry
Brainware University
Kolkata, West Bengal, India

Ajay Deshpande
Ashank Desai Centre for Policy Studies (ADCPS)
Indian Institute of Technology, Bombay (IITB)
Mumbai, Maharashtra, India

Brajesh Dubey
School of Environmental Science and Engineering
Indian Institute of Technology Kharagpur
Kharagpur, West Bengal, India

Sanjay Garg
Medicar Environmental Management Pvt Ltd
Ghaziabad, Uttar Pradesh, India

Anasua GuhaRay
Department of Civil Engineering
BITS Pilani Hyderabad Campus
Secunderabad, Telangana, India

Alana L. Hayshida-Boyles
College of Science & Engineering
James Cook University
Townsville, Queensland, Australia

Shaliza Ibrahim
Institute of Ocean and Earth Sciences (IOES)
University of Malaya
Kuala Lumpur, Malaysia

Meenu Krishnan
Department of Civil Engineering
BITS Pilani Hyderabad Campus
Secunderabad, Telangana, India

Jitesh Lalwani
Department of Civil Engineering
Vardhaman College of Engineering, Kacharam
Shamshabad, Hyderabad, India

Sarah Lebu
Department of Environmental Sciences and Engineering,
University of North Carolina
Chapel Hill, North Carolina, USA

Esha Mandal
Department of Biotechnology
National Institute of Technology Durgapur
Durgapur, India

Sanjeeb Kumar Mandal
Department of Biotechnology
Chaitanya Bharathi Institute of Technology
Hyderabad, India

Musa Manga
The Water Institute at UNC, Department of Environmental Sciences and Engineering
University of North Carolina
Chapel Hill, North Carolina, USA

Breda McCarthy
College of Business, Law & Governance
James Cook University
Townsville, Queensland, Australia

Oualid Meddour
Institute of Urban Technical Management
University Constantine
Constantine, Algeria

Slimane Merouani
Department of Chemical Engineering
University Constantine
Constantine, Algeria

Federico Morales
Polymer Science and Engineering Group
Institute of Materials Science and Technology (INTEMA), University of Mar Del Plata
Mar Del Plata, Argentina

Chimdi C. Muoghalu
Department of Environmental Sciences and Engineering
University of North Carolina
Chapel Hill, North Carolina, USA

N. Chandana
Center for Emerging Technologies for Sustainable Development (CETSD)
Indian Institute of Technology Jodhpur
Rajasthan, India

Anne Nakagiri
Department of Civil and Environmental Engineering
Kyambogo University
Kyambogo, Uganda

Charles B. Niwagaba
Department of Civil and Environmental Engineering
School of Engineering; College of Engineering, Design, Art and Technology, Makerere University
Kampala, Uganda

Sankar Ganesh Palani
Department of Biological Sciences
Birla Institute of Technology and Science, Pilani, Hyderabad Campus,
Telangana, India

Valeria Pettarin
Polymer Science and Engineering Group
Institute of Materials Science and Technology (INTEMA), University of Mar Del
Mar Del Plata, Argentina

Arumugam Poornima
Department of Biochemistry, Biotechnology and Bioinformatics
Avinashilingam Institute for Home Science and Higher Education for Women
Coimbatore, Tamil Nadu, India

Carlos Ramirez
Institute of Polymer Technology and Nanotechnology (ITPN)
University of Buenos Aires
Buenos Aires, Argentina

Bakul Rao
Centre for Technology Alternatives for Rural Areas (CTARA), Ashank Desai Centre for Policy Studies (ADCPS),
Indian Institute of Technology, Bombay (IITB)
Powai, Mumbai, Maharashtra, India

Omar Redjal
Institute of Urban Technical Management
University Constantine
Algeria

Caren Rosales
Polymer Science and Engineering Group
Institute of Materials Science and Technology (INTEMA), University of Mar Del Plata
Mar Del Plata, Argentina

Atun Roy Choudhury
Cube Bio Energy Pvt Ltd
Hyderabad, Telangana, India

S. Hemapriya
Department of Biological Sciences
Birla Institute of Technology and Science, Pilani, Hyderabad Campus
Telangana, India

Subhasmita Sahoo
Department of Biotechnology
National Institute of Technology Warangal
Hanamkonda, Telangana, India

Sumithra Salla
Department of Biotechnology
Chaitanya Bharathi Institute of Technology
Hyderabad, India

Aaron Salzberg
Department of Environmental Sciences and Engineering,
University of North Carolina
Chapel Hill, North Carolina, USA

Swaib Semiyaga
Department of Civil and Environmental Engineering
School of Engineering; College of Engineering, Design, Art and Technology, Makerere University
Kampala, Uganda

Neha Singh
Chadwick's FSM Laboratory
Banka BioLoo Limited
Secunderabad, India

Vijai Singhal
Rajasthan State Pollution Control Board and Director
M/s Greenhub Systems Pvt Ltd
Rajasthan, India

Balraj Sudha
Department of Biochemistry, Biotechnology and Bioinformatics
Avinashilingam Institute for Home Science and Higher Education for Women
Coimbatore, Tamil Nadu, India

Kanagaraj Suganya
Department of Biochemistry, Biotechnology and Bioinformatics
Avinashilingam Institute for Home Science and Higher Education for Women
Coimbatore, Tamil Nadu, India

Sundaravadivelu Sumathi
Department of Biochemistry, Biotechnology and Bioinformatics
Avinashilingam Institute for Home Science and Higher Education for Women
Coimbatore, Tamil Nadu, India

Subramani Yashini Vidhya
Department of Biochemistry, Biotechnology and Bioinformatics
Avinashilingam Institute for Home Science and Higher Education for Women
Coimbatore, Tamil Nadu, India

1 The 'Greening' of an Australian University

Onsite Composting of Residential Food Waste

Alana L. Hayshida-Boyles, Breda McCarthy, and Adam Connell

1.1 INTRODUCTION

In recent years, food waste has received increased attention from scholars, policy makers, and business leaders all over the world. An oft-quoted statistic is that one-third of all food produced in the world is lost or wasted, according to the Food and Agriculture Organization of the United Nations (Gustavsson et al., 2011). The second of the 17 United Nations Sustainable Development Goals (SDGs), which were officially adopted in 2015, is to 'end hunger, achieve food security and improved nutrition, and promote sustainable agriculture' (United Nations, 2024a). Reducing food waste in institutional settings would ultimately help achieve the SDG target on food waste, which is to reduce per capita food waste globally by 50% at the retail and consumer levels by the year 2030 (target 12.3) (United Nations, 2024b). Furthermore, reducing food waste should have positive economic, social, and environmental consequences and it is estimated that if food waste were its own nation, it would be the third highest producer of greenhouse gases behind the United States and China. (Hanson et al., 2015). In addition, the collection, transportation, and disposal of food waste is expensive for governments, institutions, and others, and landfills, which are quickly reaching maximum capacity, can pose risks to public health and contribute to groundwater contamination (El Fadel et al., 1997).

Academic institutions recognise that they have a role to play in promoting sustainability (Baldwin & Dripps, 2012; Smyth et al., 2010; de Vega et al., 2008; Zen et al., 2016). Universities, as centres of learning and research, are well positioned to develop sustainable behaviour among students and faculty, which may have an influence on society as a whole (Zhang et al., 2011). The special role of universities as role models and catalysts of change has been emphasised by the United Nations in its 'greening universities' program (UNEP, 2014). Several case studies of sustainability programs in the educational sector exist, such as a paper recycling pilot program at Rhodes University, South Africa (Amutenya et al., 2009). The advantages of eco-friendly

DOI: 10.1201/9781003364467-1

waste management include a better image, lower carbon emissions from waste transportation, and greater health and safety (Ball & Taleb, 2011).

This study focuses on food waste in a university setting since food service providers are often large generators of food waste. A study by Verghese et al. (2013) on food waste in Australia concluded that the food service sector is the largest single contributor to food waste in Australia (which includes businesses such as hotels, pubs, restaurants, cafes, and commercial caterers). The food service sector recycles only 2% of the food waste they generate and sends approximately 661,000 tonnes to landfill each year (Verghese et al., 2013). Furthermore, food waste minimisation strategies have benefits such as cost-effective operations and smaller consumption of resources used to produce food (Pirani & Arafat, 2016). Researchers have addressed the topic of environmental and waste management in universities (Sarjahani et al., 2009) and there have been several scholarly articles (Balwin & Dripps, 2012; Epinosa et al., 2008; Mason et al., 2003; Mbuligwe, 2002; Smyth et al., 2010; de Vega et al., 2008) that report the total waste stream generated on campus. To the best of our knowledge, no study has specifically focused on food waste and conducted a food waste audit within Australian universities.

This paper describes a case study of an innovative food recovery program in a university and offers recommendations on how to minimise food waste in an institutional setting. The methodology, results of the audit, and presented recommendations should prove applicable and beneficial to other universities.

1.2 FOOD WASTE IN THE HOSPITALITY SECTOR

Although there are several definitions of food waste in the literature, food waste can be defined quite generally as 'any food that is not consumed by humans and can be generated at any level within the food chain (farms, processing plants, manufacturers, commercial establishments, and households)' (Okazaki et al., 2008, p. 2483). The Waste and Resource Action Program (WRAP) in the UK describes four types of food waste that occur in food service, which are as follows:

(1) spoilage: food that is not safe to eat, is damaged or past the expiry date;
(2) preparation: waste produced during preparation, for example vegetable peelings;
(3) plate waste: uneaten food that has come back from the customer;
(4) over-production: surplus food that was uneaten after service that could not be used for future service.

(WRAP, 2013)

Kitchen waste from food preparation and processing is typically referred to as 'pre-consumer food waste' (Pirani & Arafat, 2016). A growing number of scholars are studying food waste, spanning people's cognitions and behaviours around food, household demographics, and socio-economic drivers (McCarthy et al., 2016; Priefer et al., 2016; Stancu et al., 2016; Thyberg & Tonjes, 2016). In contrast to the large literature on household food waste, scholarly publications on food waste management in the food service sector are sparse, but growing (Okazaki et al., 2008; Papargyropoulou et al., 2016; Pirani & Arafat, 2014; Pirani & Arafat, 2016; Silvennoinen et al., 2015;

TABLE 1.1
Food Waste in Food Service

Factor
Menu planning, meal composition and presentation
Procurement of food, type of food (fresh, prepared and frozen), delivery size and frequency
Food storage and stock management
Oversized dishes
Staff behaviour (i.e., awareness of waste generation; cooking mistakes)
Offer of buffets at fixed prices encouraging people to take more than they can eat
Consumer behaviour (i.e., not offering 'sides', large portion sizes, no doggy bags)
Use of individual portion packs (e.g. for jams, cereals, juice and milk) that do not meet the customer's needs
Difficulties in assessing the demand (number of customers)
Food safety and hygiene rules (e.g. two-hour guarantee on unrefrigerated products in the EU)
Untouched leftover food

Sonnino & McWilliam, 2011; Sundt, 2012). There have been several reports on food waste in relation to food service (Parfitt et al., 2013; Verghese et al., 2013; WRAP, 2013), showing that scholars, policy makers, and practitioners are seeking to identify the reasons for food waste and design effective policies for food waste prevention. Table 1.1 summarises the many and varied reasons why food waste occurs in the food service sector. Food waste may be explained by numerous factors, such as inaccurate forecasting of consumer demand; the style of service (buffet style); the type of food served, such as rice, or other carbohydrates, that are generally served as meal fillers or as a side-dish; large portions and plate size (Pirani & Arafat, 2016). Catering and procurement practices, such as bulk preparation, the need for standardisation, along with budget concerns, result in high levels of plate waste (Sonnino & McWilliams, 2011). Other explanatory factors include the skill level of employees, culture, and sensitivity to environmental issues (Dhir et al., 2020).

1.2.1 The Circular Economy and Food Waste Management

The vision of a circular economy is gaining traction in academic and practitioner-oriented literature. However, there is no commonly accepted definition of the circular economy (Yuan et al., 2006), and scholars have in fact identified 114 definitions of circular economy (Kirchherr et al., 2017). The concept of a circular economy, in general terms, promotes resource minimisation and is most frequently depicted as a combination of reduce, reuse, and recycle activities (Kirchherr et al., 2017). In a linear economic model, the physical environment is treated as a receptacle for waste products from the economy, and design-for-disassembly, recycling, and reuse are not fundamental parts of the system. This is inefficient since resources (i.e., materials, energy, water, etc.) flow out of the system. A circular economy, on the other hand, is an industrial system that is restorative or regenerative by intention and design. It is founded on the principle of the earth as a closed economic system and the circular

Circular Economy	Circularity Strategies	Examples
↑	Avoid or reduce waste	Education campaigns. Staff training. More efficient and effective food procurement and cooking methods. Menu planning. Cold chain management. Packaging initiatives to improve shelf life. Food rescue/charity donations.
	Repurpose	Composting, anaerobic digestion, soil enrichers, worm farms, community gardens. Animal feed (for farmed fish, chickens, livestock etc.) and biotechnology solutions for animal feed.
	Rethink / reprocess / redesign	Use parts, or all, of discarded product in a new, improved product (juices, fermented products, fruit platters) with a different function.
	Process / value-add	Use lower quality (low grade) materials for food processing – jams, pickles, sauces, etc.
↓	Recover	Incineration of food waste (waste-to-energy).
Linear Economy	Disposal	Produce left to rot in bins, going to sewer or landfill.

FIGURE 1.1 Moving food service towards a circular economy.

nature refers to materials flowing within a closed loop, to be reused again and again (Jackson et al., 2014). Figure 1.1 offers examples of how a linear food service system can be turned into a circular system.

Food waste management strategies should follow waste hierarchy principles with waste prevention (Papargyropoulou et al., 2016), such as education campaigns, or new forms of packaging to prevent food from spoiling (Verghese et al., 2013) being the preferred options. However, evaluating the relative merits of waste management alternatives is a complex task (Garcia-Garcia et al., 2017). Life cycle analysis shows that some actions (such as cold chain management) should always be prioritised since they avoid a high environmental impact at a low cost (Cristóbal et al., 2018). It is possible to convert food waste into energy through processes such as incineration and biomethanation, although it is not the most sustainable and cost-effective option for dealing with food waste. Problems relate to the capacity of treatment infrastructure, difficulty in separating food waste from other waste streams, and production of excessive emissions and hazardous by-products (Angelidaki et al., 2011; Kibler et al., 2018; Sridhar et al., 2021). Another solution to the food waste problem is to divert it to animal feed. While there is a long tradition of giving surplus food from kitchens to animals, health and safety regulations (i.e., concern about mad cow disease) prevent this from happening in many countries (Alexander et al., 2013).

Food waste can be turned into compost, which serves as an input into the food production process, which helps turn a linear system into a circular system (Kirchheer et al., 2017). Composting has proven to be an attractive measure to deal with waste since it is cost effective as well as being environmentally friendly (Wei et al., 2017). It also produces a useful, final product; for instance, horticulturists can use high quality compost to enhance plant growth, increase vitamin C and soluble sugar content, and reduce nitrate content (Guo et al., 2018). Compost is attractive to governments since it helps lower the cost of managing landfills, pollution impacts, and public health threats. In addition, it supports land management practices, since long-term, inappropriate chemical fertiliser application can lead to degraded soil health (particularly in China) and compost can improve soil fertility (Guo et al., 2018).

Similar to composting, the process of biological fermentation, either aerobically or anaerobically through the use of microorganisms, can be utilised to generate bio-surfactants, bio-plastics, and bio-fertilisers from diverted food waste (Sharma et al., 2020). Sharma et al. (2020) posits that this use of microorganisms could be one of the most eco-friendly and sustainable solutions to managing food waste. With an end product such as a bio-fertiliser, the opportunity for circular economy practices in food production is evident. Moreover, bio-fertilisers not only increase soil nutrients, and consequently crop production, but also neutralise the negative effects of chemical fertilisers. (Areeshi, 2022). According to Areeshi (2022), food waste enhances the production of bio-fertilisers because it is already in a biodegradable form, accelerating the activity of the microbial process and thus requiring less energy inputs. However, food waste reclamation is not standard practice in the hospitality sector. In Australia, this sector recycles only 2% of the food waste generated (Verghese et al., 2013). It has been argued that legislation makes dumping 'too easy and too cheap' (Gooch et al., 2010, p.7). Okazaki et al. (2008) offer insights into why establishments do not recycle food waste. The most common response to their survey was that the institution created little or no food waste. The second most popular reason was that separating food waste from the general waste stream was too time consuming or costly.

In summary, options for dealing with waste in the food service sector are likely to be a mix of the linear and the circular economy.

1.3 MATERIALS AND METHODS

This study has several components:

(1) A case study examining the drivers of, and barriers to, food waste minimisation in an institutional setting.
(2) A food waste recovery (microbial fermentation) program.
(3) A food waste audit.
(4) Cost-benefit analysis of the food waste recovery program along with a quantification of carbon emissions avoided.

This study adopts a case study approach since this is common in the literature that deals with sustainability programs undertaken by universities (Zen et al., 2016). Semi-structured interviews were conducted with a catering manager and an executive chef to shed light on the food service operation on campus. Both individuals were trained chefs and had worked in the 'for-profit' and 'not-for-profit' hospitality sectors. Interviews were conducted with the General Manager of a private company, VRM, and an employee who handled the food waste; VRM is a company that has pioneered a system for the biological treatment of food waste that helps return nutrients to the earth, and hence enriches the soil. All interviews were carried out in person and each interview lasted approximately one hour. The questions asked in the interviews were developed after a thorough review of publications on food waste recovery and food waste in the hospitality sector. The questions assessed food waste minimisation; reasons for participating in the food recovery program; reasons for variations in the food waste data; barriers and keys to success; measures of performance; future directions and general interest in matters related to sustainability. Ethics approval was secured from the Human Ethics Committee in the authors' university (H6601).

1.3.1 The University, Its Food Waste Recovery Program, and Business Partners

The selected university, James Cook University, has an enrolment of 12,831 students and employs 3,483 staff at its main campus in Townsville, Australia. The university has a long-standing commitment to sustainability and has won awards for its sustainability efforts under the 'Green Gown Awards Australasia' program (https://ggaa.acts.asn.au/), which aims to inspire, promote, and support change towards best practice sustainability within the tertiary education sector (ACTS, 2021). The university is also a signatory to the 'Talloires Declaration', an international agreement of university leaders for sustainability (University Leaders for a Sustainable Future, 2015). The university runs a refurbished bike program where abandoned and donated bikes are repaired by paid student mechanics and sold back to students at low cost. The university has a 'Sustainability Internship' program and engages between 30–50 students each year in on-campus sustainability projects. Staff and students are responsible for recycling and all staff rooms and public areas have recycling bins with separate bins for food waste. The university's 'Waste Reduction Management Plan' (internal report/unpublished) revealed that the university had little to no data on the food waste stream (internal report, unpublished). Beginning in 2017, the Estate Directorate worked with a student intern to begin a systematic characterisation of the residential food waste stream to address this lack of data. There are three university-owned residential halls, hosting approximately 700 students, and the vast majority, if not all, of the residents at the halls, eat all of their meals in the dining hall or in the food court, rather than in their dorm rooms.

The food recovery program was conducted with the support of two parties, a private company (VRM) and the university's contract caterer (Compass Group). Compass Group is a global company and one of Australia's largest food service providers. The organisation purchases around 50,000 tonnes of fresh food every year, making it a very large food buyer (Compass Group, 2024a). In Australia, its operations are extensive, including schools, universities, hospitals, and aged care facilities. In relation to the university sector, it provides a contract food service that covers cafes, university food halls, live-in college school meals, and special event catering. Its food waste reduction strategy includes tracking food waste on-site, donating any surplus food to charity, and implementing education and awareness-raising campaigns in partnerships with suppliers, employees, and customers (Compass Group, 2024b).

The university decided to utilise a commercial solution from a private company, VRM, to process its food waste. VRM is an Australian company that has a patented system (Bio-Regen®) that captures organic material (such as produce, dairy, meat, seafood, eggs, left overs, spoilages, pulps, and liquid products) and converts it into a liquid formulation containing key biological elements to enrich soil health (VRM BioLogik, 2020). Markets for the final product are agricultural and horticultural markets as well as local councils that manage gardens. The General Manager describes the system as a 'food-to-food' model and posits that the company is 'exceptionally environmentally conscious' (VRM, personal communication, 19 November 2018). The fermentation solution is also put forward as more sustainable than the prevailing option of composting. Discussions with VRM have indicated the microbial process sequesters carbon and nitrogen and produces water as a by-product. This is due to the microbial population containing photosynthetic bacteria (carbon sequestration) and nitrogen-fixing bacteria amongst many others. After learning about the technology, the Environment Manager from the university secured the support of VRM to conduct a pilot food waste recovery program on campus. Participation in the university's food waste recovery program was voluntary and not contractual in nature. Due to the quantity and composition characteristic of the food waste generated by the university, VRM determined that the food waste would be suitable for a reclamation process that paired their Bio-Regen® process with that of Groundswell®.

1.3.2 Primary (Bio-Regen®) Process

The Bio-Regen® processing unit acts like a commercial grinder, grinding up the food while adding water and self-dosing microbes. After food is macerated, it is pumped into one of six 1,000 L IBC tanks. Once a tank is filled, the liquid slurry then undergoes a 28-day fermentation process driven by the microbial inoculants. After 28 days, the product reaches the end point and is ready for collection by the company supplying the technology (VRM). The tanks are emptied by the contractor and the liquid is taken away for final processing to create a quality-controlled end product.

1.3.3 Secondary (Groundswell®) Process

A secondary processing option was utilised in any case where the Bio-Regen® process was unavailable, either due to blockages, fermentation time, etc. In these instances, the Groundswell® process was utilised to compost food waste. The Groundswell® process places food waste in a bay made up of pine wood sleepers and gravel, lined with a heavy-duty tarpaulin to retain moisture. After each addition of food waste to the bays, the mounds are sprayed with a microbial solution and wetted down to ensure the microbes spread throughout the mound. Other organic wastes (i.e., garden wastes, mulch, grass clippings) can be added to the mounds until each mound reaches a height of approximately 2 m. The mound is then re-sprayed with microbes and covered for a 6-month period to undertake the microbial process, with the end product being a nutrient- and humus-rich soil.

1.4 DATA ANALYSIS

With regard to the food waste audit, all data was analysed by the same student intern over the course of the eight-month study, ensuring consistency and continuity in raw data interpretation, formulation, and expression. Excel was used to sort the data, run all formulas, and determine and identify any trends present in the rate or quantity of the food waste recovered. Data was sorted into categories based on college dining hall, the week the waste was collected, and if the waste was processed by the Bio-Regen® unit or diverted to Groundswell®. In addition, residential college capacity and student demographics (i.e., presence of domestic or international students in the hall, general year of study) were taken into consideration when attempting to identify trends. A cost-benefit analysis was also undertaken in order to determine the benefits of this new method of food waste diversion, and in addition, to determine the cost efficiency of the primary Bio-Regen® processing system.

All interviews were recorded and the data was transcribed verbatim. The qualitative data was analysed for themes and simple meanings. Due to the limited number of interviews, the data was analysed manually without the assistance of computer software.

1.5 RESULTS

1.5.1 The Food Recovery Program: Collection System, Data Recording, and Processing

A fixed collection system was applied in all three residential colleges, which had 646 students in total. Food waste was collected from the kitchens in buckets each weekday and the weight of waste from each kitchen was measured using commercial grade scales. Food service employees were guided to dispose of all food waste, including preparation waste, spoiled food, and uneaten leftovers in the 15 L buckets provided by the Estate Directorate. The bins had numbers to identify their collection location and were lidded, with instructions on the top of each bucket indicating what foods should, and should not, be included. A single university maintenance employee was instructed to collect the food waste buckets from each of the three kitchens on weekday mornings, over the eight-month period. The food waste buckets were transferred to a vehicle and transported to the central processing site. Clean buckets were provided to the kitchens to replace those that were collected. After collection of the liquid end product by VRM, the product was tested to ensure a quality-controlled product prior to sale.

Data was recorded in a systematic way. The number of each bucket was recorded against the kitchen it was collected from, and each bucket was weighed at the central processing site. Data was recorded by the university maintenance employee for the student intern, who would later compile the raw data into a spreadsheet for later analysis. Inoculant amount used and fermentation tank volume was also recorded. Any food waste that could not be processed through the Bio-Regen® system was re-weighed before it was sent to the Groundswell® system.

Collected food was processed through one of two methods: processing through the Bio-Regen® unit, or through the Groundswell® process in the instance where the Bio-Regen was blocked, broken, the storage tanks were full and awaiting pickup, or the food scraps were not suitable for Bio-Regen® processing due to their size or physical attributes (e.g., bones, hard seeds, fibrous materials, etc.). As this was new technology, the university staff worked closely with VRM throughout the program to ensure the unit functioned correctly and any issues encountered were quickly resolved.

1.5.2 Food Waste Audit

Understanding the amount of food wasted is a critical first step towards developing a carbon footprint and cost-benefit analysis. The amount of food waste, by weight and number of buckets, differed vastly by college over the length of the study, with George Roberts consistently having more buckets, and Western Courts consistently having the fewest buckets and the lightest buckets. Over the course of the eight-month study, this translated to George Roberts producing nearly half of the total waste collected (Table 1.2). Since this was an audit on a new process, data was recorded for how much of the total food waste collected was able to be processed by the primary Bio-Regen® process, and how much had to be diverted to the secondary Groundswell® process. The Bio-Regen® was found to process 7,644.80 kg of the total 12,229.14 kg processed, or 62.5% (Table 1.3).

TABLE 1.2
Total Food Weight (kg)

Residential Hall	Total Food Weight (kg)	Number of residents
University Hall	4,476.33	289
Western Courts	2,378.37	107
George Roberts	5,678.33	250
Total	**12,533.03**	646

TABLE 1.3
Total Food Weight (kg) Processed by the Bio-Regen® or Diverted to Groundswell®

Process	Total Food Weight (kg)
Bio-Regen®	7,644.80
Groundswell®	4,584.34
Total	**12,229.14**

When compared to the university's academic calendar, a clear trend in the food waste produced became evident. As expected, there was a substantial drop in food waste during the winter holidays. During this period, one dining hall is closed, and students eat their meals out of one of the two larger dining halls still in operation. For the 15 weeks leading up to the winter holiday, the colleges were producing an average of 363 kg of food waste each week. Over the four weeks of winter holiday, only 197 kg of food waste were produced on average each week across all the colleges, indicating a 54% decrease in the amount of food produced (Table 1.A1, Appendix). Additionally, when looked at by college, George Roberts (250 students) produced the most amount of food waste on all but five of the 36 weeks in the study, with Western Courts (107 students) consistently producing the least throughout the study. Furthermore, there appeared to be an increase in total food waste on particular weeks that was well above average, such as week 3 (453.1 kg), week 6 (454.34) and week 12 (465.18), as well as weeks 23 (485.81 kg), 25 (453.18 kg) and 26 (457.93 kg).

1.5.3 Cost-Benefit Analysis

The analysis in Table 1.4 outlines the cost/benefit analysis of the food recovery program. Capital costs for the processing unit and installation were modest in

TABLE 1.4
The Cost-benefit Analysis of the Composting Treatment

Expenditure	Unit	AU $	Totals
	Capital costs		
Processing unit (Bio-Regen®)		14,000	
Secondary processing site (Groundswell®, cost per bay)		1,841	
Other installation expenses		2,500	
Total capital costs			18,341
	Operational costs		
Maintenance		Nil (provided as service by company)	
Employee labour		18,000/annum	
Utility bills (water, electricity)		185/annum	
Microbial inoculants		1,276/annum	
Total operational costs			19,461
Total expenditure			37,802
	Income		
Reduction in waste disposal costs from kitchens	@$30.12 per bin lift	12,469/annum	
Fertiliser buy back (first year)	@ $0.20 per litre	8,010	
Groundswell® humisoil value	5 tonnes	1,000	
Total savings			12,528
Cash inflow			9,010
Net cash loss			–10,451

comparison to other food waste processing systems (like those dependent upon the re-capture of gases to produce energy) and the operating costs were also reasonable. The major operating expense was labour, which was required for the collection and processing of the food waste, and this contributed to a net cash loss of AU$10,451. In terms of income, the liquid bio-fertiliser held the most value and buy-back from the provider over the first eight months totalled AU$5,340. Extrapolated to 12 months, this equated to AU$8,010 per year. It is expected that additional liquid bio-fertiliser production will occur in the second year, due to planned collections from two additional large residential kitchens on campus. Minimal additional labour time will be required to achieve this additional processing. Improvements in processing efficiencies and a reduction in breakdowns is expected in the second year as well, helping to decrease overall labour costs. This is likely to result in more food being processed through the primary (Bio-Regen®) process and less being diverted to the lower value secondary (Groundswell®) process, thus generating a higher return on investment.

1.5.4 Carbon Footprint Analysis

In order to illustrate the environmental benefits of the food recovery program, the weight of diverted food waste was converted to tonnes of CO_2 emissions. The total food waste diverted from landfill over the course of the study was 12,533 kg, which equates to preventing 23.81 tonnes of CO_2 from being emitted into the atmosphere (Table 1.5).

1.5.5 Case Study

The food recovery program required a minor change in practice for the food service provider. The respondents indicated that they had some concerns initially with working with an outside party, such as the food bins taking up too much space in the kitchens or not being managed properly. However, they were very pleased with how well organised the food recovery program was:

> I can't think of any major negatives. It's been the best one I've been involved in. This has probably been the easiest one . . . It's very difficult to get into a routine . . . all the buckets are different shapes and sizes, so things get too big and too difficult to weigh, so you have people coming along with your own

TABLE 1.5
Carbon Footprint Conversion for the Total Food Waste Diverted from Landfill over the Course of the Eight-month Study

Variable	Food waste diverted from landfill	Emission factor (t CO2-e/t waste)	Carbon emissions (t CO2-e emissions)
Amount	12,533 kg	1.9	23.81 tonnes

Note: Conversion factor taken from Department of Environment and Energy (2017).

> buckets, picking them up, that's been key, and weighing it . . . it's been great. Every week, you monitor it.

Educating kitchen staff on how to dispose of the food waste and avoid contamination was not difficult and instructions became part of the staff induction program. There was some variation in terms of level of food waste between the three university residential halls and the catering staff suggested that this was due to the number of residents at each hall. The general manager of VRM saw the pilot as successful in terms of a 'practical operation process' as they learned to operate the Bio-Regen® unit without 'having it break down, without having smells, leaks and so on'. The main driver of success from VRM's perspective was the enthusiasm and commitment of the university's Environment Manager:

> Let's be honest about it. If you're driven by this as a people, then it tends to be contagious. So, I think the fact that Adam and his team have been so proactive in that respect has allowed the program to grow and to have success.

In addition, VRM saw an opportunity for the university to 'be a leader in that learning environment' in terms of sharing their experience with food waste with local schools and other universities. A need for new thinking about waste was emphasised:

> . . . and shift that mindset from waste to resource because I think that's the big thing. Getting people to understand that waste is not a burden. Waste is actually a resource that can be used.

The interviews with the catering managers indicated that several strategies were used to minimise the amount of food wasted in the kitchens. The details of the food waste management process are shown in Figure 1.2. The first stage, ordering and preparing the right amount of food, was seen as the most critical control point in the food service process that helped minimise food waste. The amount of food ordered and prepared was based on the number of people expected to eat the meal and previous waste levels (i.e., meal counts). The waste generated depended heavily on the type of produce ordered. For instance, if watermelon was prepared, then the chef knew that this would lead to a very high level of wastage in terms of weight (e.g., rind, seeds). Repurposing of food was commonplace, for instance if potatoes were served for dinner, then leftovers would be used in a salad the following day, or tomatoes would became tomato soup. The menu choices were based on the experience of the chef, but they had to adhere to corporate guidelines in relation to cost. The catering manager emphasised that staff were highly cost-conscious, remarking that, 'any time you waste food, you are wasting money'. The chefs knew the price of each item on the menu, and food waste awareness was instilled in the kitchen staff through corporate culture and peer pressure. In addition, Compass Group placed a strong focus on measuring waste. In the words of the catering manager:

> As an entire business, we look at trends and we look at our rates, and it goes up onto a big chart, we can break it down by sector, we're the educational sector . . . so we can see if particular sectors are wasting more than other sectors are. So maybe one sector is doing really, really well, and we can say – why are you

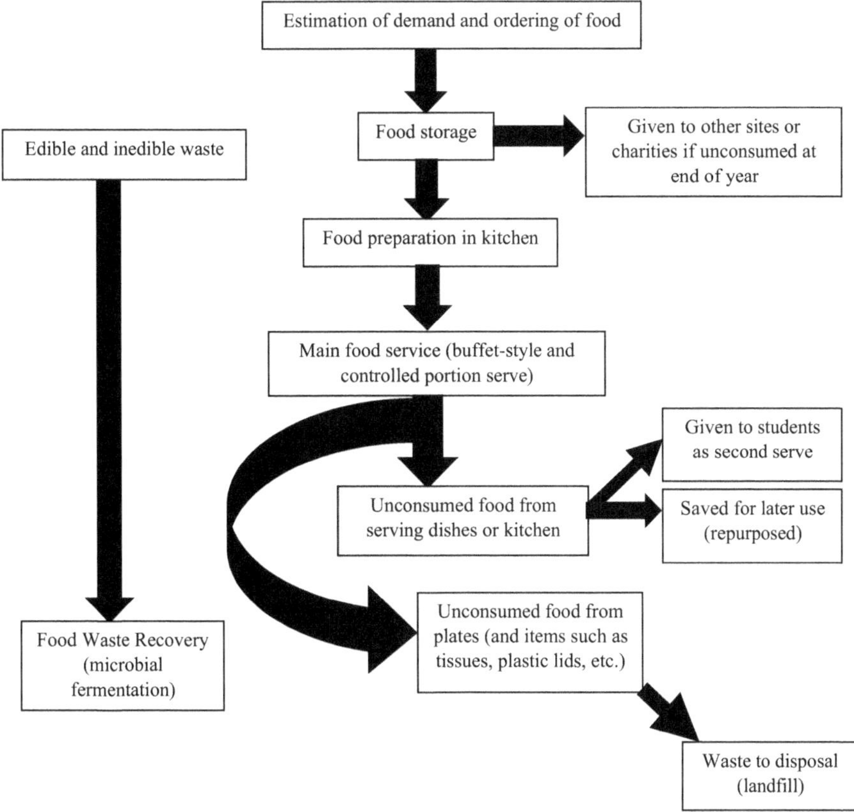

FIGURE 1.2 The flow chart of activities in food service that helped minimise food waste.

> guys doing so much better than everyone else? What are you doing differently that we are not? All the data is helpful for us; also it goes to the entire company and it allows us to analyse that data.

The canteen used the regular, buffet style of service, where consumers line up, serve themselves and take food on their plates. The only exception to this was the meat dishes, wherein the canteen workers served the meat to people in order to control portion size, and hence minimise cost and wastage. A second serving of food took place after the main meal to enable students to use up food left in the serving trays. In addition, a decision was taken to serve take-away food on plates rather than in large containers, which not only reduced waste, but also helped convey abundance in the mind of the diner: 'that's a whole plate of food'. The catering staff were mindful of health and safety restrictions which specify how long food can be kept and under what conditions before it must be thrown away (i.e., is unfit for human consumption). Moreover, the chefs and kitchen staff used strategies to minimise food waste generated due to spoilage or expiry while in storage. This included cold chain management, ensuring that a first-in, first-out system was applied for food inventory, such as using up perishable ingredients, and at the end of year, using up all frozen or stored food before buying new provisions. They also knew when students would be away on holidays. If food was unconsumed, then they had the option of giving away excess food to other sites or charities, although that was considered to be a sign of poor stock management. Plate waste, which was estimated to be around 15% of total food waste, was seen as uncontrollable, and the chefs realised that they had to satisfy the students and allow them to take large servings even though they did not always eat it all. They also suggested that food waste tended to increase around exam times, when students were likely to be stressed or preoccupied, and hence inclined to eat more, take larger servings and waste more food.

1.6 DISCUSSION

It is evident from the results that the food recovery program delivered cost savings and environmental benefits. Landfill diversion resulted in waste disposal savings of AU$12,528, but arguably the most important outcome was avoiding environmental harm whilst building a positive image for the university. Roughly 12,533 kg of food waste was diverted from landfill, which represented an elimination of 23.81 tonnes of carbon emissions over the span of eight months. Previous studies confirm the benefits of sustainability to the university sector (Amutenya et al., 2009). The cost-benefit analysis shows that installing the Bio-Regen® technology to recover food waste could be a cost-efficient option in the long term for large food generators, particularly given the introduction of new waste levies (AU$75 per ton) by the state government (Queensland Government, 2018).

The food waste audit showed that the colleges were producing an average of 363 kg of food waste each week. At certain periods, food waste levels increased to over 450 kg, although the interviews with the catering staff did not reveal clear explanations for these patterns, apart from the type of food served (i.e., watermelons leading to more preparation waste). There was some variation in terms of level of food waste between the three university residential halls. Over the course of the eight-month study, George Roberts (250 students) produced nearly half of the total waste collected, at 5,678.33 kg, whereas Western Courts (107 students) consistently produced the least throughout the study, at 2,378.37 kg. Interviews with catering staff suggested that this was due to the number of residents at each hall. Previous research has found a link between food establishment size (measured by the number of meals served per day or the number of employees) and the amount of food waste recycled (Okazaki et al., 2008). The food waste audit revealed that food waste dropped substantially over the winter holidays, with a 54% decrease recorded, and not surprisingly, the catering contractors limited the amount of food ordered and served when holiday periods arose. Based on the interviews, the catering staff appeared to be doing an acceptable job in estimating demand based on seasonality, and meal counts, but having exact figures on food waste each week might raise awareness of kitchen waste further and help them adjust catering practices to accurately meet demand.

Broadly speaking, the case study showed that catering practices for dealing with food waste conformed to circular economy thinking and the waste hierarchy, such as that developed by the European Commission under its 2008/9 Directive on Waste (European Commission, 2016). The waste hierarchy refers to a ranked series of preferential measures for dealing with waste, moving down from reducing waste, to recovery, to landfill (European Commission, 2016). A concerted effort was made to diminish dining hall food waste, such as ordering and cooking just enough food; portion control of protein; having a second serving of food and serving food on plates rather than in large take-away containers. The data findings are in line with previous research, such as avoiding excessively large portions (Alexander et al., 2013); controlling servings of meat, an element that is often central to the meal; accurate ordering (Maguire, 2016) and correctly estimating consumption (Silvennoinen et al., 2015). The kitchen staff, as skilled cooks, knew how to re-use leftovers or ingredients in order to avoid throwing out edible food. Hence, a continued focus on training for how to re-use and repurpose food is critical, which is in line with industry reports (Bloom, 2010; Gunders, 2012; WRAP, 2013). The interviews revealed that the culture of the parent organisation, Compass Group, with its strong focus on avoiding waste, helped set the tone for all staff and foster commitment to reducing kitchen waste –which represented a waste of money. However, the overwhelming focus on cost may indicate a lack of awareness of the environmental impacts of food waste and need for sustainability training. Prior study reveals that culture, goal setting, and social pressure can distinguish recyclers from non-recyclers. (Bolaane, 2006). The interviews highlighted that the measurement of food waste and dissemination of information throughout the company was useful in driving waste reduction. Likewise, the WRAP report in the UK (Parfitt et al., 2013) highlights the importance of waste monitoring.

Plate waste, or post-consumer waste (Pirani & Arafat, 2016), was not considered by the chefs to be under their control, and there was a reluctance to limit servings since this would decrease customer satisfaction, which is in line with previous research (Maguire, 2016). It was estimated that plate waste was around 15% of total waste, which is quite high. Hence, there may be opportunities to reduce plate waste through a change in catering practices, such as serving food in measured portions, rather than a buffet, and offering multiple servings to cater for diners who need more food. Investing in education and social marketing interventions would also be worthwhile. Whitehair et al. (2013), for example, achieved a 15% reduction in food waste generation from university canteens by using written messages such as 'Eat what you take. Don't waste food'. Interventions, such as showing a weekly pile of plate waste to the students, or running competitions between the colleges, might engage students by linking them to the waste they produce and promoting mindfulness of habitual behaviours.

The food recovery program, which entailed the microbial fermentation of food waste, represented a novel process for dealing with food waste. Over the course of the eight-month study, 40,050 L of bio-fertiliser were produced from the recovered food waste, all of which were top quality and could be resold by VRM to their other customers (VRM, personal communication, 19 November 2018). It was evident from the interviews that the role of the university's Environment Manager was crucial to the success of the recovery program. This study provides support for the work of scholars on how to avoid failure when implementing a waste management strategy. For instance, if composting is left to residents or staff in kitchens, it is likely to have its challenges, such as the time and inconvenience of putting waste into bins, need for student education regarding proper usage of the composting units, odour control in the vicinity of the compost vessels, and proper nutrient balance within the vessels (Balwin & Dripps, 2012). The case study showed that giving information to kitchen staff on the food recovery program helped promote compliance. Research in the recycling field shows that pro-recycling attitudes are the major contributor to recycling, along with facilities and knowledge, but situational factors or deterrents to physically recycling, such as time, space, and inconvenience, should be avoided (Clarke & Maantay, 2006; Tonglet et al., 2004). Recommendations for other universities that wish to implement similar programs are to ensure that the process is as seamless as possible, remove any inconvenience to kitchen staff, work closely with business partners, and be prepared to invest time and energy into solving practical problems.

1.7 CONCLUSION

In conclusion, universities are central to promoting sustainability since they are generators of waste and serve as role models for students, staff, and the community. The results of a food waste audit, conducted at a regional Australian university, showed that diverting food waste from landfill had economic and environmental benefits. Amongst its success factors were the commitment of the university's Environment Manager to the program, an effective internship program, and the support of the business partners. The design of an efficient food waste recovery program requires not only an assessment of the food waste stream, but also a thorough understanding of the catering practices and student behaviours responsible for generating the waste. Given that most studies on food waste deal with households, this study adds to the literature on food waste in an institutional setting.

APPENDIX

TABLE 1A.1
Total Food Weight (kg) Produced at JCU Colleges Per Week over the Course of the Eight-month (36-week) Study. All Weights Reported Have Already Been Adjusted to Not Include the Weight of the Buckets in which the Food Waste was Weighed

Hall	Week 1	Week 2	Week 3	Week 4	Week 5	Week 6	Week 7	Week 8	Week 9	Week 10	Week 11	Week 12
UH	96.4	121.66	152.5	84.85	152.72	149.14	136.28	156.66	104.68	169.96	115.98	164.42
WC	51.54	72.88	122.78	102.72	81.78	80.28	0	46.66	64.16	64.82	43.18	102.88
GR	85.02	177.8	177.82	185.64	151.32	224.92	168.32	143.14	102.34	171.36	151.06	197.88
Total	**232.96**	**372.34**	**453.1**	**373.21**	**385.82**	**454.34**	**304.6**	**346.46**	**271.18**	**406.14**	**310.22**	**465.18**

Week 13	Week 14	Week 15	Week 16	Week 17	Week 18	Week 19	Week 20	Week 21	Week 22	Week 23	Week 24
130.08	144.2	112.02	81.71	100.36	92.32	59.72	86.68	111.16	124.92	153.8	142.26
83.56	112.2	72.86	0	0	0	0	50.58	76.52	84.22	134.46	38.4
138.52	124.52	147.6	121.16	108.86	134.78	88.24	147.92	154.75	164.56	197.55	178.97
352.16	**380.92**	**332.48**	**202.87**	**209.22**	**227.1**	**147.96**	**285.18**	**342.43**	**373.7**	**485.81**	**359.63**

Week 25	Week 26	Week 27	Week 28	Week 29	Week 30	Week 31	Week 32	Week 33	Week 34	Week 35	Week 36
162.54	149.82	145.7	172.8	131.34	77.24	117.47	68.28	140.8	121.3	143.54	101.02
90.04	112.4	111.66	88.3	98.48	38.04	28.76	37.46	69.04	84.68	80.25	52.78
200.6	195.71	176.06	221.66	163.65	139.52	155.86	108.34	166.19	193.3	161.92	151.47
453.18	**457.93**	**433.42**	**482.76**	**393.47**	**254.8**	**302.09**	**214.08**	**376.03**	**399.28**	**385.71**	**305.27**

REFERENCES

Alexander, C., Gregson, N., & Gille, Z. (2013). Food waste. In A. Murcott, P. Jackson, & W. Belasco (Eds.), *The Handbook of Food Research* (pp. 471–285). Bloomsbury. http://doi.org/10.13140/2.1.2239.2964

Amutenya, N., Shackleton, C. M., & Whittington-Jones, K. (2009). Paper recycling patterns and potential interventions in the education sector: A case study of paper streams at Rhodes University, South Africa. *Resources, Conservation and Recycling*, *53*(5), 237–242. https://doi.org/10.1016/j.resconrec.2008.12.001

Angelidaki, I., Karakashev, D., Batstone, D. J., Plugge, C. M., & Stams, A. J. (2011). Biomethanation and its potential. *Methods in enzymology, 494*, 327–351. https://doi.org/10.1016/B978-0-12-385112-3.00016-0

Areeshi, M. Y. (2022). Recent advances on organic biofertilizer production from anaerobic fermentation of food waste: Overview. *International Journal of Food Microbiology*, *374*, 109719. https://doi.org/10.1016/j.ijfoodmicro.2022.109719

Australasian Campuses Towards Sustainability. (2021). *Home*. www.acts.asn.au

Baldwin, E., & Dripps, W. (2012). Spatial characterization and analysis of the campus residential waste stream at a small private liberal arts institution. *Resources, Conservation and Recycling, 65*, 107–115. https://doi.org/10.1016/j.resconrec.2012.06.002

Ball, S., & Taleb, M. A. (2011). Benchmarking waste disposal in the Egyptian hotel industry. *Tourism and Hospitality Research, 11*(1), 1–18. https://doi.org/10.1057/thr.2010.16

Bloom, J. (2010). *American Wasteland: How America Throws away Nearly Half of Its Food (and What We Can Do about It)*. Da Capo Lifelong Books.

Bolaane, B. (2006). Constraints to promoting people centred approaches in recycling. *Habitat International*, *30*(4), 731–740. https://doi.org/10.1016/j.habitatint.2005.10.002

Clarke, M. J., & Maantay, J. A. (2006). Optimizing recycling in all of New York City's neighborhoods: Using GIS to develop the REAP index for improved recycling education, awareness, and participation. *Resources, Conservation and Recycling*, *46*(2), 128–148. https://doi.org/10.1016/j.resconrec.2005.06.008

Compass Group. (n.d.-a). *Sustainable Impact*. www.compass-group.com.au/why-compass/corporate-social-responsibility/#reducing-waste

Compass Group. (n.d.-b). *Compass Group targets food waste with international day of action*. www.compass-group.com.au/compass-group-targets-food-waste-with-international-day-of-action/

Cristóbal, J., Castellani, V., Manfredi, S., & Sala, S. (2018). Prioritizing and optimizing sustainable measures for food waste prevention and management. *Waste Management, 72*, 3–16. https://doi.org/10.1016/j.wasman.2017.11.007

Department of the Environment and Energy. (2017). *National Greenhouse Accounts Factors*. Commonwealth of Australia. www.environment.gov.au/system/files/resources/5a169bfb-f417-4b00-9b70-6ba328ea8671/files/national-greenhouse-accounts-factors-july-2017.pdf

Dhir, A., Talwar, S., Kaur, P., & Malibari, A. (2020). Food waste in hospitality and food services: A systematic literature review and framework development approach. *Journal of Cleaner Production, 270*, 122861.

El-Fadel, M., Findikakis, A. N., & Leckie, J. O. (1997). Environmental impacts of solid waste landfilling. *Journal of Environmental Management*, *50*(1), 1–25. https://doi.org/10.1006/jema.1995.0131

Espinosa, R. M., Turpin, S., Polanco, G., De laTorre, A., Delfín, I., & Raygoza, I. (2008). Integral urban solid waste management program in a Mexican university. *Waste Management, 28*, S27-S32. https://doi.org/10.1016/j.wasman.2008.03.023

European Commission. (2016). *Directive 2008/98/EC on waste*. EUROPA. http://ec.europa.eu/environment/waste/framework/

Garcia-Garcia, G., Woolley, E., Rahimifard, S., Colwill, J., White, R., & Needham, L. (2017). A methodology for sustainable management of food waste. *Waste and Biomass Valorization*, *8*(6), 2209–2227. https://doi-org.elibrary.jcu.edu.au/10.1007/s12649-016-9720-0

Gooch, M., Felfel, A., & Marenick, N. (2010). *Food Waste in Canada: Opportunities to Increase the Competitiveness of Canada's Agri-food sector, while Simultaneously Improving the Environment*. Value Chain Management Centre. http://vcm-international.com/wp-content/uploads/2013/04/Food-Waste-in-Canada-112410.pdf

Gunders, D. (2012). Wasted: How America is losing up to 40 percent of its food from farm to fork to landfill. *Natural Resources Defense Council, 26*, 1–26.

Guo, W., Zhou, Y., Zhu, N., Hu, H., Shen, W., Huang, X., … & Li, Z. (2018). On site composting of food waste: A pilot scale case study in China. *Resources, Conservation and Recycling, 132*, 130–138. https://doi.org/10.1016/j.resconrec.2018.01.033

Gustavsson, J., Cederberg, C., Sonesson, U., Van Otterdijk, R., & Meybeck, A. (2011). Global food losses and food waste: extent, causes and prevention, Food and Agriculture Organisation of the United Nations (FAO), Rome (2011), p. 29.

Hanson, C., Lipinski, B., Friedrich, J., & O'Connor, C. (2015). *What's Food Loss and Waste Got to Do with Climate Change? A Lot, Actually*. World Resources Institute. www.wri.org/blog/2015/12/whats-food-loss-and-waste-got-do-climate-change-lot-actually

Jackson, M., Lederwasch, A., & Giurco, D. (2014). Transitions in theory and practice: Managing metals in the circular economy. *Resources*, *3*(3), 516–543. http://dx.doi.org/10.3390/resources3030516

Kibler, K. M., Reinhart, D., Hawkins, C., Motlagh, A. M., & Wright, J. (2018). Food waste and the food-energy-water nexus: A review of food waste management alternatives. *Waste Management*, *74,* 52–62. https://doi.org/10.1016/j.wasman.2018.01.014

Kirchherr, J., Reike, D., & Hekkert, M. (2017). Conceptualizing the circular economy: An analysis of 114 definitions. *Resources, Conservation & Recycling,* 127, 221–232. https://doi.org/10.1016/j.resconrec.2017.09.005

Maguire, M. A. (2016). *Wasted potential: A food waste reduction strategy for Toronto restaurants* (Master's thesis). Retrieved September, 2020 from https://yorkspace.library.yorku.ca/xmlui/handle/10315/34820

Mason, I. G., Brooking, A. K., Oberender, A., Harford, J. M., & Horsley, P. G. (2003). Implementation of a zero waste program at a university campus. *Resources, Conservation and Recycling*, *38*(4), 257–269. https://doi.org/10.1016/S0921-3449(02)00147-7

Mbuligwe, S. E. (2002). Institutional solid waste management practices in developing countries: A case study of three academic institutions in Tanzania. *Resources, Conservation and Recycling*, *35*(3), 131–146. https://doi.org/10.1016/S0921-3449(01)00113-6

McCarthy, B., Liu, H. B., & Chen, T. (2016). Innovations in the agro-food system: Adoption of certified organic food and green food by Chinese consumers. *British Food Journal*, *118*(6), 1334–1349.

Okazaki, W.K., Turn, S.Q., & Flachsbart, P.G. (2008). Characterization of food waste generators: A Hawaii case study. *Waste Management, 28,* 2483–2494. http://dx.doi.org/10.1016/j.wasman.2008.01.016

Papargyropoulou, E., Wright, N., Lozano, R., Steinberger, J., Padfield, R., & Ujang, Z. (2016). Conceptual framework for the study of food waste generation and prevention in the hospitality sector. *Waste Management, 49*, 326–336. https://doi.org/10.1016/j.wasman.2016.01.017

Parfitt, J., Eatherley, D., Hawkins, R., & Prowse, G. (2013). *Waste in the UK Hospitality and Food Service Sector (Technical Report No. HFS001-00 6)*. Waste and Resources Action Programme (WRAP).

Pirani, S. I., & Arafat, H. A. (2014). Solid waste management in the hospitality industry: A review. *Journal of Environmental Management, 146*, 320–336. https://doi.org/10.1016/j.jenvman.2014.07.038

Pirani, S. I., & Arafat, H. A. (2016). Reduction of food waste generation in the hospitality industry. *Journal of Cleaner Production, 132*, 129–145. https://doi.org/10.1016/j.jclepro.2015.07.146

Priefer, C., Jörissen, J., & Bräutigam, K. R. (2016). Food waste prevention in Europe: A cause-driven approach to identify the most relevant leverage points for action. *Resources, Conservation and Recycling, 109*, 155–165. https://doi.org/10.1016/j.resconrec.2016.03.004

Queensland Government. (2018). *Waste disposal levy*. www.qld.gov.au/environment/pollution/management/waste/recovery/disposal-levy

Sarjahani, A., Serrano, E. L., & Johnson, R. (2009). Food and non-edible, compostable waste in a university dining facility. *Journal of Hunger, Environment & Nutrition, 4*, 95–102. http://dx.doi.org/10.1080/19320240802706874

Sharma, P., Gaur, V. K., Kim, S. H., & Pandey, A. (2020). Microbial strategies for bio-transforming food waste into resources. *Bioresource technology, 299*, 122580.

Silvennoinen, K., Heikkilä, L., Katajajuuri, J. M., & Reinikainen, A. (2015). Food waste volume and origin: Case studies in the Finnish food service sector. *Waste Management, 46*, 140–145. https://doi.org/10.1016/j.wasman.2015.09.010

Smyth, D. P., Fredeen, A. L., & Booth, A. L. (2010). Reducing solid waste in higher education: The first step towards 'greening'a university campus. *Resources, Conservation and Recycling, 54*(11), 1007–1016. https://doi.org/10.1016/j.resconrec.2010.02.008

Sonnino, R., & McWilliam, S. (2011). Food waste, catering practices and public procurement: A case study of hospital food systems in Wales. *Food Policy, 36*(6), 823–829. https://doi.org/10.1016/j.foodpol.2011.09.003

Sridhar, A., Kapoor, A., Kumar, P. S., Ponnuchamy, M., Balasubramanian, S., & Prabhakar, S. (2021). Conversion of food waste to energy: A focus on sustainability and life cycle assessment. *Fuel, 302*, 121069.

Stancu, V., Haugaard, P., & Lähteenmäki, L. (2016). Determinants of consumer food waste behaviour: Two routes to food waste. *Appetite, 96*, 7–17. https://doi.org/10.1016/j.appet.2015.08.025

Sundt, P. (2012). *Prevention of Food Waste in Restaurants, Hotels, Canteens and Catering*. Nordic Council of Ministers.

Thyberg, K. L., & Tonjes, D. J. (2016). Drivers of food waste and their implications for sustainable policy development. *Resources, Conservation and Recycling, 106*, 110–123. https://doi.org/10.1016/j.resconrec.2015.11.016

Tonglet, M., Phillips, P. S., & Read, A. D. (2004). Using the Theory of Planned Behaviour to investigate the determinants of recycling behaviour: A case study from Brixworth, UK. *Resources, Conservation and Recycling, 41*(3), 191–214. https://doi.org/10.1016/j.resconrec.2003.11.001

United Nations. (2024a). *The 17 Goals*. https://sdgs.un.org/goals/goal2

United Nations. (2024b). *Targets and Indicators*. https://sdgs.un.org/goals/goal12#targets_and_indicators

United Nations Environment Program. (2014). *Greening Universities Toolkit V2.0. Transforming Universities into green and sustainable campuses: A toolkit for implementers*. www.unep.org/resources/toolkits-manuals-and-guides/greening-universities-toolkit-v20

University Leaders for a Sustainable Future. (2015). *Talloires Declaration*. http://ulsf.org/talloires-declaration/

de Vega, C. A., Benítez, S. O., & Barreto, M. E. R. (2008). Solid waste characterization and recycling potential for a university campus. *Waste Management, 28*, S21–S26. https://doi.org/10.1016/j.wasman.2008.03.022

Verghese, K., Lewis, H., Lockrey, S., & Williams, H. (2013). The role of packaging in minimising food waste in the supply chain of the future. Prepared for: CHEP Australia.

VRM BioLogik. (2020). *The Bio-Regen®*. VRM BioLogik. Retrieved October, 2020 from https://72d3b2da-4e4f-492e-8fc3-1e4d0224ddfc.filesusr.com/ugd/eca595_14ba4ec7b07a47e5ad6a18fde919cbc9.pdf

Wei, Y., Li, J., Shi, D., Liu, G., Zhao, Y., & Shimaoka, T. (2017). Environmental challenges impeding the composting of biodegradable municipal solid waste: A critical review. *Resources, Conservation and Recycling, 122*, 51–65. https://doi.org/10.1016/j.resconrec.2017.01.024

Whitehair, K. J., Shanklin, C. W., & Brannon, L. A. (2013). Written messages improve edible food waste behaviors in a university dining facility. *Journal of the Academy of Nutrition and Dietetics, 113*(1), 63–69. https://doi.org/10.1016/j.jand.2012.09.015

WRAP. (2013). *Overview of waste in the UK hospitality and food service sector*. www.wrap.org.uk/food-drink/business-food-waste/hospitality-food-service

Yuan, Z., Bi, J., Moriguichi, Y. (2006). The circular economy: A new development strategy in China. *Journal of Industrial Ecology, 10*(1–2), 4–8. https://doi.org/10.1162/108819806775545321

Zen, I. S., Subramaniam, D., Sulaiman, H., Saleh, A. L., Omar, W., & Salim, M. R. (2016). Institutionalize waste minimization governance towards campus sustainability: A case study of Green Office initiatives in Universiti Teknologi Malaysia. *Journal of Cleaner Production, 135*, 1407–1422. https://doi.org/10.1016/j.jclepro.2016.07.053

Zhang, N., Williams, I. D., Kemp, S., & Smith, N. F. (2011). Greening academia: Developing sustainable waste management at Higher Education Institutions. *Waste Management, 31*(7), 1606–1616. https://doi.org/10.1016/j.wasman.2011.03.006

2 Evaluating the Bacterial Concrete as a Solution to Construction Debris Waste

N. Chandana and Vishal Akula

2.1 INTRODUCTION

Rapid urbanization, modernization, and industrialization have increased the load on resources. Concrete remains the world's top consumed resource, which causes server load on natural resources and is also one of the high contributors of greenhouse gases. Concrete will eventually crack due to shrinkage with age and volume reduction from hydration products, which causes poor tensile strength and material deterioration. Internal fractures impair the durability and shorten the structure's service life by allowing harsh substances to enter [1,2]. The general and conventional methods for closing the cracks are epoxy injection, stitching, drilling and plugging method, grouting and surface treatments, etc., which may incur heavy maintenance costs [3]. On-time maintenance of the structure is of higher importance. Suppose the structures are not treated during their lifespan. In that case, there comes the issue of construction debris and waste occurring within the life cycle of buildings during the construction, modification, and demolition phases. These construction wastes have become severe environmental problems in many countries [4]. And also, without proper control or immediate repair, the propagation of cracks may extend further by deteriorating the concrete, risking the safety of concrete infrastructure [5]. The situation worsens when the structures are built in harsh environments where corrosive elements, such as chloride ions, may enter the concrete matrix through cracks and cause degradation of the internal reinforcement that is implanted, eventually lowering the durability and service life of the concrete [6]. In practice, cracks in concrete also reduce the durability, permeability, and strength of the concrete. The extreme winter situation also deteriorates as water seeps through these cracks and freezes, causing a widening of gaps [7]. To overcome this problem, self-healing concrete, unlike conventional concrete, has been known for decades. Several investigations have been conducted to understand this phenomenon of repairing concrete by itself [8–11].

One of the self-healing concretes is produced by adding bacteria during concrete preparation. To better understand the self-healing mechanism, various researchers have conducted numerous experiments on concrete fracture repair over many years

 DOI: 10.1201/9781003364467-2

[12]. Techniques for self-healing fall into two categories: autonomous and autogenous healing [13]. There are two ways that autogenous healing might take place: (i) first, calcium carbonate ($CaCO_3$) crystals are formed via a direct reaction between the carbonate and calcium ions in the water and cement, respectively and (ii) second, exposure to humid environments causes some pozzolanic materials or anhydrous cement particles to continue to hydrate [12–14]. Concrete cracks are then repaired using the $CaCO_3$ that was created. In biotechnological healing techniques, bacteria heal the cracks by precipitating calcium in the heals [14–18]. In self-healing concrete, bacteria are used along with calcium nutrients known as calcium lactate. The calcium lactate is added while preparing the concrete mix in wet conditions. On the substrate, the bacteria multiply, and microbial activity produces $CaCO_3$, which closes the pores and fills the gaps [18–20]. In addition to extending a structure's useful life, microbial self-healing is known to use less cement, lowering CO_2 emissions. Due to its environmentally friendly approach to crack treatment and concrete preservation, this method has attracted much attention and may be used in the building sector. Thus, this paper has systematically analyzed the previous research available on bacterial concrete to understand bacterial concrete as self-healing concrete. The study sought to: (i) identify the commonly used bacteria for producing a self-healing bacterium; (ii) evaluate the efficiency of self-healing bacterial concrete as a modern concrete material; (iii) identify the mechanism and strength of each type of bacterial concrete; and (iv) evaluate the strength properties of bacterial concrete.

2.2 METHODOLOGY

2.2.1 Research Protocol

A qualitative content analysis was conducted to provide a systematic literature review on bacterial concrete as a self-healing technology. The systematic review was conducted by using PSALSAR (Protocol, Search, Appraisal, Synthesis, Analysis, and Reporting) framework, as shown in Table 2.1 and Preferred Reporting Items for Systematic reviews and Meta-Analyses (PRISMA) Framework [21,22].

2.2.2 Search

The search commenced in October 2022 with relevant data in different databases sources like Scopus, science direct, PubMed, and Google scholar. To ensure every literature is fetched different search strings are employed in the databases as given in Table 2.2.

2.2.3 Appraisal

The appraisal was conducted based on the identification and assessment of articles based on the research objectives from the search of databases. The articles were screened and selected for analysis using the inclusion and exclusion criteria, as shown in Table 2.3.

TABLE 2.1
PSALSAR Framework for the Systematic Literature Review

Steps	Outcome	Method
Protocol	Scope of study defined	Only studies on bio concrete as a self-healing technology/microbial self-healing concrete/ Immobilizing bacteria/Self healing in concrete
Search	Define the search strategy search studies	Selecting studies Conducting a quality assessment of studies
Appraisal	Selecting studies Quality assessment of studies	Defining & employing inclusion and exclusion criteria Employing quality criteria
Synthesis	Extract data Categorize the data	Developing extraction template Categorizing the data based on the defined criterion Organizing the data for further analysis
Analysis	Data analysing Result and discussion	Analysing the quantitative data Comparing and contrasting the studies, Identifying trends and research gaps
Reporting	Conclusion	Developing conclusion Providing recommendations
	Report writing	Applying the Preferred Reporting Items for Systematic Reviews and Meta-Analysis (PRISMA) methodology for qualitative Analysis
	Journal article production	Summarizing the report into a journal manuscript

A total of articles (41 from Google Scholar, 63 from PubMed, 14 from Science Direct, and 8 from Scopus) from the search strings in the source databases. The appraisal of 126 processes eliminated 29 duplicates resulting in a total of 97 articles. Titles were accessed using Inclusion and Exclusion eligibility criteria. The title assessment process eliminated 18 articles, and 79 articles had their abstracts screened. A total 11 articles were excluded because they were based mainly on literature review papers. The full publications of the resulting articles were accessed and the main body reading identified 59 articles for further analysis. The main body skimming eliminated articles that were not of primary research and lacked clear assessment methods. A total of 54 articles were considered for the systematic literature review.

2.2.4 Synthesis and Analysis

A criteria extraction template used for categorizing and analyzing the data has been provided in Table 2.4. The data is extracted into an excel sheet for data analysis and processing into meaningful information to build knowledge about bacterial concrete. The analysis addressing each research objective is developed to narrate the results.

TABLE 2.2
Search Strings for the Selected Databases

Database	Search string	Number of studies
Scopus	(Self-healing concrete) or (bacterial concrete) or (mechanical strength of bacterial concrete) or (microbial induced calcite precipitation concrete) or (self-healing agents) or (self-healing bioconcrete) or (microbial self-healing concrete) or (self-healing technologies) or (biological concrete) or (bioconcerte) (Microbially Induced Calcium Carbonate Precipitation (MICP) and their role in bioconcrete) or (MICP concrete) or (MICP bio concrete) or (bio concrete) or (bio concrete) or (calcifying bacteria) or (Immobilizing bacteria) or (self-healing in concrete)	
Science Direct	(Self-healing concrete) or (bacterial concrete) or (mechanical strength of bacterial concrete) or (microbial induced calcite precipitation concrete) or (self-healing agents) or (self-healing bioconcrete) or (microbial self-healing concrete) or (self-healing technologies) or (biological concrete) or (bioconcerte) (Microbially Induced Calcium Carbonate Precipitation (MICP) and their role in bioconcrete) or (MICP concrete) or (MICP bio concrete) or (bio concrete) or (bio concrete) or (calcifying bacteria) or (Immobilizing bacteria) or (self-healing in concrete)	
Google scholar	(Self-healing concrete) or (bacterial concrete) or (mechanical strength of bacterial concrete) or (microbial induced calcite precipitation concrete) or (self-healing agents) or (self-healing bioconcrete) or (microbial self-healing concrete) or (self-healing technologies) or (biological concrete) or (bioconcerte) (Microbially Induced Calcium Carbonate Precipitation (MICP) and their role in bioconcrete) or (MICP concrete) or (MICP bio concrete) or (bio concrete) or (bio concrete) or (calcifying bacteria) or (Immobilizing bacteria) or (self-healing in concrete)	
PubMed	(Self-healing concrete) or (bacterial concrete) or (mechanical strength of bacterial concrete) or (microbial induced calcite precipitation concrete) or (self-healing agents) or (self-healing bioconcrete) or (microbial self-healing concrete) or (self-healing technologies) or (biological concrete) or (bioconcerte) (Microbially Induced Calcium Carbonate Precipitation (MICP) and their role in bioconcrete) or (MICP concrete) or (MICP bio concrete) or (bio concrete) or (bio concrete) or (calcifying bacteria) or (Immobilizing bacteria) or (self-healing in concrete)	

2.2.5 Statistical Analysis

The results are reported as mean values or ±standard error of duplicates and are analyzed using IBM SPSS 21.0 software. The strength values of lightweight coconut shell concrete for different time zones and macro structures are analyzed using the

TABLE 2.3
Eligibility Using Inclusion and Exclusion Criteria

Criteria	Decision
Where the predefined study keywords were identified in the title, in whole or part	Inclusion
Where the predefined study keywords were identified in the abstract	Inclusion
Where the paper was published in a scientific peer-reviewed journal	Inclusion
Where the paper was written in English	Inclusion
Paper is duplicated within the different databases	Exclusion
Paper is not primary research	Exclusion
Paper is studying other self healing technologies other than bio concrete	Exclusion

TABLE 2.4
Summary of Extraction Template

Criteria	Justification
Year of publication	The year of publication offered the trends and the state-of-art regarding self-healing bio concrete
Name of journal	Studies published in peer-reviewed journals to demonstrated reliability of the information
Location of study	Studies conducted throughout the world were considered in order to establish the geographical concentration of the composting studies
Types of data sources	To ensure the reliability of the articles, and therefore of the review
Type of self healing technologies of concrete	To establish which type of self healing technology has been used and excluding the technologies except microbial inclusion
Type of bacteria used	To establish which type of bacteria has been used
Efficiency of each bacteria	To establish the self healing capacity of bacterial species on various micro and macro cracks
Strength enhancement	To establish the strength enhancement by different bacterial species
Calcium precipitation–strength relationship	To establish the calcium precipitation–strength criteria
Other mechanisms for strength increase and crack reduction	To determine other mechanisms responsible for strength increase and crack reduction

non-parametric Friedman test. Only 95% confidence level with a $p= 0.05$ value was considered for analyzing the results.

2.3 BACTERIA AS A SELF-HEALING AGENT FOR CRACKS IN CEMENT

Previous studies [1–20] claim that the bacteria heal the cracks in concrete by mineral precipitation. The bacteria can stay inactive for up to 200 years and become active as soon as it encounters water seeping through the cracks in concrete. Any cracks that emerge in self-healing concrete cause microorganisms to awake from their hibernation state. When bacteria encounter air and water, it initiates the germination of bacterial spores, which feeds on calcium lactate-consuming oxygen. Previous studies have used different types of bacteria, as shown in Table 2.5, as self-healing concrete agents. These bacteria produce a calcium precipitate that is insoluble in water, and the crack will be filled, increasing the strength of the structure [12–20]. Previous studies have shown that autogenous healing is most effective when the cracks are between 0.01-0.1 mm wide when adequate humidity is present. Autonomous self-healing is incorporating non-traditional engineering modifications into the concrete matrix to achieve crack closure. Cracks up to 0.35 mm generally demonstrate satisfactory healing for early-age cracks. For cracks between 0.3 mm and 0.5 mm in width, specimens with supplemental cementitious materials (SCMs) exhibit significant crack healing of about 32% [13–20]. Figure 2.1 provides the healing characteristics of concrete of various types of bacteria different [20–42].

Autonomous healing, however, has been regarded as a different approach to a higher level of crack closure. [12]. Calcium carbonate precipitates into the cracks during the self-healing process, repairing them due to the metabolic activities of bacteria. Bacteria resume their hibernating stage once the calcium carbonate has filled the cracks. Future activation of the bacteria fills up any fractures that may appear [20]. Further previous researcher [25–30] has specified specific bacteria for specific applications, as shown in Table 2.5. Previous researchers have observed that the

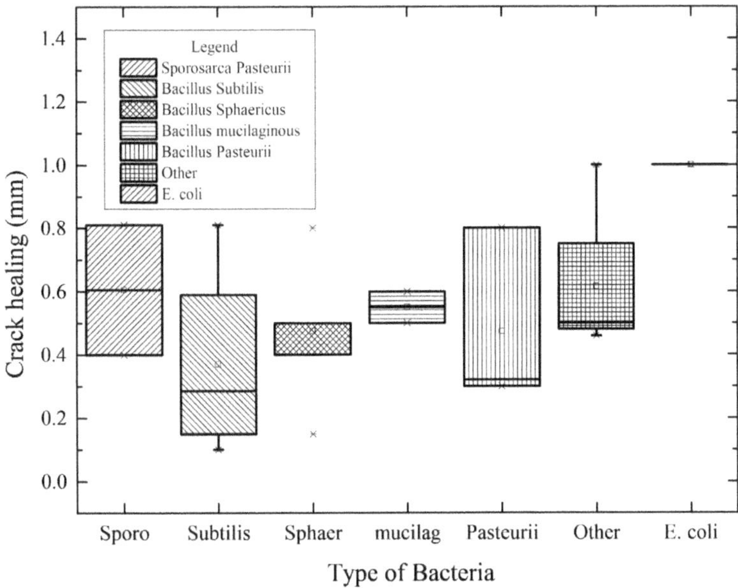

FIGURE 2.1 Healing characteristics of concrete of various bacterial self-healing concrete presented by previous literature for various curing times.

TABLE 2.5
Bacterial Self-healing Concrete Applications Presented by Previous Literature for Various Bacterial Species

Type of bacteria	Application	Reference
Bacillus Subtilis	Enhancement of concrete properties	[25]
Bacillus Sphaericus	For surface treatment	[26]
Bacillus Pasteurii/ Sporosarcina Pasteurii	As a crack healer	[27]
Escherichia Coli	Higher strength and more durable concrete	[28]
Bacillus Pseudifirmus	Self-healing ability of concrete	[29]
Bacillus Alkalinitrilicus	Self-healing concrete	[30]

addition of E. coli to concrete resulted in improvements in compressive strength test results [20].

2.4 BACTERIAL SELF-HEALING CONCRETE AS A STRUCTURAL MATERIAL

Reduction in lifetime maintenance and repair of concrete by itself leads to an increase in the durability of concrete structures, reduces the long-term maintenance of structures, and can reduce the construction debris reaching landfills. In addition, self-healing concrete has the potential to accomplish a significant decrease in costs associated with the health monitoring, damage detection, and maintenance of concrete structures while guaranteeing that the structure will have a secure service life. However, the concrete basic need is to take care of loads coming on it, and it can be measured by understanding its strength properties.

2.4.1 Compressive Strength

A detailed analysis of compressive strength reported by the previous researcher on self-healing bacterial concrete is presented in Figure 2.2. It can be inferred from Figure 2.2 that strength of bacterial concrete ranges from 2.53 to 37.96 MPa for three days of curing time, 13.7 to 44.23 MPa for seven days of curing time, 9.57 to 52 MPa for 14 days of curing time, 16.98 to 60 MPa for 28 days of curing time, 19.95 to 55.85 MPa for 58 days of curing time, and 36.13 to 45.45 MPa for 120 days of curing time. As the curing time of the cubes increases, the compressive strength of the bacterial self-healing concrete also increases. However, it is also observed that for 120 days of curing time, the maximum strength has reduced to 45.45 MPa. Compared to the control concrete samples, the strength at later ages is lower, which may be due to the consumption of large amounts of water at an early age during hydration [31,32]. Their decline in strength after 90 days is probably caused by the type of crystals produced by microbial interaction. Instead of firm crystals, the created calcite crystals may have resembled powder and might have impacted the concrete's compressive strength [30–34]. With the addition of bacteria, it was shown that the compressive strength

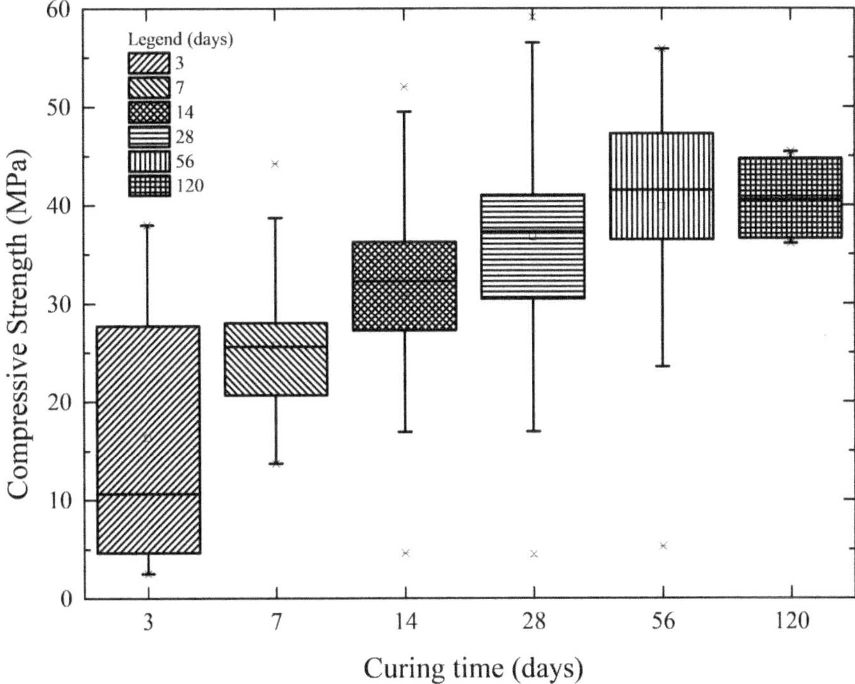

FIGURE 2.2 Compressive strength values of bacterial self-healing concrete presented by previous literature for various curing times.

increased. This increase is mostly due to the deposition of microbially induced calcium carbonate precipitation on the microorganism cell surfaces and inside the mortar pores.

Though the strength of bacterial concrete has a good value compared with normal concrete, it depends on the mixed grade and concentration of bacteria added. Further detailed analyses on the concentration effect of compressive strength were studied, as shown in Figure 2.3. Bacteria concentration has a major influence on mechanical properties, especially over the compressive strength of concrete. The compressive strength of 50 Mpa was achieved for 10^6 cells/ml concentration, which is the maximum value observed. The literature shows that the usual bacterial concentration of 10^3–10^8 cells/ml has been utilized in self-healing concrete.

Figure 2.4 shows that any grade of conventional concrete at target mean strength obtained at 28 days ranges from 10^3–10^8 cells/ml. The increase in strength is mainly due to the deposition of calcium carbonate precipitate on the cell surfaces of microorganisms. With an increase in bacteria concentration up to 10^6 cells/ml the average compressive strength has increased, and with further increase of cell concentration, the strength reduced but not lesser than the conventional concrete mix. In addition, Figure 2.4 shows that the addition of bacteria to the concrete mix has obtained results better than the conventional mix, and the mean value of compressive strength increases with an increase in bacterial concentration and has reduced from 10^7 cells/

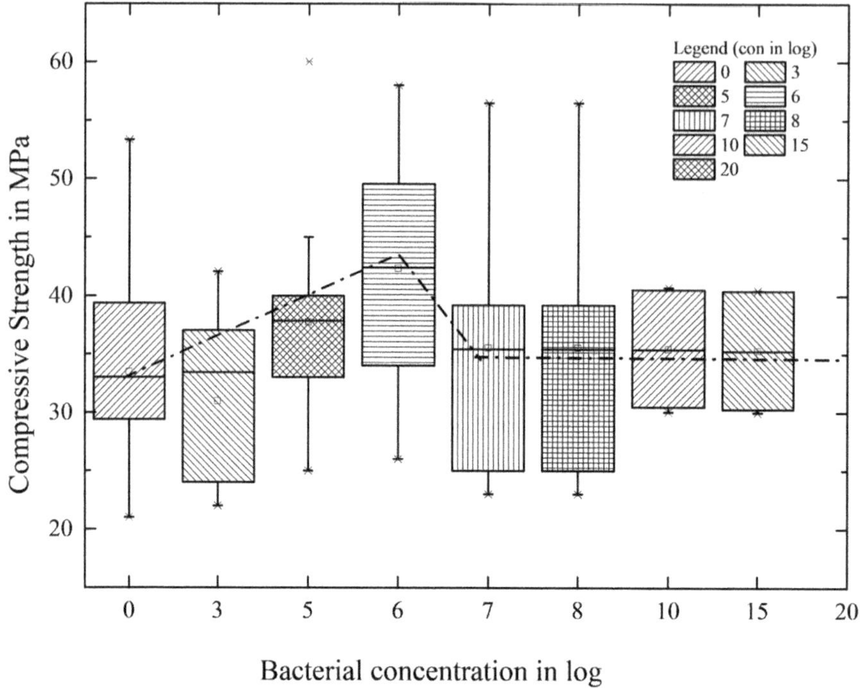

FIGURE 2.3 Compressive strength values of bacterial self-healing concrete presented by previous literature for different bacterial concentration.

ml. Table 2.6 gives a comparison of strength increase by different bacterial species and their concentration.

At higher cell concentrations of bacteria (i.e., 10^7 cells/ml) the matrix integrity may have been disrupted due to excessive bacterial activity, thus decreasing the compressive strength of concrete, as shown in Figure 2.3.

The application of different bacteria species affects the strength properties. In addition, the application of bacteria may vary depending on its availability, and for the purpose, it is used. Previously reported values about the effect of type of bacteria on compressive strength have been studied for 28 days of strength was analyzed as shown in Figure 2.5. It can be inferred from Figure 2.5 that various types of bacteria are used in concrete to enhance its properties. It is evident from Figure 2.5 that maximum compressive strength is obtained for the Bacillus subtilis bacteria, which is a Gram-positive, catalase-positive bacterium and is usually found in soil and the gastrointestinal tract of ruminants, humans, and marine sponges.

In a few cases, 28 days strength has reached a peak value of 60 Mpa. The increase in strength is due to the deposition of calcium carbonate ($CaCO_3$) in the pores. The bacterial consortium, a combination of a group of different microorganisms that could act together in a community, has a positive impact on compressive strength with a minimum strength of 30 Mpa and a maximum of 40 Mpa. Other bacteria like

TABLE 2.6
Effect of Different Bacterial Species and Its Concentration on Compressive Strength Values of Bacterial Self-healing Concrete Presented in Previous Literature

Bacteria type	Bacteria Properties	Concentration of bacteria	Inferences	Ref
Bacillus Subtilis	Gram-positive, catalase-positive bacterium.	10^5 cell/ml	12% increase in compressive strength	[42]
Bacillus Sphaericus	Aerobic gram-positive rod-shaped bacterium Pore forming bacterium, dormant for several years and would be able to withstand extreme temperature	10^6 cell/ml	7% increase in compressive strength	[43]
Bacillus Pasteurii/ Sporosarcina Pasteurii	Gram positive bacterium with the ability to precipitate calcite	10^5 cell/ml	37% increase in compressive strength	[44]
Escherichia coli	Gram-negative, facultative anaerobic, rod-shaped	10^6 cell/ml	50% increase in compressive strength	[45]
Bacillus Pseudifirmus	Gram positive	10^5 cell/ml	25.93% increase in compressive strength	[46]
Bacillus megaterium	Rod-like, Gram-positive, mainly aerobic spore forming bacterium	10^6 cell/ml	22.58% more than control concrete	[47]

E. coli, Sporosarcina Pasteurii, Ureolytic Bacteria, and Yeast have a positive impact when added to concrete by increasing the compressive strength and providing an ability to repair wide cracks in a short time. Previous researchers have observed that the addition of Escherichia Coli to concrete resulted in improvements in compressive strength test results [35]. The compressive strength of concrete-containing different bacteria is the result of the bacteria forming colonies and calcite precipitating in the concrete, which may have filled in the gaps and increased the density and strength of the concrete [36]. The minimum compressive strength value for Bacillus bacterium is nearly 35 Mpa, except for that with Bacillus megaterium, a lower-grade concrete, and the maximum value ranges between 40–50 Mpa. Thus, a detailed analysis of the effect of grade on the concrete mix was evaluated in bacterial concrete. Figure 2.6 provides the strength obtained by bacterial concrete in different grades of concrete mix, and it is evident that more work has been carried out using the M20 grade of concrete. It can be inferred from Figure 2.6 that M20 grade of concrete has gained more strength than M40. However, bacteria type, concentration, mode of application, cement grade, cement content, water/cement (W/C) ratio, and admixtures may influence compressive strength.

FIGURE 2.4 Excess bacterial activity observed at higher bacterial concentration addition in bacterial self-healing concrete.

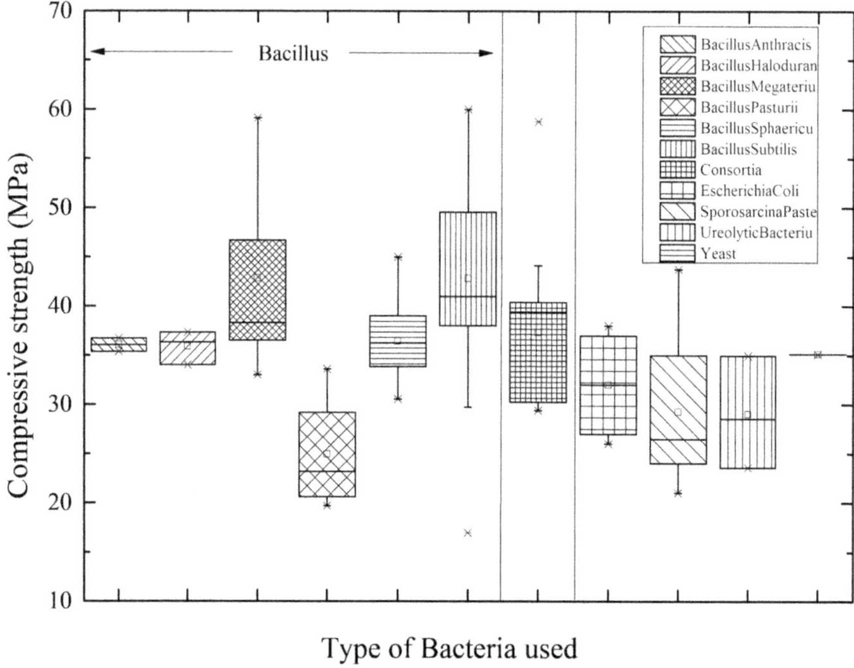

FIGURE 2.5 Compressive strength values of bacterial self-healing concrete presented in previous literature for different bacterial species.

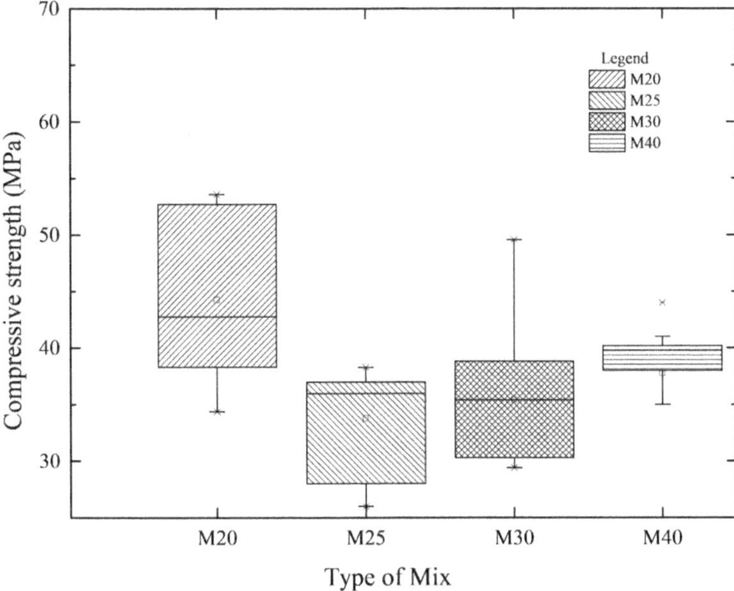

FIGURE 2.6 Compressive strength values of bacterial self-healing concrete presented in previous literature for different mix designs.

2.4.2 Flexural Strength

Flexural strength determines the resistance of the structure to bending stress. Considering compressive and tensile stress at a failure section, one can determine the flexural strength of bacterial self-healing concrete. Figure 2.7 provides the flexure strength values reported by previous literature on self-healing bacterial concrete. The increase in flexural strength of bacterial concrete is because of the deposition of a coating of calcite on the samples' surfaces and the calcite precipitation by bacteria that filled the concrete's pores [36–39]. As the curing age increases, the flexural strength has also increased with an average value of 4.5, 4.6, 4.9, 5.2,5.8 N/mm^2 for 3,7,14, 28, and 56 days. A maximum peak value of 12.4 N/mm^2 was achieved for 28 days, as indicated in Figure 2.7. Compared to compression strength (refer to Figure 2.1), fewer studies are conducted on the flexural strength of self-healing bacterial concrete.

The factors such as type of bacteria, Concentration of bacteria, Mode of application of bacteria, grade of cement, Cement & Water Content, and admixtures may influence flexural strength [40–47]. The effect of the concentration of bacteria addition on flexural strength was evaluated, as shown in Figure 2.8. Similar to that compressive strength, even flexural strength increases with an increase in bacteria concentration up to 10^5 cells/ml.

The other mixes have also shown an increase in flexural strength at various concentrations in comparison with conventional concrete. The mean value of flexural strength of all concentrations is greater than that of the conventional mix. It is inferred that use of bacterial concrete with a concentration of 10^5 cells/ml increases

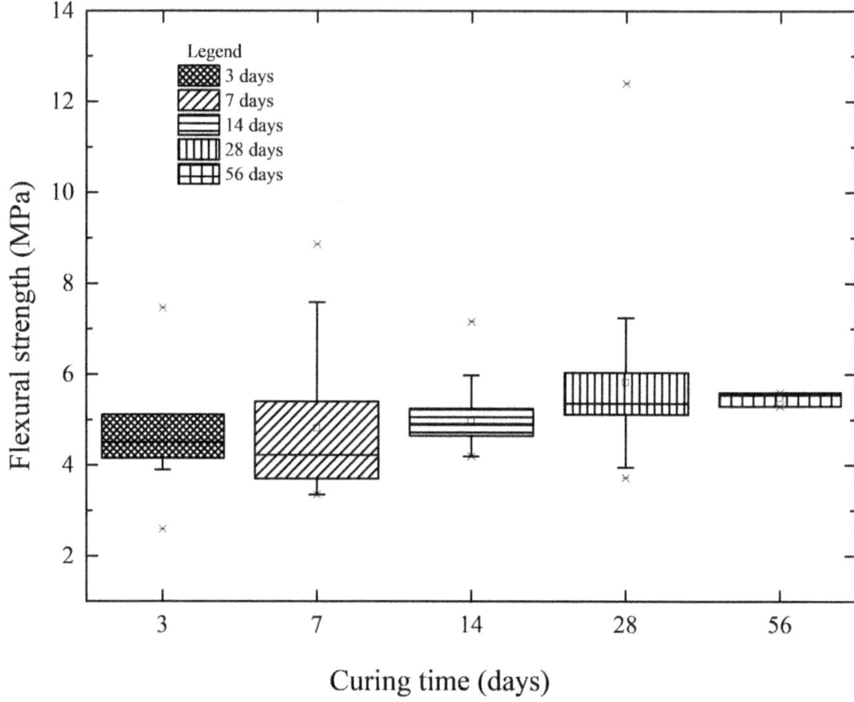

FIGURE 2.7 Flexural strength values of bacterial self-healing concrete presented in previous literature for various curing time.

9.8% in flexural strength compared with conventional concrete. And comparing the strengths of 10^5 cells/ml with 10^6 cells/ml there is a slight decrease in flexural strength of 7.14%. Further comparing 10^6 cells/ml with 10^7 cells/ml, there is a further decrease of 3.85%. But it is observed that the use of bacteria has enhanced the properties of the mix when compared to conventional concrete. The factors such as type of bacteria, Concentration of bacteria, Mode of application of bacteria, grade of cement, Cement & Water Content, and admixtures may influence flexural strength. Thus, the effect of the type of bacteria addition on flexural strength was evaluated, as shown in Figure 2.9.

Studies on the flexural strength of self-healing bacterial concrete are shown in Figure 2.9, few Bacilli, one study using Ureolytic Bacteria and Yeast have been used for understanding the flexural strength of bacterial concrete. Bacterial Pasturi and Ureolytic bacteria have shown higher flexure strength than other species. It can be inferred from Figure 2.9 that, 11%–20% increase in flexural strength was seen by using different types of bacteria when compared to conventional concrete. The effect of the mix design on flexural strength was studied and presented in Figure 2.10.

From Figure 2.10, it is evident that a lot of research has been done using M20-grade of concrete. The mean value of flexural strength for M20, M30, and M40 grades of concrete is found to be above 5 N/mm^2. The average strength has been increasing with the increase in the grade of concrete.

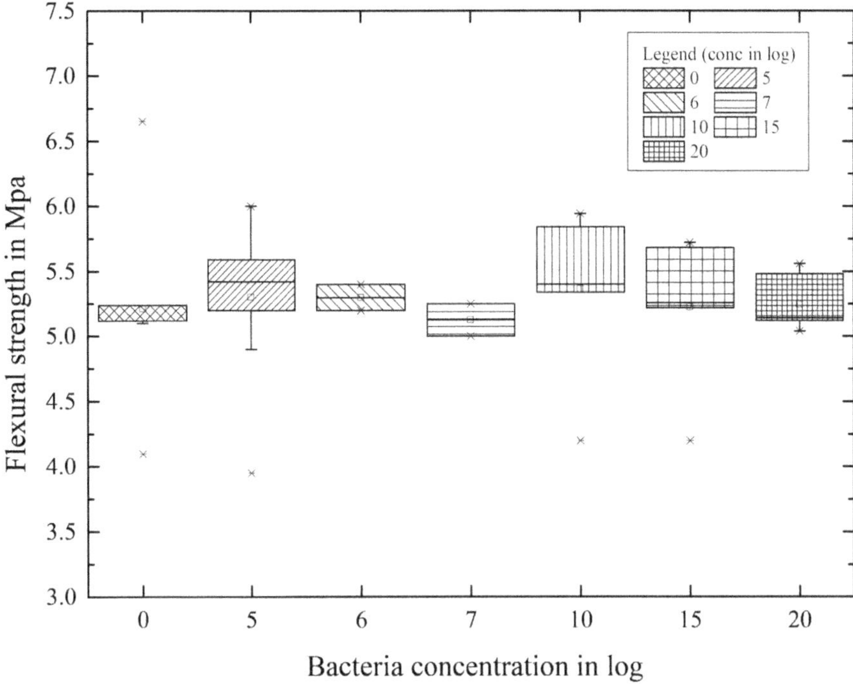

FIGURE 2.8 Flexural strength values of bacterial self-healing concrete presented by previous literature for different bacterial concentration.

2.4.3 Split Tensile Strength

The split tensile strength affects the extent of cracks that will be formed in structures, in addition, concrete is weak in tensile due to its brittle nature. The addition of bacteria effect on split strength has been analyzed by considering split tensile values reported by previous researchers, as shown in Figure 2.11.

It can be inferred from Figure 2.11 that with the increase in curing time, the split tensile strength has increased, and very few studies have been conducted on split tensile strength properties. The strength of bacterial concrete ranges from 1.6 to 5.2 Mpa for three days of curing time, 2 to 6.2 Mpa for 14 days of curing time, and 3.5 to 6.8 Mpa for 28 days of curing time. The increase in strength is mainly observed due to the complete hydration of cement particles. As the specimens are exposed to oxygen dissolved in the water, microbiologically induced calcium carbonate precipitation continues to occur. The calcium salts and $CaCO_3$ generated precipitate fill the concrete pores, strengthening the bio-concrete matrix integrity and leading to increased density. Thus, greatly increases the split tensile strength at a later stage [13]. Split tensile affects the type of bacterial addition, the concentration of bacterial addition, and the mix design [40–49]. The effect of the type of bacteria has been analyzed, as shown in Figure 2.12.

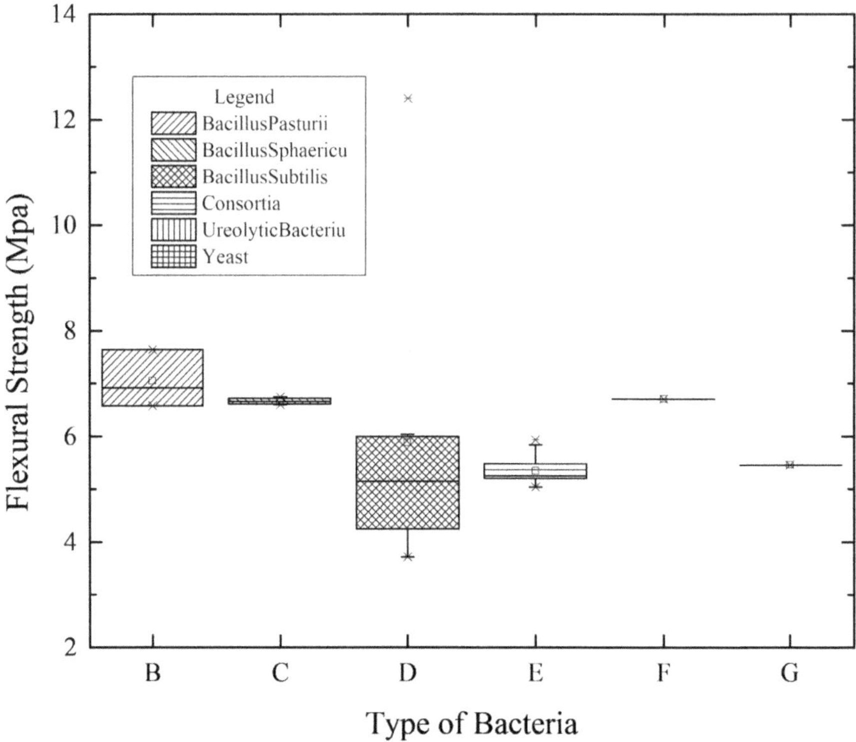

FIGURE 2.9 Flexural strength values of bacterial self-healing concrete presented in previous literature for different bacterial species.

The addition of bacteria to the concrete mix has a positive impact on the split tensile strength of concrete. As previously mentioned, only a few studies are conducted on flexural strength, and only four types of bacterial species have been used, and only the bacillus family. The fact that concrete is weak in tension is important because utilizing the Bacterium Megaterium, Bacillus subtilis, and their consortia, respectively, increased the tensile strength of concrete by 18.49%, 25.3%, and 19.58% [36–49]. As shown in Figure 2.13, only a few concentrations variation has been checked for their effect on bacterial concrete split tensile strength. The effect of bacteria concentration positively impacts the strength of concrete up to a concentration of 10^5 cells/ml. There is an increase in the split tensile strength of concrete by 20% when concentrated with 10^5 cells/ml. Further, only a little variation is observed with an increase in bacteria concentration, and the variation is around a 5%–9% increase compared to conventional concrete.

The effect of mix design on split tensile strength has been analyzed, as shown in Figure 2.14, and it was found that by an increase in the grade of concrete from M30 to M40 the average value of tensile strength has increased from 4.8 Mpa to 6.2 Mpa, however, it should be noted that only very few grade mix properties.

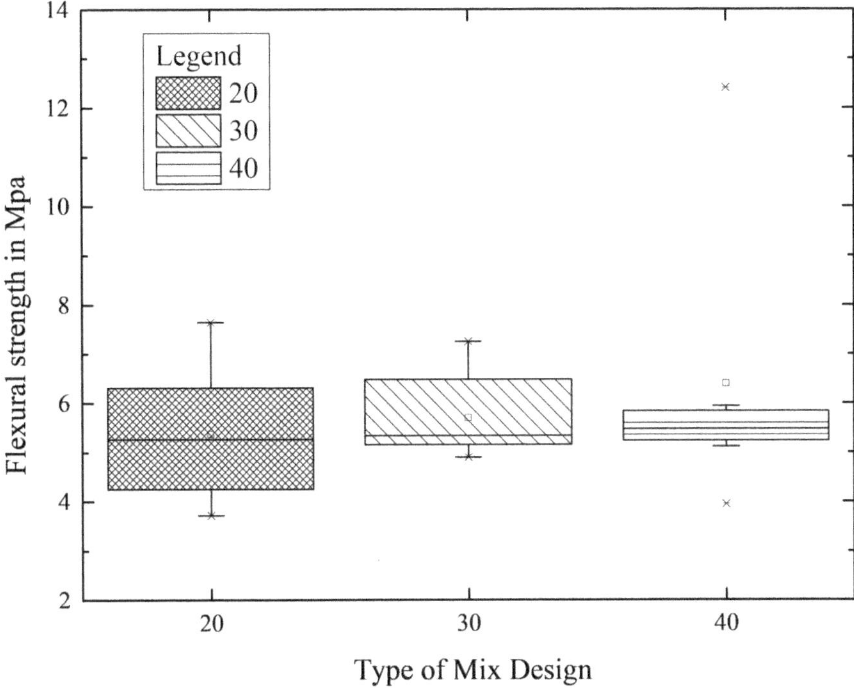

FIGURE 2.10 Flexural strength values of bacterial self-healing concrete presented in previous literature for different mix designs.

It can be inferred from the above discussion that bacterial concrete has a higher ability to be used as special concrete in modern society. Previous literature estimates that the cost of bacterial concrete will be close to 30% higher than ordinary concrete. Bacterial concrete prepared with admixtures like fly ash, silica fume, metakaolin and fibers like natural jute or artificial fibers like polypropylene, steel, etc., also gives better strength and durability. However, bacterial concrete will require much lesser maintenance than regular concrete. The calcium nutrient is the main cause of the rise in the price of microbiological concrete (calcium lactate is very costly). Calcium lactate can be replaced with wheat bran, corn steep liquor (CSL), or sugar-based nutrition to cut costs. Structures that need to retain water can benefit from bacterial concrete. By using a self-healing technology, leaks can be prevented and cracks can be closed. The future of bacterial concrete appears bright, particularly in underground constructions where restoration is challenging. When bacterial concrete is fully developed, it might replace OPC and counteract its hazardous effects on environmental contamination by closing cracks. As a result of its corrosion resistance, it can be employed in construction. The present study results recommend that bacterial concrete can be an alternative material being a high-quality concrete sealant whose initial cost may be high, but the cost incurred on maintenance during its life span is cost-effective and

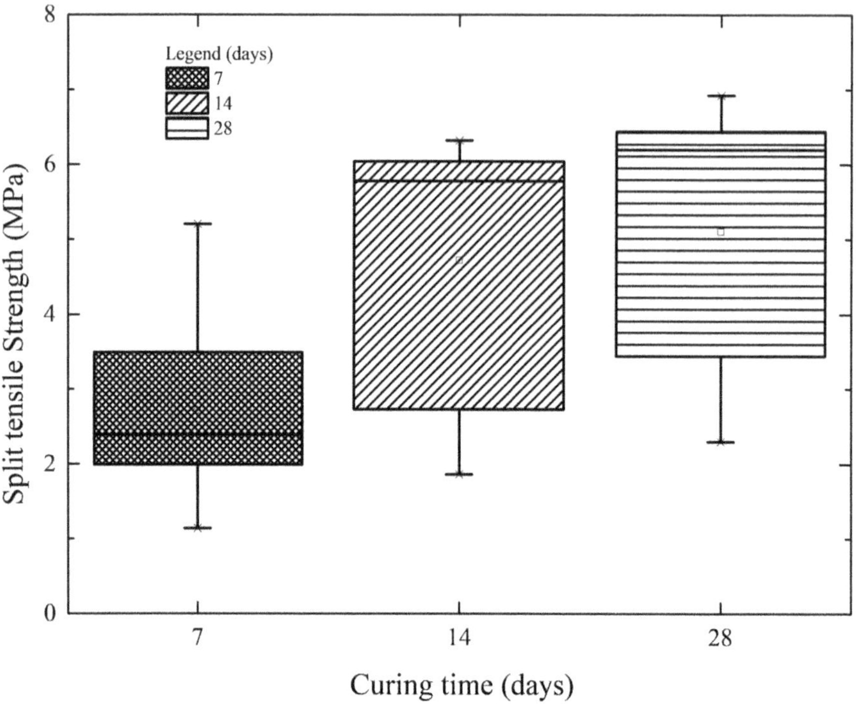

FIGURE 2.11 Split tensile strength values of bacterial self-healing concrete presented in previous literature for various curing time.

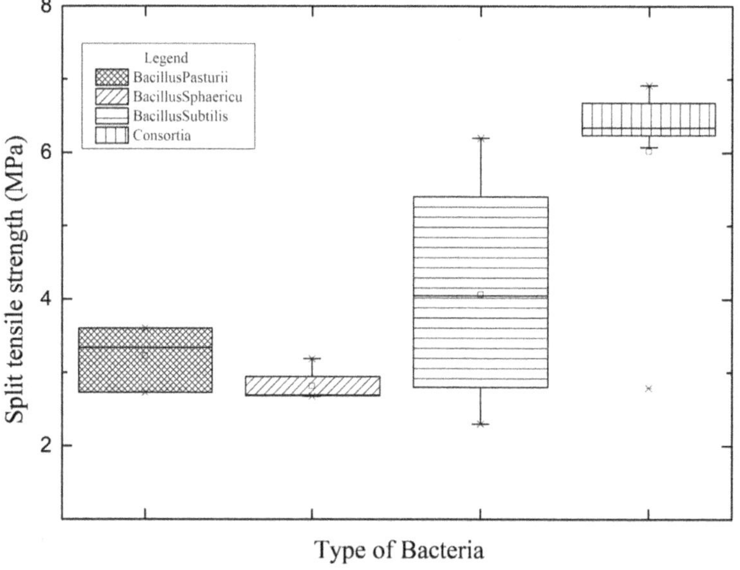

FIGURE 2.12 Split tensile strength values of bacterial self-healing concrete presented in previous literature for different bacterial species.

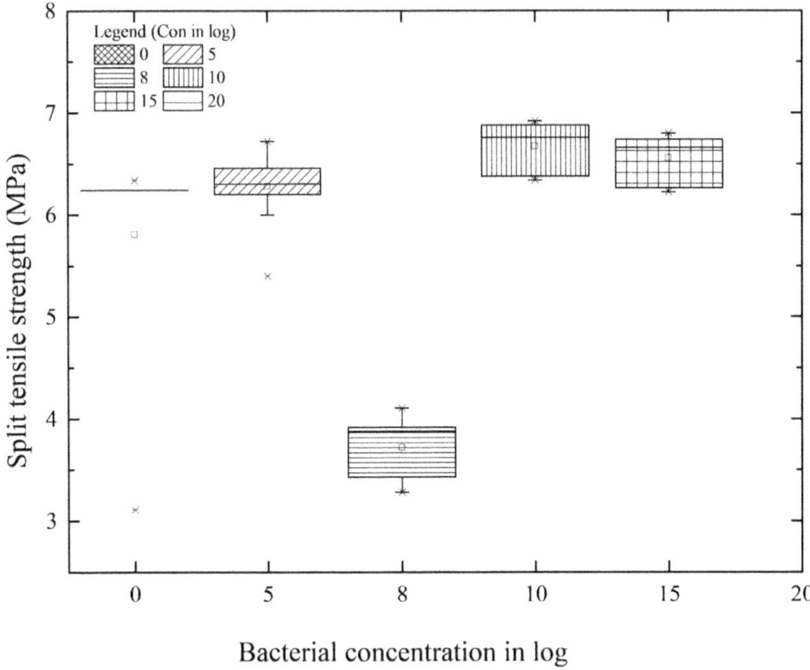

FIGURE 2.13 Split tensile strength of bacterial self-healing concrete presented by previous literature for different bacterial concentration.

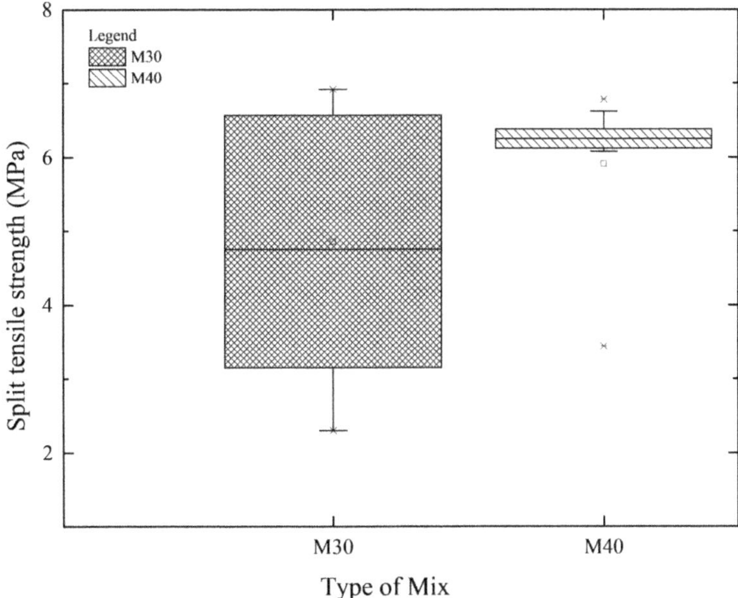

FIGURE 2.14 Split tensile strength values of bacterial self-healing concrete presented in previous literature for different mix designs.

environmentally friendly, which can reduce the construction debris reaching landfills and eventually leads to improvement in the durability of the building.

2.5 CONCLUSION

Bacterial self-healing concrete is found to be a potential replacement for conventional concrete to reduce the construction debris reaching landfills. In addition, self-healing concrete has the potential to accomplish a significant decrease in costs associated with the health monitoring, damage detection, and maintenance of concrete structures while guaranteeing that the structure will have a secure service life. Self-healing technology has proved to be better than the other conventional methods of concreting because of its eco-friendly nature, high healing capacity, and increase in the durability of various building materials. This study has also identified that the addition of bacteria positively impacts mechanical properties like concrete's compressive, tensile, and flexural strength. It is observed that the strength of bacterial concrete specimens is always greater than that of conventional concrete, irrespective of the bacterial concentration used in the mix ranging between 10^3–10^8 cells/ml. The average increase in strength with bacterial concentration 10^3 cells/ml, 10^7 cells/ml has been observed to be around 20–30%, and 5–10% only. The high strength at 10^6 cells/ml bacterial concentration corresponds to the optimum dose of self-healing bacterial concrete. The fundamental reason for the increase in mechanical strength properties is the consolidation of cement mortar pores caused by microbiologically induced calcite precipitation. Bacillus Subtilis and Bacillus Pasteurii are the most effectively used bacteria to heal cracks in concrete. Compared to standard samples, most Bacillus bacteria positively affect the compressive strength of concrete and bending strength. Using bacteria in concrete was found harmless to people living in the structure. By using a self-healing technology, leaks can be prevented, and cracks can be closed. The future of bacterial concrete appears bright, particularly in underground constructions where restoration is challenging. The deposit of $CaCO_3$ that has been formed is highly beneficial for preventing structural reinforcement corrosion, especially in coastal areas, and for enhancing the durability of such structures. The present study results recommend that environmentally friendly bacterial concrete as an alternative material for conventional concrete which can reduce the construction debris reaching landfills and eventually leads to improvement in the durability of the building.

REFERENCES

[1] Wang, J., K. Van Tittelboom, N. De Belie, W. Verstraete, Use of silica gel or polyurethane immobilized bacteria for self-healing concrete, *Constr. Build. Mater.* 26 (2012) 532–540.

[2] Xu, J., W. Yao, Z. Jiang, Non-ureolytic bacterial carbonate precipitation as a surface treatment strategy on cementitious materials, *J. Mater. Civ. Eng.* 26 (2013) 983–991.

[3] Yu, T., "The experimental study of repairing concrete cracks with epoxy resin grouting." Thesis (Master). Beijing University of Technology, Beijing, 2016.

[4] Esin, T., N. Cosgun, A study conducted to reduce construction waste generation in Turkey, *Build. Environ.* 42 (4) (2007) 1667–1674. ISSN 0360-1323.

[5] Basheer, L., J. Kropp, D. J. Cleland, Assessment of the durability of concrete from its permeation properties: a review, *Constr. Build. Mater.* 15 (2) (2001) 93–103.
[6] Bertolini, L., B. Elsener, P. Pedeferri, R. B. Polder, *Corrosion of Steel in Concrete*, vol. 392, Wiley Online Library, 2013.
[7] Vijay, K., M. Murmu, S. V. Deo, Bacteria based self-healing concrete–a review, *Constr. Build. Mater.* 152 (2017) 1008–1014.
[8] Huang, H., G. Ye, D. Damidot, Characterization and quantification of self-healing behaviors of microcracks due to further hydration in cement paste, *Cem. Concr. Res.* 52 (2013) 71–81.
[9] Jacobsen, S., E. J. Sellevold, Self-healing of high strength concrete after deterioration by freeze/thaw, *Cem. Concr. Res.* 26 (1996) 55–62.
[10] Li, V. C., Y. M. Lim, Y.-W. Chan, Feasibility study of a passive smart self-healing cementitious composite, *Compos. B Eng.* 29 (1998) 819–827.
[11] Li, V. C., E.-H. Yang, *Self-Healing in Concrete Materials, Self-Healing Materials*, Springer, 2007.
[12] Luhar, S., S. Gourav, A review paper on self-healing concrete, *J. Civ. Eng. Res.* 5 (3) (2015) 53–58.
[13] Rauf, M., W. Khaliq, R. Arsalan, I. Ahmed, Comparative performance of different bacteria immobilized in natural fibers for self-healing in concrete, *Construction and Building Materials* 258 (2020) 119578–119588, at 119578. https://doi.org/10.1016/j.conbuildmat.2020.119578
[14] Depaa, R. A. B., T. F. Kala, An experimental study on flyash as self-healing material, *Int. J. Appl. Eng. Res.* 13 (8) (2018) 5920–5925.
[15] Roig-Flores, M., P. Serna, Concrete early-age crack closing by autogenous healing. *Sustainability* 12 (11) (2020) 4476. https://doi.org/10.3390/su12114476
[16] Ratnayake, K. C., S. M. A. Nanayakkara, Effect of fly ash on self-healing of cracks in concrete, *Mercon 2018–4th International Multidisciplinary Moratuwa Engineering Research Conference* (2018) 264–269. doi: 10.1109/MERCon.2018.8421952
[17] Tomczak, K., J. Jakubowski, Ł. Kotwica, Enhanced autogenous self-healing of cement-based composites with mechanically activated fluidized-bed combustion fly ash, *Constr. Build. Mater.* 300 (2021) 124028. doi: 10.1016/j.conbuildmat.2021.124028
[18] Jonkers, H. M., H. E. J. G. Schlangen, Crack repair by concrete-immobilized bacteria, in: *Proceedings of the first international conference on self-healing materials*, 2007, pp. 18–20, *Noordwijk aan Zee, The Netherlands.*
[19] Shaheen, N., R. A. Khushnood, Bio immobilized limestone powder for autonomous healing of cementitious systems: a feasibility study, *Adv. Mater. Sci. Eng.* (2018) 7049121. doi: 10.1155/2018/7049121
[20] Jonkers, H. M., A. Thijssen, G. Muyzer, O. Copuroglu, E. Schlangen, Application of bacteria as self–healing agent for the development of sustainable concrete, *Ecol. Eng.* 36 (2010) 230–235.
[21] Tawfik, G. M., et al., A step by step guide for conducting a systematic review and meta-analysis with simulation data. *Trop. Med. Health* 47 (2019) 46.
[22] Whittemore, R., et al., Methods for knowledge synthesis: An overview. *Heart Lung.* 43 (5) (2014) 453–461.
[23] Nosouhian, F., D. Mostofinejad, H. Hasheminejad, Influence of biodeposition treatment on concrete durability in a sulphate environment. *Biosyst. Eng.* 133. (2015) 141–152. 10.1016/j.biosystemseng.2015.03.008
[24] Pacheco-Torgal, F., J. A. Labrincha, 21–Biotechconcrete: An innovative approach for concrete with enhanced durability, In *Woodhead Publishing Series in Civil*

and Structural Engineering, Eco-Efficient Concrete, Woodhead Publishing, 2013, pp. 565–576, ISBN 9780857094247, https://doi.org/10.1533/9780857098993.4
[25] Bang, S . S., K. Johnna, V. Galinat, Ramakrishnan, Calcite precipitation induced by polyurethane-immobilized Bacillus pasteurii, *Enzyme Microb.* 28 (4–5) (2001) 404–409. ISSN 0141-0229, https://doi.org/10.1016/S0141-0229(00)00348-3
[26] Sarkar, M., N. Alam, B. Chaudhuri, B. Chattopadhyay, S. Mandal, Development of improved E. coli bacterial strain for green and sustainable concrete technology, *RSC Adv.* 5 (2015) 32175–32182. doi: 10.1039/C5RA02979A
[27] Žáková, H., J. Pazderka, Z. Rácová, P. Ryparová, Effect of bacteria bacillus pseudofirmus and fungus trichoderma reesei on self-healing ability of concrete. *Acta Polytechnica CTU Proceedings* 21 (2019) 42–45. doi: 10.14311/APP.2019.21.0042
[28] Ziviloglou, E. et al., Bio-based self-healing concrete: From research to field application. In: Hager, M., van der Zwaag, S., Schubert, U. (eds) *Self-healing Materials. Advances in Polymer Science*, vol. 273. Springer, Cham, 2016. https://doi.org/10.1007/12_2015_332.
[29] Yoneyama, A., H. Choi, M. Inoue, J. Kim, M. Lim, Y. Sudoh, Effect of a nitrite/nitrate-based accelerator on the strength development and hydrate formation in cold-weather cementitious materials, *Mater (Basel)* 14 (4) (2021) 1–14. doi: 10.3390/ma14041006
[30] Basaran Bundur, Z., M. J. Kirisits, R. D. Ferron, Biomineralized cement-based materials: Impact of inoculating vegetative bacterial cells on hydration and strength, *Cem. Concr. Res.* 67 (2015) 237–245.
[31] Algaifi, H. A., S. A. Bakar, A. R. M. Sam, M. Ismail, A. R. Z. Abidin, S. Shahir, W. A. H. Altowayti, Insight into the role of microbial calcium carbonate and the factors involved in self-healing concrete, *Constr. Build. Mater.* 254 (2020) 119258.
[32] Jogi, P. K., T. V. Lakshmi, Self-healing concrete based on different bacteria: A review, *Mater. Today Proc.* 43 (2020) 1246–1252.
[33] Wang, J.-Y., N. De Belie, W. Verstraete, Diatomaceous earth as a protective vehicle for bacteria applied for self-healing concrete, *J. Ind. Microbiol. Biotechnol.* 39 (2012) 567–577.
[34] Kashif Ur Rehman, S., S. Kumarova, S. Ali Memon, M. F. Javed, M. Jameel, A review of microscale, rheological, mechanical, thermoelectrical and piezoresistive properties of graphene-based cement composite, *Nanomaterials* 10 (2020) 2076.
[35] Qian, C. X., M. Luo, L. F. Ren, R. X. Wang, R. Y. Li, Q. F. Pan, H. C. Chen, Self-healing and repairing concrete cracks based on bio-mineralization, 15 (21): 7796. *Key Engineering Materials; Trans Tech Publications*; pp. 494–503.
[36] Rao, M., V. S. Reddy, M. Hafsa, P. Veena, P. Anusha, Bioengineered concrete—a sustainable self-healing construction material, *Res. J. Eng. Sci.* 2 (2013) 45–51.
[37] Metwally, G. A., M. Mahdy, A. E.-R. H. El Abd, Performance of bio concrete by using bacillus pasteurii bacteria, *Civ. Eng. J.* 6 (2020) 1443–1456.
[38] Xu, J., X. Wang, J. Zuo, X. Liu, Self-healing of concrete cracks by ceramsite-loaded microorganisms, *Adv. Mater. Sci. Eng.* (2018) 5153041.
[39] Zheng, T., Y. Su, X. Zhang, H. Zhou, C. Qian, Effect and mechanism of encapsulation-based spores on self-healing concrete at different curing ages. *ACS Appl. Mater. Interfaces* 12 (2020) 52415–52432.
[40] Luo, M., C. X. Qian, R. Y. Li, Factors affecting crack repairing capacity of bacteria based self–healing concrete, *Constr. Build. Mater.* 87 (2015) 1–7.
[41] Wang, J., D. Snoeck, S. Van Vlierberghe, W. H. Verstraete, N. De Belie, Application of hydrogel encapsulated carbonate precipitating bacteria for approaching a realistic self-healing in concrete. *Constr. Build. Mater.* 68 (2014) 110–119.

[42] Chao Liu, Xiaoyu Xu, Zhenyuan Lv, Lu Xing, Self-healing of concrete cracks by immobilizing microorganisms in recycled aggregate, *J. Adv. Concr. Technol.* 18 (April 2020) 168–178.
[43] Njau, M. W., J. Mwero, Z. Abiero-Gariy, V. Matiru, Effect of temperature on the self-healing efficiency of bacteria and on that of fly ash in concrete, *Int. J. Latest Trends Eng. Technol.* 70 (4) (April 2022), 174–187. ISSN: 2231–5381. https://doi.org/10.14445/22315381/IJETT-V70I4P215
[44] Lakshmi L., C. M. Meera, C. Eldhose, Durability and self-healing behavior of bacterial impregnated concrete, *Int. J. Innov. Res. Technol. Sci. Eng.* 5 (8) (August 2016) 2319–8753.
[45] Wang J. Y., N. De Belie, W. Verstraete, Diatomaceous earth as a protective vehicle for bacteria applied for self-healing concrete, *J. Ind. Microbiol. Biotechnol.* 39 (2012) 567–577.
[46] Wang J., J. Dewanckele, V. Cnudde, et al., X-ray computed tomography proof of bacterial based self–healing in concrete, *Cem. Concr. Compos.* 53 (2014) 289–304.
[47] Islam, Md. M., N. Hoque, M. Islam, I. Ibney Gias, An experimental study on the strength and crack healing performance of e. coli bacteria-induced microbial concrete, *Hindawi Adv. Civ. Eng.* 2022 (3060230)13. https://doi.org/10.1155/2022/3060230
[48] Vijay, K., M. Murumu, Effect of calcium lactate on compressive strength and self–healing of cracks in microbial, *Concr. Strut. Civil. Eng.* (2019) 13 (3): 515–525.
[49] Sundharam, M., R. Jeevakkumar, K. Shankar, An experimental study on performance of bacteria in concrete, *IJIRCST* 2 (6) (November 2014): 119578. ISSN: 2347-5552.

3 Sustainable Utilization of Construction and Demolition Waste in Geotechnical Engineering
A State-of-the-Art Review

Anasua GuhaRay and Meenu Krishnan

3.1 INTRODUCTION

According to a report by the UN, India will have 675 million urban residents in 2035, ranking second only to China's one billion. According to a recent World Bank assessment, India will need to invest $840 billion, or $55 billion annually, in urban infrastructure for the next 15 years to adequately satisfy the demands of its rapidly growing urban population. This rapid urbanization and industrialization are putting pressure on the building and infrastructure sectors. According to the Building Materials and Technology Promotion Council (BMTPC), there will be a demand for 380 million tonnes of cement, 50 million tonnes of steel, 600 billion bricks, 400 million cubic metres of aggregate, and 40 million cubic metres of wood in the years 2021–2022. Thus, India suffers from a severe lack of conventional building materials. Redevelopment of existing infrastructures and construction of new ones to meet the growing demand will produce significant construction and demolition waste (CDW).

Cities produce roughly 1.3 billion tonnes of solid trash annually on a global scale. In a 2012 report, the World Bank predicted that by 2025, this volume would reach 2.2 billion tonnes. About half of all the solid waste produced worldwide is related to the construction industry. Construction and demolition waste refers to trash made of building debris and rubble left over after constructing, remodelling, repairing, or demolishing any type of civil structure. The massive debris accompanying calamities such as earthquakes and floods is added to this. Table 3.1 shows the quantity of CDW formed in different countries annually.

(1: (Duan et al. 2019) 2: (Villoria Sáez & Osmani 2019), 3: (Jain et al. 2021), 4: United States Environment Protection Agency 2018, 5: (Bergsdal et al. 2007)

 DOI: 10.1201/9781003364467-3

TABLE 3.1
Amount of CDW Produced Globally

Country	CDW (Million MT/year)
China	2500[1]
Europe	850[2]
India	431[3]
USA	600[4]
Japan	77[1]
Norway	1.25[5]

Until recently, disposing of C&D trash in landfills was the most affordable and practical option. Landfill exhaustion, a lack of land for future landfill construction, rising landfill costs, strict environmental legislation, opposition from the public, etc., have prompted administrators to adopt the reuse and recycling of CDW trash. For all parties involved, proper CDW management and processing would result in a win-win situation. In addition to making the scarce building construction materials available, it would prevent indiscriminate dumping, air and land pollution. Additionally, it would ease the strain on the environment caused by the exploitation of natural resources for use in road and building construction.

This chapter focuses on the types of CDW available and the physical and mechanical properties that determine its applications. It also discusses the geotechnical applications of recycled CDW to enhance the properties of soil under different conditions.

3.2 DIFFERENT TYPES OF CDW

Construction debris produced during building demolition and renovation includes elements including concrete, wood, brick, and metal. CDW also includes any materials that a natural disaster, such as an earthquake or flood, may generate abruptly. Environmental issues arise from the demolition of concrete structures and the subsequent disposal of trash. So, it is imperative to use innovative methods whenever possible to recycle concrete rather than producing fresh concrete. Figure 3.1 shows the typical composition of CDW.

A general classification of CDW based on the source of origin is shown in Figure 3.2.

Over half of all CDW materials produced by the demolition activities include concrete, brick, and asphalt. Different types of CDW available in the field are discussed below.

3.2.1 Recycled Concrete Aggregate

Recycled concrete aggregate (RCA) is obtained by crushing and screening concrete debris to remove impurities such as gypsum, paper, wood, and reinforcement (Wagih et al. 2013). Figure 3.3 shows typical RCA available in the market. RCA can be

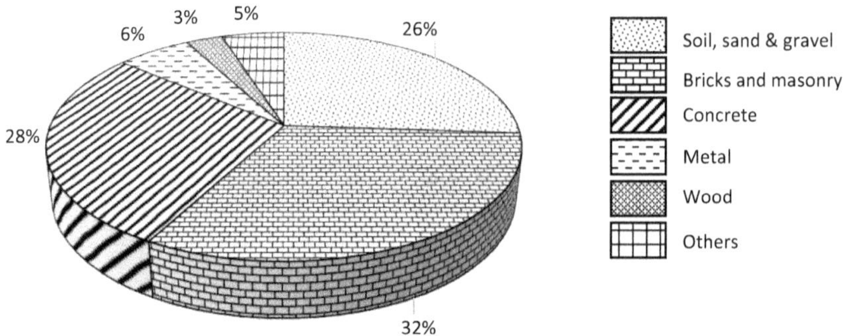

FIGURE 3.1 Composition of CDW (BMTPC).

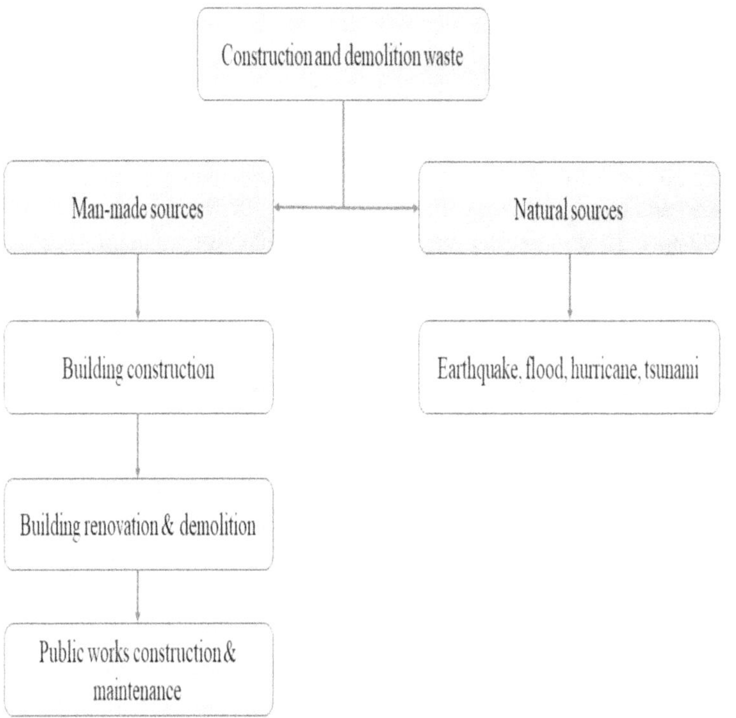

FIGURE 3.2 Classification of CDW as per its source of origin.

Source: Menegaki & Damigos 2018.

effectively utilized as an alternative to natural aggregates. The ageing and strength qualities of the concrete used to make RCA determine their characteristics, and the type of crusher controls the particle size distribution (Arulrajah et al. 2012). It has the highest potential for use in civil engineering applications, owing to its widespread

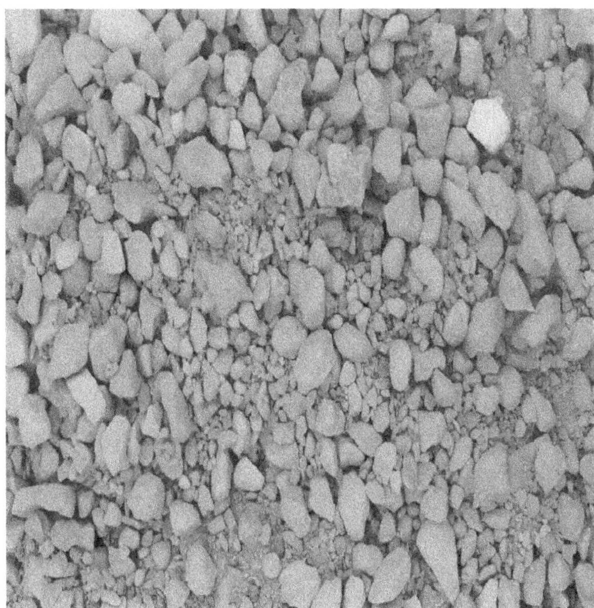

FIGURE 3.3 Recycled concrete aggregate.

Source: Bai et al. 2020.

availability, particularly in urban or industrialized locations (Serridge 2005). RCA can be used for various applications, including embankments, pavement, highways, and bridges, if it is crushed to both coarse and fine aggregates ranging from 40mm to <5mm meeting the gradation requirements of ASTM D2940 (2009) and ASTM D448 (2008) (Poon & Chan 2006).

The Los Angeles abrasion test was used to measure the percentage loss of RCA, and it was found to fall within the permissible range of 50% for structural applications. The percentage loss of recycled aggregates produced from concrete with a minimum strength of 30 MPa is within the authorized range of 12%, according to the results of the aggregate soundness test, which entailed five cycles in a saturated sodium sulphate solution (Tabsh & Abdelfatah 2009). Although the gradation of RCA and natural aggregate particles is comparable, RCA particles have a more rounded shape. The compressive strength of concrete is reduced when natural aggregate is replaced with RCA, but the splitting tensile strength is comparable. Even though RCA can be an inferior aggregate and negatively affect the material properties of concrete, large-scale testing shows that RCA can still be used to produce concrete when considering an entire structural part (McNeil & Kang 2013). Minimizing the negative impacts of RCA on concrete performance is one way to encourage and promote the use of RCA for structural concrete applications. It would be ideal for modelling an acceptable recycling process in the production of RCAs with improved characteristics. Previous research showed that the porosity of cement mortar attached on the surface of RCA resulted in a relative density that was 7 to 9% lower and a water absorption rate that was twice that of natural aggregates (Limbachiya et al. 2000). With acid treatment, RCA's characteristics including density, water absorption, and mechanical strength can be enhanced. The acid treatments can improve the physical qualities of RCA by successfully removing a sizable amount of loose materials and weak cement mortar from the surface. Immersing RCA in acids of low concentration were found to be safe and non-detrimental to RCA particles (Ismail & Ramli 2013).

3.2.2 Crushed Brick

In civil engineering applications, recycled crushed brick (CB) is a practical substitute for natural construction materials. Crushed brick will function effectively at low moisture levels since it has a stronger propensity to absorb moisture. If it is utilized in locations with high moisture content, such as pavement applications, it should be combined with other recycled materials. Figure 3.4 shows the typical construction bricks and recycled crushed bricks available. It has been claimed that the mechanical and physical qualities of recycled concrete aggregate and crushed rock are barely impacted by the addition of crushed brick. So, it was established that the mixtures of crushed rock, recycled concrete aggregate, and crushed brick could successfully satisfy the demands of the state road authority. Crushed brick can safely make up to 25% of recycled concrete aggregate and crushed rock mixtures used in pavement sub-base applications (Arulrajah et al. 2012).

The aggregate impact value (AIV) and aggregate crushing value (ACV) of crushed brick were found to be more than the maximum acceptable value of 30% recommended by the Building Research Establishment (BRE) (BRE 2000). Therefore, it is not ideal to use this alone as a filler material for stone columns in soft soil. Crushed bricks have higher water absorption and lower density compared to natural aggregates. It used to show complete disintegration during the soundness test. However, compared to a subbase using recycled concrete aggregate as the fine aggregate, the subbase using crushed clay brick as the fine aggregate was less sensitive to changes in moisture (Poon & Chan 2006). According to studies on the use of crushed brick in concrete building, it can replace up to 50% of coarse and fine materials. These concrete

FIGURE 3.4 Original waste clay brick (a) & fine aggregates formed by crushing the brick (b). **Source: Huang et al. 2021.**

mixes exhibited comparable mechanical properties to concrete prepared with natural aggregates (Debieb & Kenai 2008). When reinforcing steel is absent, it has been found that crushed bricks can replace natural aggregates without significantly reducing the durability of concrete. Crushed brick should not be used in place of natural aggregates in concrete that has steel reinforcement. This is explained by the fact that the corrosion of the reinforcing steel bars began earlier in brick-based samples than it did in samples with natural aggregates (Adamson et al. 2015).

3.2.3 Crushed Marble

Demolition of existing structures produces some amount of marble chips as shown in Figure 3.5, that have the potential to be used in several civil engineering applications. Concrete made by replacing 100% natural coarse aggregate using recycled aggregate from marble waste showed only a 10% reduction in compressive strength. Mechanical and durability-related performance of concrete made by replacing fine aggregate with crushed marble showed a loss of compressive strength between 10% and 20%. Replacing 100% natural aggregate with marble aggregate in concrete mixes showed an improvement in workability because of the low water absorption and flat smooth surface of the latter (Martins et al. 2014).

The Los Angeles Abrasion test results on marble aggregates were within the BIS 2386 (part IV) – 1963 tolerance limits. Thus, concrete pavement projects can employ marble aggregates (Kore & Vyas 2016). It was reported that marble waste's aggregate impact value (AIV) was well below 35%, indicating that it is suitable for pavement applications. Studies on its use in chemically aggressive environments showed that the AIV of marble aggregate immersed in acid is less than that of virgin samples because of the marble and salt composite formation on its surface. The shear strength of soil stabilized with marble waste showed a high value for angle of internal friction than that of virgin soil. However, exposure to H_2SO_4 leads to a significant increase in the angle of internal friction due to the formation of $CaSO_4$ salts and salt-marble composite. These greater shear values do not apply to marble aggregate alone but to a mixture of salt and marble aggregate. Therefore, it is not suggested to employ the same in a chemically aggressive environment for ground improvement (Surana et al. 2018).

FIGURE 3.5 Coarse aggregate from marble waste.

3.2.4 Recycled Glass

Ignorance of its geotechnical characteristics and associated environmental risks are the main barriers to recycled glass (RG) sustainable usage in roads and backfills. As the chemical composition and angular particle shape is comparable for both crushed waste glass (CWG) and natural sand (NS), they showed similar behaviour in geotechnical characterization tests with superior permeability and abrasion resistance (Kazmi et al. 2021). Figure 3.6 shows the crushed waste glass available. CWG is more stable under the saturated condition as shown by its higher friction angle under the soaked condition. Fine recycled glass (FRG) of maximum particle size 4.75 mm is more similar to naturally occurring sand. It was observed that FRG shows excellent drainage characteristics and comparable shear strength parameters as to natural aggregates as observed by direct and consolidated drained triaxial tests (Disfani et al. 2011).

The Los Angeles abrasion test demonstrates the endurance of recycled glass in geotechnical applications, although the repeated load triaxial test (RLT) and California bearing ratio (CBR) standards are not met. However, it can be made better using additives or mixed with aggregates to allow for its usage in applications for pavement subbase (Arulrajah et al. 2013). It was reported that recycled glass could be used in the concrete mix as recycled glass sand (RGS) without significantly affecting its compressive strength. However, alkali-silica reaction (ASR) will be a major obstacle in using RGS in concrete as it will produce alkali-silica gel that can absorb water and swell at a later stage. This will eventually cause the formation of cracks in concrete made of RGS as fine aggregate without any safety measures to reduce this risk (Taha & Nounu 2009). Concrete that can be placed and compacted by its own weight with little to no vibration effort is known as self-compacting concrete (SCC). Even though SCC made of river sand and recycled glass (0%, 15%, 30% & 45%) as fine aggregate showed an increase in slump flow and air content and a decrease in compressive strength, static modulus of elasticity and shrinkage, the overall evaluation of both the fresh and hardened characteristics showed that it

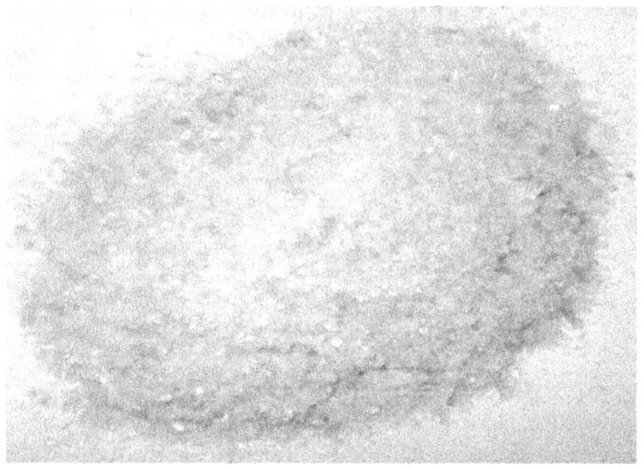

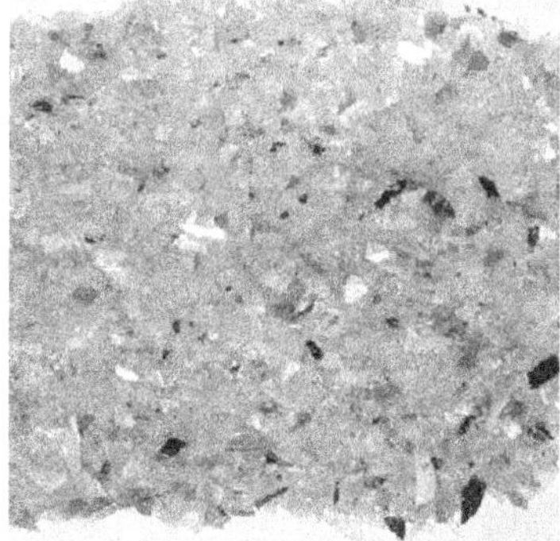

FIGURE 3.6 Recycled glass aggregate.

Source: Mohajerani et al. 2017.

is feasible to make SCC from recycled glass cullet. Also, by substituting 33% of the cement's weight with fly ash, the ASR growth of each specimen was dramatically reduced (Kou & Poon 2009).

3.2.5 Railway Track Ballast

Spent Railway Ballast is a high-quality material (usually granite or limestone) excavated from beneath train tracks when excessive fines develop as a result of repetitive railway carriage loads. Ballast degradation is typically seen under cyclic loads, which affects the track's drainage and bearing capacity. Periodic maintenance is required to maintain stability and safety, including cleaning and replacement of ballast, both of which generate a significant amount of waste. Ballast should be finer than 63 mm, with no more than 0.8% finer than 1.18 mm, and the majority of the material should fall between 50 and 28 mm, according to Railtrack (Network Rail) Line Specification RT/CE/S/00642. The stone has to be clean, robust, and durable, as well as angular in shape and dust-free. Due to these characteristics, ballast is particularly well suited for higher-value applications like stone column construction.

Recycled ballast can be mixed with fresh ballast for further application in laying rail tracks. The interlock between recycled and fresh ballast is weaker due to the loss of angularity and reduction in surface texture in used ballast. This will result in a reduction in shear strength and friction angle. However, it was observed that the reduction in shear stress is minimal when the percentage of recycled ballast is less than 30% (Jia et al. 2019).

3.2.6 Reclaimed Asphalt Pavement

Reclaimed asphalt pavement (RAP) are materials obtained by recycling asphalt and aggregates of existing pavements. RAP was found to be a well-graded material with lower optimum moisture content (OMC) than that of natural aggregates used for pavement applications. It is also observed that the maximum dry density (MDD) results from the compaction study are comparable with that of conventional aggregates (Sayed et al. 1994). Figure 3.7 illustrates typical recycled asphalt pavement aggregate. When building roads, landfill capping systems, retaining walls, drainage systems, and slope protections, reclaimed asphalt pavement can be utilized in place of natural aggregates.

The synthesis of bituminous binder from crude petroleum and the subsequent manufacturing of the bituminous mix at hot mix plants are energy-intensive steps in the traditional approach of adding bituminous surface to flexible pavements. Reclaimed asphalt pavement materials are periodically combined with new materials and a recycling agent (RA) to make hot mix asphalt mixtures, which is a process known as hot mix recycling. Recycled mixtures can perform as least as well as traditional hot mix asphalt mixtures when properly developed. Recycling agents (RA) are organic substances with physical and chemical properties chosen to restore the desired specifications to aged asphalt. In general, high-viscosity, old asphalts can be restored using lower-viscosity RA kinds, and vice versa (Pradyumna et al. 2013).

FIGURE 3.7 Reclaimed asphalt pavement aggregate.

Source: Debbarma et al. 2019.

IRC prohibits the use of California's 100% RAP bearing ratio (CBR) as the base for flexible pavement. RAP is ideal for use as the subbase or base of flexible pavement since its soaking CBR value improves from 20% to well over 100% when mixed in various ratios with crushed stone aggregates and stabilized with tiny amounts of cement (Saha & Mandal 2017).

3.3 CURRENT SCENARIO OF CDW IN INDIA

By 2 October 2019, all solid waste generated in cities and towns, including CDW rubbish, must be processed. This is one of the key objectives of the Swachh Bharat Mission. The Ministry of Urban Development (MoUD) instructed all states to establish CDW recycling facilities in all cities with a population of more than 1 million people in a circular dated 28 June 2012 (Central Pollution Control Board (CPCB) 2017). In order to support this endeavour, C&D trash reuse and recycling initiatives may also be implemented in cities with a population under one million. This is due to the fact that just 23% of urban residents live in megacities and cities with a population of one million or more (Census 2011).

Prior to 2010, municipal solid waste (MSW) included CDW as a component. Therefore, it wasn't adequately addressed. Once CDW and MSD are combined, separating the two is challenging, reducing the material's potential for recycling. Additionally, this causes issues in MSW processing facilities. The MoEF-created working committee on MSW management indicated that it is vital to produce data on CDW in 2010. Following the publication of C&D Waste Management rules, 2016 by

TABLE 3.2
C&D Generation in India: Divergent Estimates

Organization	Year	CDW (Million MT/year)
Ministry of Urban Development (MoUD)	2000	10–15
Technology Information, Forecasting & Assessment Council (TIFAC)	2001	12–15
Ministry of Environment, Forest and Climate Change (MoEF&CC)	2010	10–12
Building Materials & Technology Promotion Council (BMTPC)	2013	165–175
Centre for Science & Environment	2014	530
Central Pollution Control Board (CPCB)	2017	12

TABLE 3.3
CDW Recycling Plants Available in India

Recycling plant location	Year of establishment	Recycling capacity (Tonnes/day)
Burari	2009	2000
Ahmedabad	2014	1100
East Kidwai Nagar, New Delhi	2014	150
Shastri Park, New Delhi	2015	500
Surat	2019	300
Mulund, Mumbai	2019	600
Jeedimetla, Hyderabad	2020	500
Fathullaguda, Hyderabad	2021	500

the Ministry of Environment, Forest and Climate Change (MoEF&CC), the actions related to CDW management, processing, and re-use have become more active. The responsibilities of the many parties involved are clearly outlined in this, including those of the waste generator, service provider, local authority, State Pollution Control Board (SPCB), Central Pollution Control Board (CPCB), and State, and Central governments. The standards for choosing storage locations and recycling facilities are laid down in the regulations. It also discussed the duration for developing and enacting rules dependent based on a city's population. In determining the amount of CDW generation, it was recognized that there are uncertainties. This can be ascribed to a variety of things, such as various estimation techniques for the amount of CDW trash produced, variable rates of urban expansion, and the requirement to rebuild cities owing to fast urbanization. Table 3.2 shows the variability in the estimate of the quantity of CDW generation according to various agencies.

Table 3.3 shows the details of working recycling plants available in India. The figure shows that recycling plants have insufficient capacity to manage the CDW generated.

TABLE 3.4
IS: 383-2016 – Indian Standard for Coarse and Fine Aggregate for Concrete

CDW BIS IS: 383	Plain Concrete	Reinforced Concrete	Lean Concrete (<M15 Grade)	Extent of Utilization
Recycled Concrete Aggregate (RCA)	25%	20% (Only up to M20 grade)	100%	Coarse aggregate
Recycled aggregate (RA)	Nil	Nil	100%	Coarse aggregate
Recycled Concrete Aggregate (RCA)	25%	20% (Only up to M20 grade)	100%	Fine aggregate

The following estimates were created by the Technology Information, Forecasting and Assessment Council (TIFAC), which acknowledges that the generation of C&D waste is project-specific (Central Pollution Control Board (CPCB 2017):

- New construction: 40–60 kg/m^2
- Repairing of buildings: 40–50 kg/m^2
- Demolition of buildings: 300–500 kg/m^2

The creation of codes of conduct and standards for the use of recycled materials and byproducts of construction and demolition waste in relation to construction operations was stated to fall under the purview of the Bureau of Indian Standards (BIS) and the Indian Roads Congress (IRS) (Central Pollution Control Board (CPCB 2017). In 2016, the standard for coarse and fine aggregates that can be used in concrete, IS: 383, underwent its third amendment, allowing the use of recycled aggregates up to 25% in plain concrete, 20% in reinforced concrete of grade M 25 or below, and up to 100% in lean concretes of grade M 15 or lower (BIS: 383, 2016). Recycled aggregate (RA) contains concrete, brick, tiles, etc., whereas recycled concrete aggregate (RCA) contains more than 90% of concrete debris (BIS: 383, 2016). Table 3.4 shows the revised portion of IS:383-2016.

The Central Public Works Department (CPWD) and National Building Construction Company (NBCC) have advised using recycled CDW in their construction activities in line with IS:383-2016, or if the same is accessible within 100 kilometres of the construction site. Delhi is at the forefront of recycling CDW and converting it into usable construction products. Municipal corporation of Delhi (MCD) and IL&FS Environmental Infrastructure & Services Ltd (IEISL) set up a pilot project in Burari to show the possibilities of a scientifically regulated process of recycling CDW. Karolbagh, Sadar-Paharganj, and the City are the three designated zones of Delhi from which IEISL collects 500 tonnes of CDW every day (TPD). The C&D trash is then recycled into aggregates at the waste management facility, which are then used to make Ready Mix Concrete (RMC), pavement blocks, kerb stones, and concrete bricks (Centre for Science and Environment, CSE).

3.4 APPLICATIONS OF CDW IN GEOTECHNICAL ENGINEERING

3.4.1 Filler Materials for Stone Column

The increasing value of land and scarcity of suitable sites for construction force engineers to build many structures in problematic soft soil. Stone columns can enhance soft soil properties by reducing the settlement and increasing the bearing capacity. They primarily function as rigid inclusions that are stiffer, having better shear strength, and permeable than the surrounding soil thus accelerating the rate of consolidation of soft clays by providing drainage paths. A material that can withstand the impact forces of a vibrating poker and preserve structural integrity over time under the applied foundation loads should be used to make stone columns. Figure 3.8 shows a schematic diagram of stone column installation.

Generally, materials used as filler for stone column construction are stone aggregates between 20 to 75 mm, gravel and sand. However, recycled CDW such as crushed concrete and brick can substitute the depleting natural aggregates. Crushed concrete has the highest potential for use in stone columns of all recycled aggregates, owing to its widespread availability, particularly in urban or industrialized locations.

It was reported that Aggregate Impact value (AIV) and Aggregate Crushing Value (ACV) of crushed concrete are below the limit of 30% recommended by BRE (2000) standard for the acceptance of any material to be used in stone columns (Shahverdi & Haddad 2020). Testing for slake durability on concrete debris revealed acceptable durability, even after two months of soaking. 3D numerical studies on CDW-Geosynthetic encased stone columns (CDW-GECs) observed a comparable load bearing capacity and settlement as that of stone columns made of natural aggregates. They discovered a stress concentration ratio of 3, meaning that the increased stiffness of the reinforced soil caused the total vertical stresses applied to the CDW-GECs to be three times greater than the total vertical stresses applied to the surrounding soil. (Anita et al. 2023). Load tests on clayey bed reinforced with stone columns made of natural aggregates and concrete debris observed that concrete debris provided more than 70%

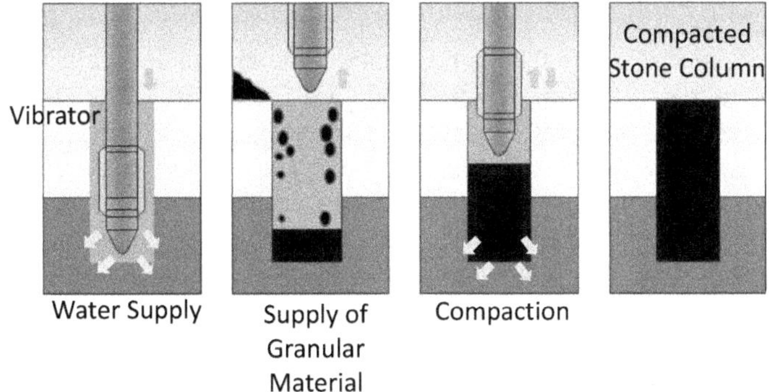

FIGURE 3.8 Stone column installation by vibro-replacement.

Source: Dheerendra Babu et al. 2013.

of the aggregate's maximum load carrying capacity. However, crushed bricks have very low compressive strength and poor durability (Sanjeewa & Nawagamuwa 2019). Particle size distribution of building debris predominantly consisted of crushed brick before and after a direct shear test revealed a sizable amount of particle breaking due to shearing (Mckelvey et al. 2002). Due to their extreme fragility, strict monitoring is necessary to assure that older bricks must not be present and must have a fine content below 5% (Serridge 2005).

Recycled railway track ballast is another aggregate that can be effectively utilized as a filler material because of its high strength and friction properties. However, spent railway track ballast usually contains hydrocarbon contaminants which will affect the performance of stone columns. Therefore, a suitable grading for stone columns should be met by solvent cleaning and screening before it is utilized in the field (Tranter et al. 2008).

3.4.2 Pavement Construction

Recycled construction and demolition waste can be employed in constructing sub-base, base and asphalt layers of pavements. Recycled concrete aggregates had a lower California Bearing Ratio (CBR), a higher Maximum Dry Density (MDD), and a higher Optimum Moisture Content (OMC) when compared to natural materials. As a result, it is more difficult to compact recycled CDW aggregates in the field since more water is needed (Poon & Chan 2006). However, CDW have a higher angle of internal friction (~40^0) that improve shear resistance and impart higher values of resilient modulus. It was claimed that a mixture of 75% natural aggregate and 25% recycled concrete aggregate produced the same resilient reaction and permanent deformation as a dense-graded aggregate base course. (Bennert et al. 2000).

The physical properties of recycled CDW may change depending on the compactive effort. The compaction process has changed the particle size distribution and increased the percentage of grains in recycled CDW by crushing and breaking the material. This physical alteration helps the aggregate become denser, which enhances its bearing capacity, resilience modulus, and resistance to permanent deformation (Leite et al. 2011). Although the abrasion resistance of concrete mixes containing recycled aggregate was lower than that of mixes containing natural aggregate, the properties of these mixes' abrasion resistance were nevertheless suitable for paving concrete (Kumar 2017). Studies on the feasibility of using recycled glass in base and subbase layers using mixtures of recycled concrete aggregates (RCA), 20% crushed brick (CB) and 10% to 40% RG stabilized with 3% cement showed a satisfactory result in accordance with the local road authority's sub-base specifications in grading, CBR, and major geotechnical requirements. The unconfined compressive strength, resilient modulus and permanent deformation properties support the use of 15% RG, 20% CB, and RCA as supplemental materials in cement-stabilized pavements with moduli between 500 MPa and 3500 MPa (Senanayake et al. 2022).

As per the guidelines stipulated by Indian Roads Congress (IRC) 121: 2017, to adhere to the water-bound macadam (WBM) gradations, recycled concrete aggregate (RCA) should be crushed and compacted after a mild sprinkling of water with tandem or vibrating rollers if it is used as a granular sub-base layer. Each compacted layer must be less than 75 mm thick (Congress 2017). RCA can be used as a component for the construction of granular base courses if it satisfies aggregate requirements as per Ministry of Road Transport and Highways (MORTH) Specifications in cases where design traffic is more than 2 MSA (million standard axles) or Ministry of Rural Development (MORD) Specifications in cases where design traffic is less than 2 MSA. For grades up to M40 for the construction of concrete pavement, up to 30% of natural mineral aggrègates can be substituted with RCA without affecting the strength properties of the concrete.

3.4.3 Embankment Construction

A huge volume of soil will be required for use as a filler in the construction of embankments. A substantial volume of CDW wastes will be used in this application if it is qualified as fill material. It was observed that the substitution of embankment clayey soil with recycled CDW reduced the rutting damage due to the highly resilient moduli and lower accumulated permanent deformation of CDW (Zhang et al. 2019). Different combinations of CDW materials contributed differently to the bearing capacity of embankments. At the completion of construction and after a certain amount of time, it was claimed that an embankment field created with 50% reclaimed asphalt pavement (RAP) and 50% concrete as fill material was the stiffest. The CDW materials that were examined revealed that stiffnesses considerably increase over time. This was ascribed to the CDW materials' self-cementing ability, which is its key mechanical advantage over pure natural unbound aggregates (Sangiorgi et al. 2015).

The quality ratio of the fine materials to coarse materials in CDW was satisfactory after rolling, according to the results of laboratory tests and field rolling tests. It shows that following a simple compaction treatment, the content of fine material has significantly increased, and the performance of the road embankment has improved due to reduced porosity and permeability coefficient (Li et al. 2017). It is crucial to comprehend how CDW infill creeps since doing so will enable you to foresee and regulate how the CDW embankment will settle. Under a simulated load, a steady creep was discovered to be the creep type of the CDW samples. The main reasons for the deformation of recycled CDW embankment filler are the fragmentation, slippage, and rearrangement of recycled particles. The rate of slippage and the filling of the particles are both significantly accelerated by water. By appropriately adjusting the filler ratio, we can reduce the embankment's tendency to deformation (Li et al. 2020).

The coefficient of interaction between the fill material and the reinforcements is one of the most important factors to take into account when designing geosynthetic reinforced structures. The maximum direct shear stress achieved with the interface material to the maximum direct shear stress achieved with the fill material, both under the same normal load, can be used to compute the coefficient of interaction. It was reported that the value of this ratio for CDW material/geosynthetic interface compares well with that of soil/geosynthetic interface, making it feasible as a filler material in the geosynthetic reinforced embankment (Vieira & Pereira 2018).

As per the guidelines stipulated by Indian Roads Congress (IRC) 121: 2017, The maximum size of coarse material used in earth embankments should not exceed 75 mm. The maximum coarse size permissible in rockfill embankments is 300 mm. A good earth cover should be constructed to protect the side slopes of CDW-built embankments. To operate as a sealing layer over the CDW embankment, a subgrade made of good earth shall be placed at the top of the embankment. It is not suggested to use CDW for a mechanically stabilized earth wall.

3.4.4 Soil Stabilization

Black cotton soil occupies around 20% of India's land area. It contains montmorillonite clay minerals, which have the propensity to alter volume when moisture is applied to them; as a result, these materials have various construction and performance issues on the site, causing structural deformities. CDW waste can stabilize soil with very high swelling potential and stability problems. It was observed that the plasticity index, cohesion and optimum moisture content of black cotton soil decreased with CDW due to the reduced percentage of silt and clay with the addition of CDW resulting in lesser surface area. However, maximum dry density and angle of internal friction increased with CDW, thus causing an improved bearing capacity of clayey soil (Abhijith et al. 2014). The CDW mixtures' intrinsic aluminosilicate components and calcium silicate hydrate (CSH) species benefit the binder qualities necessary for sustainable improvement in the compressive strength of expansive clay (Bassani & Tefa 2018).

Powdered waste glass can effectively improve the geotechnical characteristics of high-plasticity clay. Experimental results showed that adding glass powder could reduce the plasticity and linear shrinkage and improve the unconfined compressive strength of high-plasticity clay (Ibrahim et al. 2021). Studies on the effect of adding recycled glass powder as a precursor of geopolymers on the improvement in split tensile strength and durability against wet-dry cycles of cement-stabilized soil found that glass powder improves the microstructure of soil-cement-glass powder blend and reduces the porosity due to its pozzolanic activity. This resulted in improved split tensile strength and durability (de Jesús Arrieta Baldovino et al. 2020).

A comparison study of the stabilization of clayey soil using fly ash, CDW and lime showed that the optimum moisture content of soil decreased only with the addition of CDW. Seven-day unconfined compressive strength value was more prominent in the case of lime and CDW. However, lime is costlier compared to CDW and fly ash. Therefore, CDW is cost-effective when early strength gains are crucial (Sharma & Hymavathi 2016). Stabilization of expansive soil with CDW observed an increase in peak deviator stress and reduction in volumetric strain. The stiffness of unstabilized and CDW-stabilized soil was expressed in terms of the secant modulus of the stress-strain curve. An increase in secant modulus with the addition of CDW was observed for a particular confining pressure and strain level (Sharma & Sharma 2019). The presence of fine sand particles in CDW helped to mobilize the angle of shearing resistance, improving the strength of the composite, and thus causes an improvement in the California bearing ratio (CBR) of expansive soil. Due to the coarser CDW particles and its higher permeability as compared to virgin soil, it also aids in raising the coefficient of permeability of clayey soil. Thus, the stabilized soil possesses good drainage properties (Sharma & Sharma 2020).

3.4.5 Construction of Retaining Walls

Any built wall that restricts soil or other materials in sites from sliding with an abrupt elevation change is referred to as a retaining wall. Recycled aggregate (RA) from CDW can substitute a portion of natural aggregate in the structural concrete to construct slabs and walls of retaining structures. In comparison to conventional RCC retaining walls, mechanically stabilized earth retaining (MSE) walls are more frequently employed due to their lower construction and maintenance costs. The interaction of the reinforcing and fill components makes the MSE walls stable. Only crushed concrete waste can be used as a backfill for an MSE retaining wall out of all the CDW debris because of the interaction requirement (Venkatachalam & Balu 2022). Due to the high quantity of fines and scattering of its ingredients, the fine grain fraction of recycled CDW is typically not regarded as suitable for use in concrete or roadway base layers. Recycled materials can be classified as inert backfill and, when appropriately compacted, exhibit direct shear and pull-out behaviour at interfaces with various geosynthetics that are comparable to conventional earth material (Vieira & Pereira 2018).

For reinforced retaining walls, hollow segmental concrete units are utilized in the facing columns, and their cavities are filled with granular infill to improve mechanical interlocking between the courses of the facing units. Studied showed that the interface shear capacity (peak) of blocks filled with recycled concrete aggregate is almost identical to that of blocks filled with natural coarse aggregate (Bhuiyan et al. 2015).

A gabion wall is a retaining wall constructed of wire-tied stacks of gabions filled with stone. The wire's lifespan, not the contents of the basket, determines how long gabions will last. When the wire breaks, the structure will also. The life expectancy of gabions depends on the durability of the wire, not on the contents of the basket. Horizontal and vertical displacements of gabions filled with CDW bring the possibility of using the same for the construction of gabion walls up to a height of 5.5m with a factor of safety of 2 (Paschoalin Filho et al. 2020). It was found that the wrapped-face geogrid reinforced wall constructed using recycled CDW had satisfactory wall deformations, reinforcement durability, and reinforcement strains comparable to the wall performance anticipated of structures made with conventional granular backfills (Santos et al. 2013).

For a reinforced concrete cantilever retaining wall built on a seismic zone, the overturning moment and total active earth pressure under seismic and static loads will decrease with an increase in the angle of internal friction of backfill materials. Although a reduction in the angle of internal friction was observed in CDW compared to natural aggregates, its impact on the performance of the wall was not severely affected to refrain from using the same as backfill material (Saribas & Ok 2019).

3.5 CONCLUSION

Unplanned development to accommodate the need for rapid urbanization will increase the pollution load that will adversely affect the environment. Construction and demolition waste (CDW) and its batching activities are one of the world's major contributors to solid waste. So, it is inevitable to recycle the CDW for sustainable development. Recycling and reusing CDW has the dual benefits of protecting the environment by avoiding waste disposal and cutting down the costs related to the disposal process. By lowering the requirement for extracting raw materials and long-distance transportation of the resources, it reduces the production and emission of greenhouse gases and other pollutants. The need for additional landfills and the associated expenses will decline as a result of recycling and reuse of the same. It creates employment opportunities in the recycling industry.

The chapter draws on the problems associated with the handling of CDW and the prospects of using it wisely in the civil engineering construction sector.

- Different types of CDW such as recycled concrete aggregate, crushed brick, crushed marble, recycled glass, recycled railway track ballast and reclaimed asphalt pavement are potential construction materials in different applications such as pavements, embankments, soil stabilization and replacement of fine aggregate in concrete.
- Existing rules and codes such as the Guidelines on Environmental Management of Construction & Demolition (c & d) Wastes (CPCB 2017) and IS:383-2016 provides guidelines for the recycle and reuse of CDW.
- Superior mechanical properties of recycled concrete aggregate and recycled railway track ballast make it an ideal sustainable substitute for natural aggregate in the construction of stone columns, pavements, and embankments.
- Pozzolanic nature of recycled crushed glass make it a suitable material for the stabilization of soft clay.
- Different types of CDW can be effectively utilize in different geotechnical applications after the thorough study of its mechanical, physical and chemical properties.

ACKNOWLEDGMENT

This work is part of an ongoing project supported by the Science and Engineering Research Board (SERB), Govt. of India for supporting the present study through the Power Grant (Project ID: SPG/2021/001552).

REFERENCES

Abhijith, B. S., Murthy, V. S., & Kavya, S. P. (2014). Study of the effectiveness in improving montmorillonite clay soil by construction and demolition waste. *Journal of Civil Engineering and Environmental Technology*, *1*(5), 1–4. www.krishisanskriti.org/jceet.html

Adamson, M., Razmjoo, A., & Poursaee, A. (2015). Durability of concrete incorporating crushed brick as coarse aggregate. *Construction and Building Materials, 94*, 426–432. https://doi.org/10.1016/j.conbuildmat.2015.07.056

Anita, A., Karthika, S., & Divya, P. V. (2023). Construction and Demolition Waste as Valuable Resources for Geosynthetic-Encased Stone Columns. *Journal of Hazardous, Toxic, and Radioactive Waste*, *27*(2), 1–13. https://doi.org/10.1061/jhtrbp.hzeng-1175

Arulrajah, A., Piratheepan, J., Bo, M. W., & Sivakugan, N. (2012). Geotechnical characteristics of recycled crushed brick blends for pavement sub-base applications. *Canadian Geotechnical Journal*, *49*(7), 796–811. https://doi.org/10.1139/T2012-041

Arulrajah, A., Piratheepan, J., Disfani, M. M., & Bo, M. W. (2013). Geotechnical and geoenvironmental properties of recycled construction and demolition materials in pavement subbase applications. *Journal of Materials in Civil Engineering*, *25*(8), 1077–1088. https://doi.org/10.1061/(asce)mt.1943-5533.0000652

Bai, G., Zhu, C., Liu, C., & Liu, B. (2020). An evaluation of the recycled aggregate characteristics and the recycled aggregate concrete mechanical properties. *Construction and Building Materials, 240*, 117978. https://doi.org/10.1016/j.conbuildmat.2019.117978

Bassani, M., & Tefa, L. (2018). Compaction and freeze-thaw degradation assessment of recycled aggregates from unseparated construction and demolition waste. *Construction and Building Materials, 160*, 180–195. https://doi.org/10.1016/j.conbuildmat.2017.11.052

Bennert, T., Papp, J., Maher, A., & Gucunski, N. (2000). Utilization of construction and demolition debris under traffic-type loading in base and subbase applications. *Transportation Research Record*, *2*(1714), 33–39. https://doi.org/10.3141/1714-05

Bergsdal, H., Bohne, R. A., & Brattebø, H. (2007). Projection of Construction and Demolition Waste in Norway. *Journal of Industrial Ecology,* 11(3), 27–39. https://onlinelibrary.wiley.com/doi/abs/10.1162/jiec.2007.1149

Bhuiyan, M. Z. I., Ali, F. H., & Salman, F. A. (2015). Application of recycled concrete aggregates as alternative granular infills in hollow segmental block systems. *Soils and Foundations*, *55*(2), 296–303. https://doi.org/10.1016/j.sandf.2015.02.006

BRE (2000). Specifying Vibro Stone Columns. *CRC Ltd, London, UK, BR391*. Available at: www.thenbs.com/publicationindex/documents/details?Pub=BRE&DocId=250675

Census 2011. List of urban agglomerations in India. Available at: https://en.wikipedia.org/wiki/List_of_urban_agglomerations_in_India

Central Pollution Control Board (CPCB). (2017). Guidelines on environmental management of C&D waste management in India. *Prepared in Compliance of Rule 10 Sub-Rule 1(a) of C & D Waste Management Rules, 2016*, *1*(February), 1–39.

Debbarma, S., Ransinchung, G. D., & Singh, S. (2019). Feasibility of roller compacted concrete pavement containing different fractions of reclaimed asphalt pavement. *Construction and Building Materials, 199*, 508–525. https://doi.org/10.1016/j.conbuildmat.2018.12.047

Debieb, F., & Kenai, S. (2008). The use of coarse and fine crushed bricks as aggregate in concrete. *Construction and Building Materials*, *22*(5), 886–893. https://doi.org/10.1016/j.conbuildmat.2006.12.013

Dheerendra Babu, M. R., Nayak, S., & Shivashankar, R. (2013). A critical review of construction, analysis and behaviour of stone columns. *Geotechnical and Geological Engineering*, *31*(1), 1–22. https://doi.org/10.1007/s10706-012-9555-9

Disfani, M. M., Arulrajah, A., Younus Ali, M. M., & Bo, M. W. (2011). Fine recycled glass: A sustainable alternative to natural aggregates. *International Journal of Geotechnical Engineering*, *5*(3), 255–266. https://doi.org/10.3328/IJGE.2011.05.03.255-266

de Jesús Arrieta Baldovino, J., dos Santos Izzo, R., Rose, J. L., & Avanci, M. A. (2020). Geopolymers based on recycled glass powder for soil stabilization. *Geotechnical and Geological Engineering, 38*(4), 4013–4031. https://doi.org/10.1007/s10706-020-01274-w

Duan, H., Miller, T. R., Liu, G., & Tam, V. W. Y. (2019). Construction debris becomes growing concerns of growing cities. *Waste Management*, *83*, 1–5. https://doi.org/10.1016/j.wasman.2018.10.044

EPA, Construction and Demolition Debris: Material-Specific Data (2018) Available at: www.epa.gov/facts-and-figures-about-materials-waste-and-recycling/construction-and-demolition-debris-material

Huang, Q., Zhu, X., Xiong, G., Wang, C., Liu, D., & Zhao, L. (2021). Recycling of crushed waste clay brick as aggregates in cement mortars: An approach from macro–and micro-scale investigation. *Construction and Building Materials, 274*, 122068. https://doi.org/10.1016/j.conbuildmat.2020.122068

Ibrahim, H. H., Mawlood, Y. I., & Alshkane, Y. M. (2021). Using waste glass powder for stabilizing high-plasticity clay in Erbil city-Iraq. *International Journal of Geotechnical Engineering*, *15*(4), 496–503. https://doi.org/10.1080/19386362.2019.1647644

IRC:121-2017. Guidelines for use of construction and demolition guidelines for use of construction and demolition. *Indian Road Congress*, *1*(November), 1–28. https://law.resource.org/pub/in/bis/irc/irc.gov.in.121.2017.pdf

IS 383:2016. Coarse and fine aggregate for concrete–specification. *Bureau of Indian Standards (BIS) New Delhi,* 1–21. https://icikbc.org/docs/IS383-2016.pdf

Ismail, S., & Ramli, M. (2013). Engineering properties of treated recycled concrete aggregate (RCA) for structural applications. *Construction and Building Materials, 44*, 464–476. https://doi.org/10.1016/j.conbuildmat.2013.03.014

Jain, S., Singhal, S., & Jain, N. K. (2021). Construction and demolition waste (C&DW) in India: generation rate and implications of C&DW recycling. *International Journal of Construction Management*, *21*(3), 261–270. https://doi.org/10.1080/15623599.2018.1523300

Jia, W., Markine, V., Guo, Y., & Jing, G. (2019). Experimental and numerical investigations on the shear behaviour of recycled railway ballast. *Construction and Building Materials, 217*, 310–320. https://doi.org/10.1016/j.conbuildmat.2019.05.020

Kazmi, D., Serati, M., Williams, D. J., Qasim, S., & Cheng, Y. P. (2021). The potential use of crushed waste glass as a sustainable alternative to natural and manufactured sand in geotechnical applications. *Journal of Cleaner Production, 284*, 124762. https://doi.org/10.1016/j.jclepro.2020.124762

Kore, S. D., & Vyas, A. K. (2016). Impact of marble waste as coarse aggregate on properties of lean cement concrete. *Case Studies in Construction Materials, 4*, 85–92. https://doi.org/10.1016/j.cscm.2016.01.002

Kou, S. C., & Poon, C. S. (2009). Properties of self-compacting concrete prepared with recycled glass aggregate. *Cement and Concrete Composites*, *31*(2), 107–113. https://doi.org/10.1016/j.cemconcomp.2008.12.002

Kumar, R. (2017). Influence of recycled coarse aggregate derived from construction and demolition waste (CDW) on abrasion resistance of pavement concrete. *Construction and Building Materials, 142*, 248–255. https://doi.org/10.1016/j.conbuildmat.2017.03.077

Leite, F. D. C., Motta, R. D. S., Vasconcelos, K. L., & Bernucci, L. (2011). Laboratory evaluation of recycled construction and demolition waste for pavements. *Construction and Building Materials*, *25*(6), 2972–2979. https://doi.org/10.1016/j.conbuildmat.2010.11.105

Li, Y., Zhou, H., Su, L., Hou, H., & Dang, L. (2017). Investigation into the application of construction and demolition waste in urban roads. *Advances in Materials Science and Engineering,* 2017, 1–11. https://doi.org/10.1155/2017/9510212

Li, Z., Yan, S., Liu, L., & Yang, J. (2020). Investigation into creep characteristics and model of recycled construction and demolition waste used in embankment filler. *Sustainability (Switzerland)*, *12*(5), 1–22. https://doi.org/10.3390/su12051924

Limbachiya, M. C., Leelawat, T., & Dhir, R. K. (2000). Use of recycled concrete aggregate in high-strength concrete. *Materials and Structures/Materiaux et Constructions*, *33*(9), 574–580. https://doi.org/10.1007/bf02480538

Martins, P., De Brito, J., Rosa, A., & Pedro, D. (2014). Mechanical performance of concrete with incorporation of coarse waste from the marble industry. *Materials Research, 17*(5), 1093–1101. https://doi.org/10.1590/1516-1439.210413

Mckelvey, D., Sivakumar, V., Bell, A., & Mclaverty, G. (2002). Shear strength of recycled construction materials intended for use in vibro ground improvement. *Proceedings of the Institution of Civil Engineers–Ground Improvement*, *6*(2), 59–68. https://doi.org/10.1680/grim.2002.6.2.59

McNeil, K., & Kang, T. H. K. (2013). Recycled concrete aggregates: A review. *International Journal of Concrete Structures and Materials*, *7*(1), 61–69. https://doi.org/10.1007/s40069-013-0032-5

Menegaki, M., & Damigos, D. (2018). A review on current situation and challenges of construction and demolition waste management. *Current Opinion in Green and Sustainable Chemistry, 13*, 8–15. https://doi.org/10.1016/j.cogsc.2018.02.010

Mohajerani, A., Vajna, J., Cheung, T. H. H., Kurmus, H., Arulrajah, A., & Horpibulsuk, S. (2017). Practical recycling applications of crushed waste glass in construction materials: A review. *Construction and Building Materials, 156*, 443–467. https://doi.org/10.1016/j.conbuildmat.2017.09.005

Paschoalin Filho, J. A., Camelo, D. G., de Carvalho, D., Guerner Dias, A. J., & Marcondes Versolatto, B. A. (2020). Use of construction and demolition solid wastes for basket gabion filling. *Waste Management and Research, 38*(12), 1321–1330. https://doi.org/10.1177/0734242X20922591

Poon, C. S., & Chan, D. (2006). Feasible use of recycled concrete aggregates and crushed clay brick as unbound road sub-base. *Construction and Building Materials*, *20*(8), 578–585. https://doi.org/10.1016/j.conbuildmat.2005.01.045

Pradyumna, T. A., Mittal, A., & Jain, P. K. (2013). Characterization of reclaimed asphalt pavement (RAP) for use in bituminous road construction. *Procedia–Social and Behavioral Sciences, 104*, 1149–1157. https://doi.org/10.1016/j.sbspro.2013.11.211

Saha, D. C., & Mandal, J. N. (2017). Laboratory investigations on reclaimed asphalt pavement (RAP) for using it as base course of flexible pavement. *Procedia Engineering*, *189*(May), 434–439. https://doi.org/10.1016/j.proeng.2017.05.069

Sangiorgi, C., Lantieri, C., & Dondi, G. (2015). Construction and demolition waste recycling: An application for road construction. *International Journal of Pavement Engineering*, *16*(6), 530–537. https://doi.org/10.1080/10298436.2014.943134

Sanjeewa, K. W. D., & Nawagamuwa, U. P. (2019). Utilization of building debris as aggregates in stone columns. *Engineer: Journal of the Institution of Engineers, Sri Lanka*, *52*(1), 43. https://doi.org/10.4038/engineer.v52i1.7329

Santos, E. C. G., Palmeira, E. M., & Bathurst, R. J. (2013). Behaviour of a geogrid reinforced wall built with recycled construction and demolition waste backfill on a collapsible foundation. *Geotextiles and Geomembranes, 39*, 9–19. https://doi.org/10.1016/j.geotexmem.2013.07.002

Saribas, I., & Ok, B. (2019). Seismic performance of recycled aggregate–filled cantilever reinforced concrete retaining walls. *Advances in Mechanical Engineering*, *11*(4), 1–11. https://doi.org/10.1177/1687814019838112

Sayed, B. S. M., Pulsifer, J. M., & Schmitt, R. C. (1994). Project, R 31 E R32 E Alligator Project Concrete v FIG. 2. *Typical Pavement Cross Section*, *5*(3), 321–338.

Senanayake, M., Arulrajah, A., Maghool, F., & Horpibulsuk, S. (2022). Evaluation of rutting resistance and geotechnical properties of cement stabilized recycled glass, brick and concrete triple blends. *Transportation Geotechnics*, *34*(November 2021), 100755. https://doi.org/10.1016/j.trgeo.2022.100755

Serridge, C. J. (2005). Achieving sustainability in vibro stone column techniques. *Proceedings of the Institution of Civil Engineers: Engineering Sustainability*, *158*(4), 211–222. https://doi.org/10.1680/ensu.2005.158.4.211

Shahverdi, M., & Haddad, A. (2020). Use of recycled materials in floating stone columns. *Proceedings of Institution of Civil Engineers: Construction Materials*, *173*(2), 99–108. https://doi.org/10.1680/jcoma.18.00086

Sharma, A., & Sharma, R. K. (2019). Effect of addition of construction–demolition waste on strength characteristics of high plastic clays. *Innovative Infrastructure Solutions*, *4*(1), 1–11. https://doi.org/10.1007/s41062-019-0216-1

Sharma, A., & Sharma, R. K. (2020). Strength and drainage characteristics of poor soils stabilized with construction demolition waste. *Geotechnical and Geological Engineering*, *38*(5), 4753–4760. https://doi.org/10.1007/s10706-020-01324-3

Sharma, R. K., & Hymavathi, J. (2016). Effect of fly ash, construction demolition waste and lime on geotechnical characteristics of a clayey soil: a comparative study. *Environmental Earth Sciences*, *75*(5), 1–11. https://doi.org/10.1007/s12665-015-4796-6

Surana, S. R., Kar, A., Guharay, A., Ashok, ;, Suluguru, K., & Jayatheja, M. (2017). Experimental Investigations on Building Derived Materials in Chemically Aggressive Environment as a Partial Replacement for Sandy Soil in Ground Improvement. *ASCE India Conference 2017*, 244–254. https://ascelibrary.org/doi/10.1061/9780784482032.026

Tabsh, S. W., & Abdelfatah, A. S. (2009). Influence of recycled concrete aggregates on strength properties of concrete. *Construction and Building Materials*, *23*(2), 1163–1167. https://doi.org/10.1016/j.conbuildmat.2008.06.007

Taha, B., & Nounu, G. (2009). Utilizing waste recycled glass as sand/cement replacement in concrete. *Journal of Materials in Civil Engineering*, *21*(12), 709–721. https://doi.org/10.1061/(asce)0899-1561(2009)21:12(709)

Tranter, R., Jefferson, I., & Ghataora, G. (2008). The Use of Recycled Aggregate in Vibro-Stone Columns – A UK Perspective. *GeoCongress 2008: Characterization, Monitoring, and Modeling of Geosystems.* 630–637. https://doi.org/10.1061/40972(311)79

Venkatachalam, M. N., & Balu, S. (2022). A review on the application of industrial waste as reinforced earth fills in mechanically stabilized earth retaining walls. *Environmental Science and Pollution Research*, *29*, 86277–86297. https://doi.org/10.1007/s11356-021-17953-x

Vieira, C. S., & Pereira, P. M. (2018). Use of Mixed Construction and Demolition Recycled Materials in Geosynthetic Reinforced Embankments. *Indian Geotechnical Journal*, *48*(2), 279–292. https://doi.org/10.1007/s40098-017-0254-6

Villoria Sáez, P., & Osmani, M. (2019). A diagnosis of construction and demolition waste generation and recovery practice in the European Union. *Journal of Cleaner Production*, *241*. https://doi.org/10.1016/j.jclepro.2019.118400

Wagih, A. M., El-Karmoty, H. Z., Ebid, M., & Okba, S. H. (2013). Recycled construction and demolition concrete waste as aggregate for structural concrete. *HBRC Journal*, *9*(3), 193–200. https://doi.org/10.1016/j.hbrcj.2013.08.007

Zhang, J., Gu, F., & Zhang, Y. (2019). Use of building-related construction and demolition wastes in highway embankment: Laboratory and field evaluations. *Journal of Cleaner Production*, *230*, 1051–1060. https://doi.org/10.1016/j.jclepro.2019.05.182

4 On the Challenge of Recycling Massively Used Polymer-Based Packaging

Federico Morales, Gabriela Campos, Carlos Ramirez, Alejandra Costantino, Caren Rosales, and Valeria Pettarin

4.1 INTRODUCTION

Since the industrial introduction of plastic materials in the middle of twentieth century, their use has been constantly increasing worldwide. Due to the wide versatility and properties of plastics, they have replaced many other materials in food and cosmetic packaging (Cabrera et al., 2021; Horodytska et al., 2018; Morris, 2017). The three main materials used for food and cosmetic packages are polymer, glass, and metal (Sahota, 2013), but the most common used materials are polymers, more specifically thermoplastics. The main advantages of using plastics include high quality, flexibility, and resistance to breakage, which is extremely useful for products distribution. Using plastic is also extremely light, relatively cheap, and odorless compared with metals or ceramics. Common polymers used for containers include polyethylene (PE), polypropylene (PP), and polyethylene terephthalate (PET), among others.

In 2017, almost 350 million tons of plastics were produced, a large part of which belongs to the packaging industry (*Plastics Europe • Enabling a Sustainable Future*, s. f.). For example, in 2017, approximately US$25 billion of the US$532 billion valued cosmetic industry was reflected in product packaging, showing just how large of an impact the packaging aspect has in this industry (Research, 2018). Regarding the global food packaging market size, it was valued at US$346.5 billion in 2021 and is expected to expand at a compound annual growth rate of 5.5% from 2022 to 2030 (*Food Packaging Market Size, Share & Growth Report, 2030*, s. f.). Changing lifestyles and alternative eating habits are expected to boost demand for convenience food products, thus positively contributing to the industry demand. Factors such as convenience and the use of high-performance materials are expected to aid the industry growth. Improved shelf-life, coupled with heightened efficiency in the prevention of content contamination, is expected to boost the growth of the industry.

DOI: 10.1201/9781003364467-4

In addition, factors such as increasing population, rising disposable income, and shrinking households have a positive impact on the industry growth.

However, the marked growth of plastic packaging industry has not been accompanied by an adequate recovery and recycling plan, currently leading to a large volume of garbage and no adequate solutions for the mismanaged plastic waste. Changing times have brought along a sense of urgency toward sustainability and a cleaner way of living. In recent years, sustainability has become a key trend, with awareness on recycling everyday products and minimizing energy consumption becoming the norm in the average consumer's household. The concept of circular economy for plastic packaging has then emerged. The European parliament defines the circular economy as a model of production and consumption, which involves sharing, leasing, reusing, repairing, refurbishing, and recycling existing materials and products as long as possible (*Circular Economy*, 2015). Despite the fact that efforts should also aim to reduce the total amount of used packaging, e.g. by reusing, recycling is essential for extending the life span of materials. The average recycling rate for plastic packaging is still quite low, for example having been 40% in the EU in 2019 (*Statistics | Eurostat*, s. f.) (Figure 4.1).

Packaging can be first classified into rigid and flexible containers. Flexible films can be molded into any shape and are resistant to tears and punctures. One advantage of flexible packaging materials is that they can be adapted to many applications. For example, food packages can be made from flexible films that can be stretched or folded to create custom shapes for products like yogurt or cereal. Alternatively, rigid containers have a very specific shape that is difficult to change or customize. Rigid packaging is generally more durable and can handle more pressure than flexible materials. This is especially important regarding products that must be shipped and stored in a safe and secure environment.

Another important classification is in single or multi-layer packaging. Single layer (one layer) consists of only one type of material, such as plastic bags, trash bags,

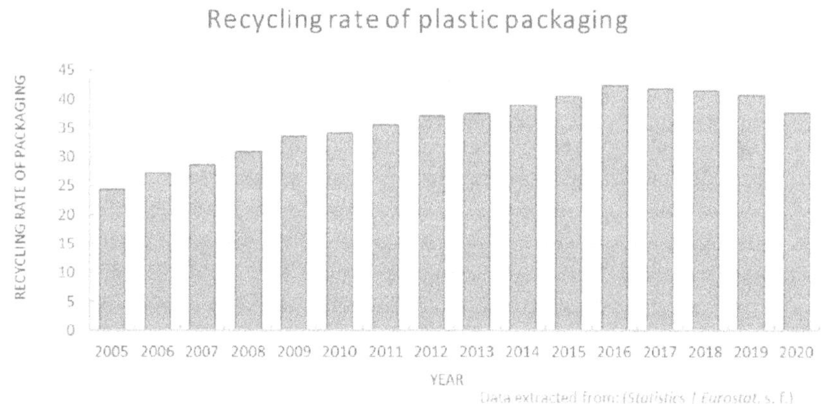

FIGURE 4.1 Recycling rate of plastic packaging in the EU.

and so on. Multi-layer (more than one layer) packaging consists of several types of materials that have several functions for each layer, including the printing layer, barrier layer, and sealing layer. Examples of multi-layer packaging include sachets (powder product packaging, packaging, snack etc.), standing pouch and others according to the designation of the product packaging to be packed. Complete classification is summarized in Figure 4.2.

Moreover, when manufacturing packaging, polymers are rarely used alone and their properties are enhanced with additives. In this way the properties of various plastics are modified depending on their final application. The use of additives in polymer packaging has multiple advantages in optimizing their performance by enhancing physical, chemical, and mechanical properties of the used polymers (Singh et al., 2012). For example, additives can extend shelf life of packaged products by preventing microbial growth, as in the case of the antimicrobial additives (Punia Bangar et al., 2021). The use of additives can also improve the processing of packages, making them more profitable by avoiding alternative methods to achieve the same properties. However, additives have some disadvantages that are worth mentioning. They may be toxic and harmful to human health, as in the case of plasticizers (Ncube et al., 2020). In addition, they can affect the recyclability of packaging, contributing to waste and damage to the environment (Kindsvater et al., 2020). Some additives can reduce the properties of the recycled material, affecting its resistance and quality. In summary, the use of additives can be an excellent alternative to improve the performance of the packaging, but it is important to keep in mind the possible problems that its use entails.

Packaging represents one of the largest fractions of plastic waste worldwide, and when discarded after only one short use, plastics maintain their properties unaltered,

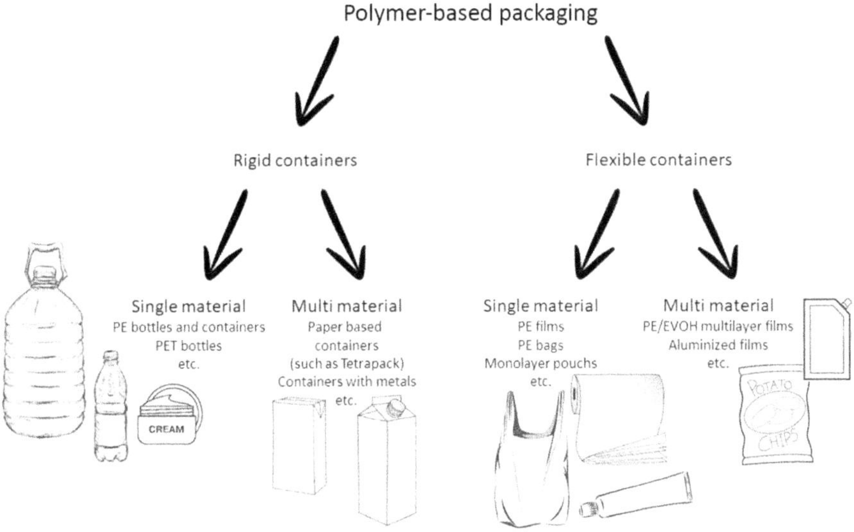

FIGURE 4.2 Classification of polymer-based packaging.

being in principle suitable for recycling. Currently, there are several alternatives for recycling plastic materials (García, 2016). The first one is mechanical recycling, which is the main route of recycling and is mainly focused on the recovery of rigid packaging such as PET bottles, while for flexible packaging, recycling rates remain very low. Chemical or feedstock recycling is another alternative that consists in breaking the polymer chains down using chemical agents or catalysts in order to produce lower molecular weight fractions to be reused as new materials for synthesis. The last alternative is energy recovery which consists in incinerating plastics materials to recover the energy stored in chemical bonds. Recycling routes for post-consumer plastic fractions that are technologically and economically feasible remain a challenge. The problem is emphasized when talking about multi-layer packaging. This fact has generated a negative view of these materials, overshadowing the positive impact that they provide due to their numerous advantages in terms of low cost and products' protection (Cabrera et al., 2021; Horodytska et al., 2018; Kaiser et al., 2018; Kulkarni AK et al., 2011; Maris et al., 2018; Morris, 2017; Vollmer et al., 2020).

Through this chapter, the current situation of rigid, flexible, single, and multi material packaging recycling, along with its problems and future challenges are discussed. Finally, the possibility of the plastics replacement by biodegradable polymers is introduced.

4.2 RIGID CONTAINERS

4.2.1 Mono Material Packaging

Single-layer packages are those that are predominantly composed of a single class of polymer, for example, polyethylene (LDPE, LLDPE, HDPE), polypropylene, PET, or paper, in at least 90% of their composition (minimum threshold when including adhesives, additives, and inks). The main advantage of this kind of packages is that they are most suitable for being recycled compared to multi-layer ones. The recycling process is at first easy as materials do not have to be separated individually as occurs with multi-layer packages during the process, diminishing the risks of contamination. In this way, the process is easy, fast, and cheap (OSP, 2022). Another advantage of single-layer packages is that their manufacture normally requires less energy consumption and economic resources in their production process because their processing is usually simpler and faster compared to packages made up of two or more materials. Single-layer packages facilitate recycling reducing its environmental impact, but also promotes socio-economic development (OSP, 2022).

Today the world produces twice as much plastic waste as it did two decades ago, most of which goes to landfills, is incinerated, or seeps into the environment, and only 9% is successfully recycled, according to a new report from the OECD (Organization for Economic Cooperation and Development) (OECD, s. f.). Fortunately, in recent years there has been a greater awareness of the development of the Circular Economy, which proposes to reuse and recycle polymeric products, and thus prevent them from becoming waste without additional useful lives. The world production of plastic from recycled plastic – or secondary – has quadrupled, from 6.8 million tons (Mt) in 2000 to 29.1 Mt in 2019, but it still represents only 6% of the volume of total plastic production (OECD, s. f.). This increase in the amount of recycled plastic percentage is due to a greater awareness of the impact of plastic waste on the environment. In addition to the waste and pollution that plastics cause, the reduction of resources is also a driving force as recycled plastic saves significant amounts of energy, carbon dioxide, and petroleum resources, compared to virgin material. As much consumer awareness evolves and becomes increasingly aware of the growing problems related to plastic waste, major brands and retailers including Adidas, Nike, Walmart, Ikea, and Coca-Cola, among others, have responded by carrying out various sustainability initiatives in relation to plastic use and waste. For example, Adidas has announced that they will use 100% recycled polyester in its garments by 2024; Ikea has committed to the same goal by 2030; and Coca-Cola has announced that it will recycle one bottle for every bottle it sells by 2030. Unilever too committed to use 100% recyclable plastic packaging by 2025 (Davis, s. f.).

There are many types of plastics, although the single-layer packaging market is dominated by Polyethylene (PE) (plastic bags, plastic sheets and films, bottles, microspheres for cosmetics, and abrasive products) and Polyester (PET) (bottles, containers, clothing, X-ray films etc.). Despite the general advantages that these materials present because they are single-layer, these two types of plastics present some difficulties when it comes to being recycled due to their chemical structure.

PET, for example, is susceptible to absorb environmental humidity, due to its hygroscopic nature. At high temperatures, the presence of water may cause chemical reactions with the PET structure such as hydrolysis, in which a water molecule breaks one or more chemical bonds on the main chains of PET structure, which in turn may modify the mechanical performance of the material. For this reason, the drying of a resin with internal humidity, as is the case of PET, considers special dehumidification conditions. Then, the dew point parameters are important and large drying time is needed, since it is much slower to move the humidity from the core of the material and it is required to handle temperatures up to 120–160 °C for a minimum of 4 to 6 hours (Davis, s. f.). In other words, humidity of recycled PET must be controlled in order to obtain a material with good mechanical properties. Drying then is the limiting stage in recycling process due to the time it takes, making it a less economically viable process compared to the production of virgin resin. To overcome this problem the use of infrared drying is currently being investigated. Infrared dryers make a more efficient use of energy, reducing energy costs by up to 45% compared to conventional techniques (*Cristalizador / secador infrarrojo: Plastics Technology México*, s. f.), they require a short drying time and their operation is not limited by the point of air dew so it can be used especially in applications with high degrees of inlet humidity. Although this type of drying is of recent use and is not yet used on a massive scale in industries, it is an extremely promising method to optimize the recycling cycle of PET and hygroscopic resins (Campos et al., 2022).

There is another method of recycling PET: chemical or feedstock recycling. It is applied using different chemical methods in order to produce total or partial depolymerization of PET into monomers or oligomers (hydrolysis, glycolysis, hydrogenation etc.) (Ghosal & Nayak, 2022), which can be used to re-polymerize and regenerate the original polymer (Achilias et al., 2012). This method has the advantage of being able to transform plastic waste into new molecules and materials (Ghosal & Nayak, 2022). However, it is not the most industrially used method since it implies a huge initial investment and personnel knowledge, and more research is needed in order to successfully implement it at an industry level (Achilias et al., 2012).

Other problems in PET recycling are due to the presence of other materials (for example, aluminum or paper) or other polymers. In particular, small amounts of PVC make PET difficult to process (La Mantia & Vinci, 1994). Moreover, they promote aesthetic problems due to discoloration and opacity, which reduces the value of the recycled material.

Regarding polyethylene (PE), it also presents contamination problems, mainly with polypropylene, PP. Polypropylene is, in volume, the second most common polymer found in plastic waste. It is a semi-rigid, transparent, inexpensive, and easy-to-process polymer. PP and HDPE are both from the polyolefin family, with great similarity in their structure and behavior, and with a density of less than 1 g/cm^3, the lowest density compared to other thermoplastics (Dai et al., 2023). Those similarities make the separation process very difficult, technologically speaking. For example, separation by floating in water, which is a widely used technique for polymer separation, is not suitable to separate PE and PP, and complete separation is sometimes impossible. Their combination is especially important, so recycling them as a blend is currently the most viable industrial option, and in fact a trend to reduce the impact of plastic wastes as blends is under study (Bertin & Robin, 2002; Mourad et al., 2009; Penava et al., 2013; Strapasson et al., 2005; Tai et al., 2000). Several authors have studied PE/PP blends, including blends of PP with HDPE, LDPE, and LLDPE (Fortelný et al., 1996; Jose et al., 2004; Rosales, Bernal et al., 2020; Rosales, Brendstrup et al., 2020; Rosales, Costantino et al., 2020; Wong & Lam, 2002). The mechanical properties of the blends are determined by their two-phase morphology, which is also influenced by the thermal history of the final parts. Rosales et al. extensively studied LLDPE/PP blends in different ratio, with and without incorporation of different compatibilizers (Rosales, Bernal et al., 2020; Rosales, Brendstrup et al., 2020; Rosales, Costantino et al., 2020). The blends present a biphasic morphology, where each blend composition gives rise to a different microstructure. The tensile properties of the blends are significantly affected by both the blend composition and the type of compatibilizer used. Thus, different mechanical behaviors were observed, from ductile to brittle under quasi-static and dynamic loading conditions. The effect of composition and morphology on the fracture behavior of LLDPE/PP blends was also investigated. Different fracture mechanics approaches had to be applied to characterize the fracture behavior of the different blends, using J-instability and EWF methodology. It is well known that PE and PP are immiscible and incompatible. The same research groups have used this incompatibility and analyzed the manipulation of the morphology of these blends using conventional processing techniques to obtain microfibrillated composites, developing improved mechanical properties in both tensile test, impact resistance, and fracture toughness under quasi-static conditions (Rosales et al., 2022). However, it must be taken into account that the mixtures also have limitations, such as percentage of each phase, thermal properties etc. For example, great differences between melting temperatures may be a problem for materials systems like PP and PE. PP has a melting temperature between 160 and 170 °C, which is higher than that of HDPE of 130 °C. During the PP melting process there may be degradation of HDPE in the form of black particles that can contaminate and cause mechanical or visual failures in the final product of the recovered polymer (*Procesabilidad y sustentabilidad del polipropileno*, s. f.).

As it is evident, it is still important to continue working on the different stages of recycling of this kind of materials, either to optimize the stages in terms of time and energy consumed, as well as the supplies (as water) used in these processes.

4.2.2 Multi Material Packaging

Aluminized multilayer rigid containers are liquid-food and beverage packages widely used throughout the world. These kind of products were devised in the 1960s and revolutionized the liquid-food market, because they allow products considered perishable to be distributed and stored without refrigeration for long periods of time, preserving the nutritional properties of food. The excellent properties they have are due to the way in which they are manufactured. They are made up of six thin layers: a layer of cellulose (~75% of the total weight), another of aluminum (~5%), and four of low-density polyethylene (~20%), conformed through processes of lamination, coating, deposition, and evaporation. In Figure 4.3 the layer arrangement of these rigid multilayer containers is shown. This manufacturing process allows generating products with specific properties of durability, mechanical and humidity resistance, besides other characteristics. The cellulose layer gives the container stability and structural resistance, the aluminum one is used as a barrier against oxygen and light, while the polyethylene ones are used as protection against humidity, as an adhesive between the cardboard/aluminum and as an internal sealant. The layer arrangement and specific characteristics of each layer make this kind of containers light, resistant and hermetic, suitable for easy transport and storage for long periods of time, thus becoming one of the main food containers currently.

The worldwide acceptance of this type of rigid packaging responds not only to the good properties that they possess, but also to economic reasons. It is advantageous for producers because it allows them to easily and efficiently transport a liquid product, while for distributors and merchants it saves on storage space (Hidalgo Molina, 2013). It allows the consumer to purchase basic liquid foods in a container that preserves them for a long time and that resists impact solicitations. These factors, added to certain changes in consumer habits in society, have been decisive for the demand for these containers to grow permanently. As an example, the Tetra Pak® company marketed a total of 192 billion containers in 2021 worldwide, which is equivalent to approximately 24 containers used per person in that year (*Datos y cifras de Tetra Pak, incluidos los resultados del desempeño financiero*, s. f.).

The use of large volumes of these containers makes their recycling an issue to deal with. In 2019, only 26% of all discarded packaging was recycled. On one hand, its structure of six compacted sheets makes the natural degradation of the container impossible. On the other hand, current recycling technologies for common cardboard are not adapted to process this type of material since, when plastic and aluminum

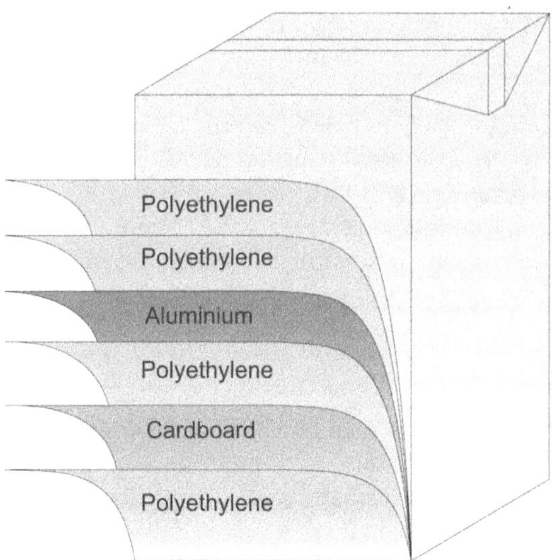

FIGURE 4.3 Structure of typical multi-layer rigid container based on polymers and aluminum.

break up, filters and meshes are clogged in subsequent processes, being this a serious problem for industrial practices. For this reason, rigid multilayer containers are mainly discarded in landfills at the end of their life cycle, after a sole use (remember that they are single-use containers). However, there are currently some alternatives for recycling these materials.

One of the alternatives to recycle multilayer rigid containers is to recover the aluminum from the container through thermochemical processes such as pyrolysis or through the separation of polyethylene with aluminum through the implementation of an acid-based wet environment (Korkmaz et al., 2009; Şahin & Karaboyacı, 2021; Zmijková et al., 2022; Zúñiga-Muro et al., 2021). Although this technique is viable for the recovery of aluminum on a laboratory scale, it is not recommended on an industrial scale. Only the metal is recovered (there is only 5% of metal in each container), in addition to the fact that it is a harmful method for the environment due to the emission of greenhouse gases and/or the use of toxic solvents.

Another alternative is the manufacture of synthetic wood by treating waste from containers, which is commercially available as Chiptek, Maplar, or Tectán®, the latter being the most widespread worldwide. These high-quality, durable synthetic woods are manufactured by hot-pressing waste from rigid multi-laminate packaging. First, the containers are crushed, washed, and dried. Subsequently, they are arranged in a heated press that, through pressure and temperature, melts the polyethylene that acts as a binder between the cellulose and the aluminum. Then, the product is allowed to cool, generating a highly resistant agglomerate with an impermeable surface. This method for recycling containers has a wide range of good properties: products are resistant to humidity, thermoformable, flexible, and manageable, they are free of resins (they do not pollute the air, soil, and water), and they are immune to pests and fungi, they resist temperatures up to 120°C, have a great acoustic insulation capacity and good mechanical properties, they can be sawn, shaped, glued, and screwed (it can be worked in the same way as wood) and have a long useful life.

The last commercially accepted option for the recycling of this type of packaging is by using the hydropulper technique, which consists of the mechanical separation of cellulose from polyethylene and aluminum, leaving as a by-product a mixture between these last two materials, which is known as PolyAl. The cellulose can be reused for the manufacture of paper and packaging materials. PolyAl can be used as an additive in the processing of polyethylene items through injection, extrusion, hot-forming techniques, among others, achieving different types of products. Thanks to the implementation of this mixture between polyethylene and aluminum, the manufactured parts manage to have good properties, such as thermal and acoustic insulation, thermoformable, fire resistant, flexible, waterproof, resistant to fungi, bacteria and pests, and the typical silver color of aluminum, among other properties (Şahin & Karaboyacı, 2021; Zawadiak et al., 2017).

Despite the various alternatives mentioned, there are still many significant challenges to overcome that limit the full recycling of rigid multilayer packaging that contain aluminum, such as the lack of adequate infrastructure in most parts of the world, high cost of recycling methods, high energy consumption, in addition to the low demand for recycled final products, little commitment and lack of education of society.

4.3 FLEXIBLE CONTAINERS

Post-consumer flexible packaging waste is considered one of the main contributors to plastic environment pollution. It is a major problem worldwide, and is caused by the inappropriate disposal and treatment of plastic waste, due to cost of collection and separation, added to the low market value (The Plastic Leak Project Guidelines, 2020). The current treatment of packaging waste is recycled to produce plastic wood, used in park benches, garden furniture, or garbage bags (Lase et al., 2022).

In this context, it is essential to generate greater recycling capacity for flexible containers (Lase et al., 2022). To improve this, several options exist, including enhanced mechanical recycling, chemical recycling (including solvent-based recycling, such as dissolve-precipitation or delamination), and energy recovery (Horodytska et al., 2018; Kol et al., 2021; Ragaert et al., 2017; Vollmer et al., 2020). Chemical recycling is still under development although there is some promising solvent-based research (Schwarz et al., 2021; Vollmer et al., 2020).

It is essential to increase the recycling rate, therefore it is necessary to improve the currently available technologies to increase the performance of the process and the quality of the waste regranulates (Lase et al., 2022). As for improved mechanical recycling, a higher level of technological readiness on a commercial scale are being implemented, such as improved washing and extrusion (Horodytska et al., 2020; Lase et al., 2022).

The main polymers used in flexible packaging are polyolefins (PO), low-density polyethylene (LDPE), linear low-density polyethylene (LLDPE) and polypropylene (PP) (*CEFLEX | A Circular Economy for Flexible Packaging*, s. f.; Faraca & Astrup, 2019; Horodytska et al., 2018). It is estimated that between 70% and 80% (~3 million tons) are single-layer PO containers (*CEFLEX | A Circular Economy for Flexible Packaging*, s. f.; Faraca & Astrup, 2019) and around 20% (750 kilotons) of the total post-consumer packaging waste is of multiple materials (Lase et al., 2022).

4.3.1 Mono Material Packaging

Polymers can be classified into two groups, thermoplastics and thermosets. This classification is related to their heat behavior. Unlike thermosetting polymers, thermoplastics are economical and can be reprocessed, which provides recyclability. For this reason thermoplastic polymers are the main source of plastic flexible packaging. Most monolayer films are made from polyolefins (PO), i.e. polyethylene (PE) and polypropylene (PP).

Polyethylene is inexpensive, lightweight, easy to process, has good impact resistance and excellent chemical resistance. The molecular structure of PE, in terms of branching, affects crystallinity and therefore density. High-density polyethylene (HDPE) has minimal branching, so molecules can regularly link together to form crystals, giving it a stiffer structure than low-density polyethylene (LDPE). LDPE has more branched molecules than HDPE, so it is less crystalline and less brittle, and is the main component of flexible packaging, accounting for 38% of plastic waste (Salehi Morgani et al., 2021).

The obvious consequences of the excessive use of plastic films lead researchers to seek solutions to reduce the harmful environmental effects. As people are educated to be more careful with the consumption of plastics, recycling is the best option to find a second life for the plastic waste that today is discarded in landfills. In this way, the plastic waste is converted into other products of greater value (Salehi Morgani et al., 2021). Currently, all existing types of recycling, i.e. mechanical, chemical, and thermal recycling, are being investigated to find the most economical, sustainable, and industrial way of reprocessing the plastic waste generated (Soares et al., 2022).

Thermal or quaternary recycling involves the recovery of energy from the incineration of plastic waste. This method can be used as a way to replace the fossil fuels used today for heating and energy. However, the burning of petrochemical compounds generates high CO_2 emissions, being a concern for current global warming (Zallaya et al., 2023).

Chemical recycling remains the main area of study, because it would allow the possibility of infinite recyclability. This recycling method involves the application of high temperatures (pyrolysis) and the use of catalyzers (catalytic cracking) to modify the chemical structure of polymers through depolymerization (Zallaya et al., 2023). The addition of catalyzer to the pyrolysis process can save energy and reduce production costs, since the process temperature can be reduced from 450°C to 300°C approximately. It is important to point out that the use of catalytic cracking was performed only on pure polymers, since if there are any contaminants mixed with the plastic the efficiency of the process is drastically reduced (Almeida & Marques, 2016). Nonetheless, there are some disadvantages in this process: it involves high costs of implementation and the use of skilled labor; the polymer waste requires pre-treatment to clean it from all contaminants because the process is very sensitive to contamination of the feed stock; and there are some components in waste based on chloride and nitrogen that could deactivate the catalyst and cause a temperature increase (Solis & Silveira, 2020).

Due to the reasons mentioned above, mechanical recycling remains to this day the most widely used technique for processing plastic waste. A typical conventional mechanical recycling consists of shredding, washing, density separation, mechanical and thermal drying and extrusion (Ragaert et al., 2018). Using this technique, the discarded plastic films are shredded into flakes and washed to remove contaminants such as organic traces. They are then separated by flotation in a water-based medium; the polyolefins will float and other materials will mostly sink. Then, the mechanical and thermal drying stage proceeds and, finally, the flakes undergo a last stage of reprocessing by extrusion (Larrain et al., 2021; Ragaert et al., 2018). The efficiency of the mechanical recycling process for PE and PO films can range from 60% to 80%, depending on the quality of the input material (sorting) and the efficiency of the recycling equipment (Faraca & Astrup, 2019; Larrain et al., 2021).

As in the case of rigid containers, mechanical recycling, in terms of the final properties offered by the recycled materials, is highly dependent on the purity of the post-consumer waste. The main drawback of mechanical recycling is the impossibility to completely separate materials before recycling, having to deal with polymeric blends that are often incompatible and/or immiscible. In many countries, flexible packaging is still not adequately separated at source. In Europe, only a limited number of countries separate plastic waste at source, such as rigid plastic packaging, metals and beverage cartons (Barceló & Kostianoy, 2022; Brouwer et al., 2018; Picuno et al., 2021). Currently, research is focused on the development of automated methods for the identification of post-consumer polymers.

Taking into account the current limitations regarding the separation step of plastic waste, a research group investigated a step prior to waste separation (Arenas-Vivo et al., 2017). For this purpose, they incorporated a small amount of fluorescent markers (3–10%) into HDPE films to introduce the ability to identify and separate it from the waste stream once the film reached the end of its useful life. Results showed that the incorporation of the markers did not significantly affect the polymer structure. They also analyzed the effect of thermal, hygrothermal, and accelerated photochemical degradation. Despite the fact that signal intensity decreases with degradation, the fluorescent emission record remains distinguishable, allowing the marked plastics to be correctly identified even after exposure to aggressive conditions. HDPE containers can thus be clearly identified at the end of their service life.

Different applications for recycled PE are currently being investigated. As the most abundant polymer in solid plastic waste, a group studied the possibility of manufacturing a hydrophobic airgel membrane capable of separating oil/water emulsions (Gan et al., 2021). The results obtained showed that it is possible to use PE waste as raw material for a new application, forming a closed recycling cycle as a sustainable model.

4.3.2 Multi Material Packaging

Consumers' preference for consuming good-quality fresh food grows day by day, with as little treatment as possible, without additives or preservatives. That is the main reason why the packaging industry is constantly growing, promoting the development of new technologies that meet certain requirements, allowing the product's useful life to be extended while maintaining its quality, as well as improving its transportation and marketing.

Food packaging must meet several requirements to guarantee food conservation and handling. Materials used in food packaging must be resistant to the mechanical damage that they could suffer during handling; and in some cases they must also resist low temperatures if they contain products that need to be refrigerated. It is difficult for a single material to fulfill all the technical, protective, and commercial requirements. Therefore, the combination of layers of different materials may provide functionalities to the packaging that allow it to be optimized for various applications. Typically, two to five (or more) combined layers, may provide one or more of the desirable properties. These multi-layer packages are made up of layers that fulfill mainly these requirements: base, barrier, sealant, and/or adhesive. The base layers give the package structure. The barrier ones prevent the exchange of gases (nitrogen, oxygen, water vapor, carbon dioxide, aromas, and odors), vapors and liquids between the exterior and interior, thus increasing the useful life of the product it contains. The adhesive layer is responsible for the union between the different sheets, while the sealing layer is the one that allows the hermetic closure of the package.

Currently, flexible packaging is mainly made up of plastics or combinations of these with other materials. The main polymers used are: branched polyethylene (the most widely used polymer in the packaging industry), high and low density polyethylene, linear polyethylene, linear low density polyethylene, homo and copolymer polypropylene, polyethylene terephthalate, polyamide, among others. The main manufacturing processes for multilayer structures are lamination, extrusion coating and co-extrusion.

In Europe, approximately 58 million tons of multipurpose plastics are consumed, of which 23 million represent packaging. Which in turn represent more than 60% of the plastic that is thrown away and only 40% of it is recycled. With regard to multi-laminate packaging, some 333,000 tons of multi-layer waste are generated annually, which mainly ends up incinerated or in landfills. Approximately 60% of the films used in the food industry correspond to this type of packaging (Biblox, 2021).

Mechanical recycling is the most widely used alternative to reduce the amount of thermoplastic waste from multilayer films. However, due to its complex structure, it is not possible to separate the different layers of the packaging to recycle each material separately. This secondary recycling method generates a material composed of various phases, generally incompatible and immiscible. This is where the challenging aspect of recycling multilayer films arises, since the different phases need to be stabilized in order to develop optimal mechanical properties.

Multilayer films are very beneficial in the food and cosmetic industries since they combine the properties of several polymers in the same container. However, to date, they generate a large volume of garbage because their growth has not been accompanied by an adequate recovery and recycling plan. Trying to develop an adequate solution, some authors analyzed the feasibility of recycling multilayer commercial milk containers, which are currently made of LDPE and EVOH (Echeverria et al., 2022). The main objective of the study was to reprocess them as blends, in order to obtain films with good mechanical properties to be used in applications such as e-commerce envelopes. For this, different processing conditions were applied and a custom-designed compatibilizer and virgin polyethylene were added in variable content. It was found that properties of recycled films depend on their composition largely. Just by adding 2% of compatibilizer, mechanical behavior was improved. They also found that a large amount of compatibilizer has a negative effect, since it can agglomerate and act as a stress concentrator, inhibiting the deformation of the polymer.

In contrast to the previously mentioned study, another research group studies the incorporation of a compatibilizer in different proportions (Nasri et al., 2022). The rheological and morphological results showed poor compatibilization when the amount of compatibilizer added to the mixtures is low. However, in higher percentages (13%) an improved rheological response could be observed. It seems that results are dependent on the type of added compatibilizer.

As it was said above, food, cosmetic, and chemical industries are concerned about packaging meeting the expected requirements in terms of protecting the product from possible damage that may be caused by the environmental and physical conditions to which it may be subjected. Flexible aluminum laminates are one of the alternatives used to contain such products. Aluminized multilayer packages can be made up of different polymer materials (mainly polypropylene, polyethylene, and PET) and an intermediate layer of aluminum between 1 and 20 μm thick, which gives the packages protection against UV rays, aromas, grease, and vapors. The polymeric layer gives the packaging structural stability as well as contributing to the barrier against humidity, vapors, and aromas. This type of multi-layer packaging represents approximately 20% of all multi-layer packaging (Anukiruthika et al., 2020).

Most of these multi-layer films are processed by lamination or blow molding to obtain the polymeric layers. Subsequently, these polymeric films are subjected to an aluminum deposition process by evaporation under high vacuum. The main drawback of this technique is that a great adhesion between the components is needed (CETEC, 2017). Currently this type of packaging is being widely used due to the excellent properties it possesses in terms of preservation and product protection capacity. However, the main problem is its low rate of degradability and its very short useful life (one use only), so the recycling of this kind of multilayer films is still an issue to deal with. A medium-sized industrial plant that deals with food packaging produces around 8 tons of multilayer films per month. Most of this material ends up as waste. That is why the future of multilayer packaging has become a major environmental concern (Fávaro et al., 2013).

They are currently studying some alternatives to solve this problematic, these alternatives are focused on the technology to develop new packaging. First, it is intended to improve the barrier properties of packaging – a property provided by the aluminum layer – through the incorporation of additives thus simplifying its subsequent recycling. Second, adhesives that facilitate the delamination of the structure after its use are under study. However, these studies are oriented more to development of new kinds of multi-laminate packaging than to recycle the existing ones (Biblox, 2021). It is still then a big problem to treat the large amount of multi-layer packaging waste. The separation of the polymeric and aluminum layers of these packages is yet one of the main challenges for their recycling. Currently there are chemical delamination mechanisms that act by removing one or several layers, or by direct action on the adhesive (Kaiser et al., 2018). Although these techniques are effective, they use expensive and polluting solvents and a large part of the polymeric material is lost. That is why these methods are not suitable and efficient enough to recycle large volumes of material. The fraction of waste that can be treated is too low to justify an investment in sophisticated recycling technology (Kaiser et al., 2018).

An alternative that is currently being under study for the recycling of polymer/aluminum containers is to use the Al present as a pigment for the manufacture of polymeric parts. After cleaning the packaging, it is grounded using a mill until it reaches a particle size of the order of one micron. The mixture of metallic particles and polymeric material may be used as an additive in the injection process of new parts, thus giving the final product a metallic appearance similar to those of pristine metallic particles (Costantino et al., 2015).

4.4 BIODEGRADABLE PLASTIC PACKAGING

It is well known – and extensively developed in previous sections – that synthetic polymers, such as polyethylene (PE), polypropylene (PP), Polystyrene (PS), and Polyethylene terephthalate (PET), among others, are used for packaging purposes. It was also discussed that since these plastics cannot go under any kind of degradation (physical, chemical, or biological) their life cycle ends as an increase in waste, which influences directly in severe environmental and health-related problems (Foolmaun & Ramjeeawon, 2012; Vert et al., 2002). Biodegradable and eco-friendly polymers have then emerged as sustainable alternatives for many industrial applications to control the environmental risk caused by commodity polymers. These polymers are called bioplastics and according to the European standard EN 1675 are defined as "derived from biomass" (FBR BP Biorefinery & Sustainable Value Chains et al., 2017), they are produced from renewable sources and their properties are very similar and competitive to conventional plastic materials (Kirwan et al., 2011). Additionally, their production processes need 65% less energy than synthetic polymers and their emission of greenhouse gases is less (Ahvenainen, 2003).

Due to increasing environmental concerns and the initiatives to reduce plastic waste from governments, the production of bioplastics for packaging has increased since 2019 and it is estimated to grow very rapidly by the end of 2025, reaching a market value of up to US$4.66 billion. In 2019, global production was 2.11 million tons, and it is expected that by the end of 2024 the production will increase to 2.43 million tons, with biodegradable plastics constituting over 55.5% of the global bioplastic production (PLA, starch blends, PBS, PBAT, and others) (EUBIO_Admin, s. f.).

This section aims to provide general information on biopolymers and their role in packaging material, as an alternative that can help to reduce plastic pollution's environmental impact.

4.4.1 Biodegradability

The biodegradation of biodegradable polymers is defined as the chemical decomposition of substances, which is achieved through the enzymatic work of microorganisms that change chemical, mechanical, and structural properties of the polymer and form metabolic products, which are eco-friendly, such as methane, water, biomass, and carbon dioxide (Luckachan & Pillai, 2011). The process of biodegradation occurs in two stages (see Figure 4.4), the first one is the attack of extracellular enzymes and abiotic agents which produce oxidation, photo-degradation, and creating shorter chains through the mechanism of hydrolysis (Engineer et al., 2011; Karthika et al., 2019). The second stage is the biomineralization process, in which these shorter polymer chains are bio-assimilated by microorganisms and mineralized. This degradation can take place in the presence or absence of oxygen. If the reactions are in the presence of oxygen, then it is called aerobic degradation and produce CO_2, H_2O, biomass, and residue. On the other hand, if the reactions are carried out in absence of oxygen it is called anaerobic degradation and produce CO_2, H_2O, CH_4, biomass, and residue (Engineer et al., 2011).

The biodegradation process is affected not only by environmental factors but also by polymers' morphology, structure, chemical treatment, and molecular weight, as follows:

- Polymer structure: if the polymer has hydrolyzable links along the chain, then it will be easy to degrade it in the presence of microorganisms and hydrolytic enzymes. If the polymer has both hydrophobic and hydrophilic structures, then it is more degradable than polymers with only hydrophobic or hydrophilic structures (Ghanbarzadeh et al., 2013).

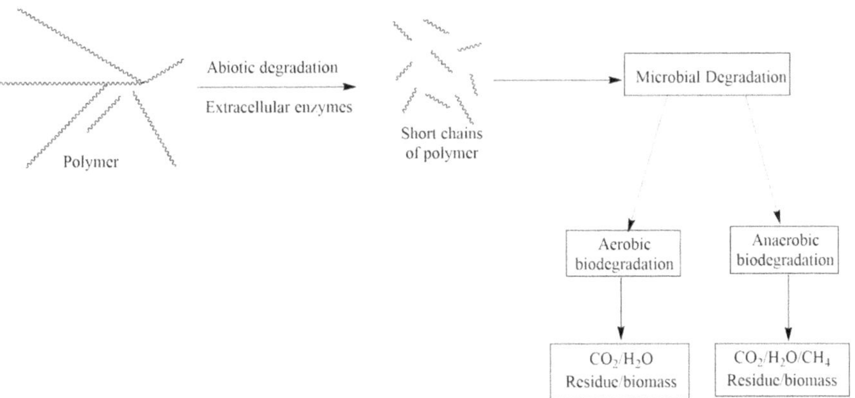

FIGURE 4.4 Biodegradation process for polymers.

- Polymer morphology: amorphous regions are more vulnerable to the attack of microorganisms than crystalline regions because the molecules are far apart from each other making them more susceptible to degradation (Saini, 2017).
- Molecular weight: as the molecular weight of the polymer increase, the biodegradability of the polymer reduces (Saini, 2017).

4.4.2 Bioplastics and Their Classification

Bioplastics could be classified according to their origin and method of production. In this section, the most important polymers from each group will be discussed, including their main properties and applications in the packaging industry.

4.4.3 Polymers Extracted/Isolated Directly from Biomass or Natural Materials

These biopolymers have biodegradability, bio-functionality, biostability, and biocompatibility making them a good candidate for food packaging (Bassyouni et al., 2022). Their main source is plant-based agricultural waste, and the most important polymers are:

- **Starch:** is composed of amylopectin (poly-α-1,4-D-glucopyranoside and α-1,6-D-glucopyranoside) and amylose (poly-α-1,4-D-glucopyranoside) (Figure 4.5a). It is extensively found and extracted from wheat, rice, potatoes, and corn. The ratio of amylose and amylopectin substantially impact the properties of starch (Ratnayake et al., 2001). Since starch has poor mechanical properties, low impact resistance, water sensitivity, and brittleness properties, it is very common to reinforce the starch matrix with fibers or modify it physically or chemically with other biodegradable polymers (Jiang & Zhang, 2017). Starch is mostly used in thermoplastic starch (TPS) or plasticized starch, this modification improves its properties and makes it suitable for packaging purposes. Corn starch and TPS are commonly used in food packaging applications. Recent studies show that the use of starch and its blends in food packaging extends the shelf life of food. They also allow to incorporate a visual indicator of food freshness to control any quality changes of seafood, meat, cooking oil, fruit, juice, and other foods (Luchese et al., 2018; Zhang et al., 2020). Companies that manufacture biodegradable starch films at the moment are: Plantic ™ from Plantic Technologies, Solanyl ™ from Rodenburg Biopolymers, Bioplast ™ from Biotec, Biopar™ from Biop and Mater Bi ™ from Novamont.
- **Cellulose:** is a polysaccharide with a molecular structure like starch, but the polymer chains made up of β (1 → 4) are connected to D-glucopyranosyl units (Figure 4.5b). This polymer is one of the main components of lignocellulosic plant cell walls, the other ones are lignin and hemicellulose (Ghanbarzadeh et al., 2013). It is crystalline, linear, biodegradable, and insoluble in organic solvents, but it has a hydrophilic nature, and it tends to absorb moisture, which causes its mechanical properties to deteriorate. That is the reason why cellulose

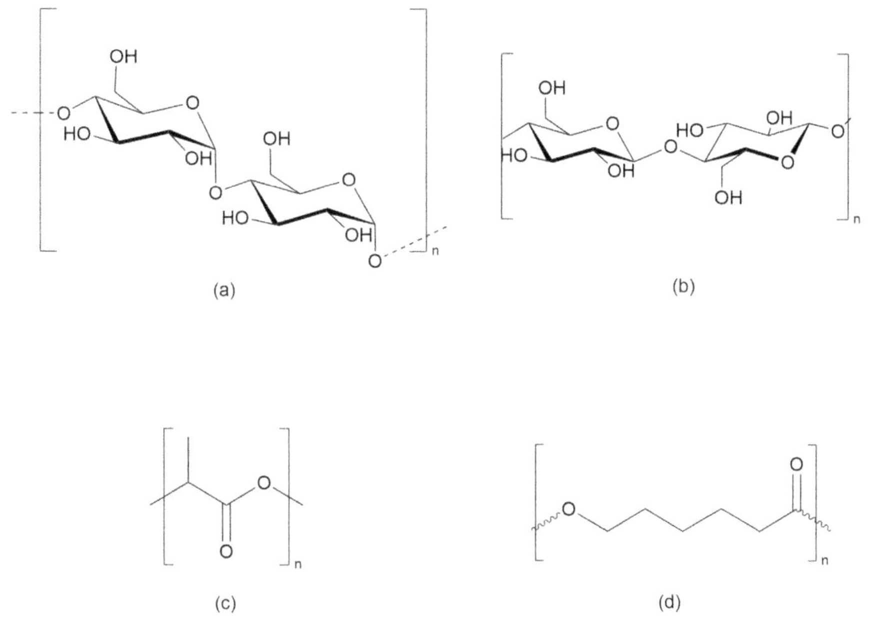

FIGURE 4.5 Chemical structure of main biopolymers: (a) starch; (b) cellulose; (c) PLA; (d) polycaprolactone.

is commonly modified to achieve the mechanical properties required to its application. These modifications involve various degree of substitution where the mechanical properties increase but the degradation rate decreases (Jiang & Zhang, 2017; Shah & Vasava, 2019). Cellulose and its derivates are edible, and lightweight thus reducing the weight of the packaging material. They also extend the shelf life of food and improves its quality. In addition, this polymer provides barrier properties to the film, making it a perfect candidate for packaging, especially for preserving fruits and vegetables whose fast deterioration is a very common issue in the industry (Klemm et al., 2002). Commercially available cellulose and its derivates are Tenite ™ from Eastman, Fasal ™ from IFA, and Natureflex ™ from UCB.

4.4.4 Synthetic Biodegradable Polymers

These polymers are obtained by conventional polymerization procedures in the form of aliphatic polyesters, polylactide, and aliphatic copolymers (Pooresmaeil et al., 2019). Degradable polyesters could be considered as the materials with higher potential to be used as substitutes for commodity polymers, since they have an appropriate degradation time, and their production processes at industrial scale are very well developed (Mangaraj et al., 2019). There are several synthetic biodegradable polymers, but the leading polymers are the ones mentioned below:

- **Polylactic Acid (PLA):** PLA is a type of aliphatic thermoplastic polyester synthesized by the ring opening of lactide monomer (Tian & Bilal, 2020) (Figure 4.5c). This method of synthesis allows controlling not only its molecular weight but also its final properties (Singla & Mehta, 2012). This monomer is usually obtained from the fermentation of natural materials like potato starch, wheat, rice bran corn, and biomass (Balla et al., 2021; Tian & Bilal, 2020). The lactic acid monomer could exist in three different diastereoisomeric structures: D-lactide, L-lactide, and meso-lactide (DL-lactide). It is possible to change the properties of PLA, even its biodegradability, by changing the monomeric ratio of these structures. This is due to the different levels of crystallinity achieved at varying monomeric ratios. It has been demonstrated that the lowest degradation rate occurs when the crystallinity of the polymer is the highest (Kale et al., 2006). PLA has very good properties that make it an excellent candidate for packaging material: it has a high molecular weight, water solubility resistance, it is easy to process by thermoforming, extrusion, injection, blow and film molding techniques, and it is biodegradable and recyclable by mechanical conversion back to lactic acid (that could be then re-polymerized). Also, its mechanical properties, especially the tensile strength, and flavor and odor barrier properties are very similar to polymers like polyethylene, PET, and PVC (Avérous & Pollet, 2012; Katiyar et al., 2014; Leja & Lewandowicz, 2010). In food packaging industry it is commonly used for short shelf-like products and forming pads and trays for serving food. Moreover, it is usually blended with other biopolymers such as TPS to improve its mechanical and barrier properties for applications such as trays, packaging, and food films (Shaikh et al., 2021; Siracusa et al., 2008). Nowadays, PLA is commercialized by different companies and commercial names, being the most popular Galacid TM by Galacid, Lacty TM from Shimadzu and Eco plasticTM by Toyota.
- **Polycaprolactone (PCL):** PCL is a cheap fossil-based and biodegradable polymer prepared by ring opening polymerization of ε-caprolactone (McKeen, 2012) (Figure 4.5d). It is a linear semi-crystalline, hydrophobic, and easy-to-process polymer, that shows flexible behavior with high elongation, and low water vapor transmission, making the food stores in PCL films to have a longer shelf-life. Even though PCL presents these excellent properties, its main issue in packaging use is its price, low barrier properties to gases, and poor mechanical properties. These are the reasons why PCL is commonly used in blends or as a compatibilizer as a property enhancer (Sarasam et al., 2006; Taherimehr et al., 2021). It is commercially available under the trade names Tone ® from Union Carbide, Celgreen ® from Daicel and CAPA ® from Solvay.

4.5 FUTURE PERSPECTIVE OF BIOPOLYMERS IN THE PACKAGING INDUSTRY

There is a need to develop alternatives for packaging purposes that could be sustainable for conventional polymers, and this interest has started many studies (Iglesias

Montes et al., 2020). This alternative could be filled by biopolymers and biomaterials, but there is still much research left to develop. The main challenge in this biopolymer development for packaging systems is that in many cases, monolayer packaging of bio-based polymers is not enough to preserve food. However, there are several types of food, such as fruit or vegetables, where this low permeability of gases could enlarge the shelf life in comparison with conventional plastic packaging (Mendes & Pedersen, 2021). In recent developments and studies, it has been reported that biocomposite materials, where you can combine a biopolymer and nanoparticles to create smart packaging which improve the shelf-life of foods and communicate when the packed items have gone bad, making them a good alternative to traditional polymers such as PET, PE, PS, etc. (Shaikh et al., 2021).

It is important to develop new industrial processes, logos, labels, and branding that consumers could identify easily to induce them to change their appreciation that biobased packaging is just a marketing trick. Also, developing new, sustainable, and industrial scaled processes and technologies could reduce the price, which is a very important matter since the costs of producing biopolymers are higher than oil-based polymers (Shaikh et al., 2021).

4.6 CONCLUSION: FUTURE AND PERSPECTIVE

Through this chapter, the current situation, problems, and future challenges of rigid and flexible packaging based on or containing polymers are discussed. It is clear that the use of these types of packaging is going to increase during the next decades due to the benefits they carry.

Ultimately, a great part of the world still lacks basic waste management and conventional recycling systems. The plastics recycling industry requires cost-effective solutions to enhance technologies. There is still a gap between established recycling technologies and the recycling targets, widening innovation opportunities for the use of more sustainable materials, such as bio-based. Hence, the use of renewable materials in packaging materials can enhance the sustainability performance where plastic packaging is essential and its recycling is not available yet. It is important to notice then that there is still much research left to develop both in recycling of non-degradable packaging and in development of degradable packaging.

REFERENCES

Achilias, D. S., Andriotis, L., Koutsidis, I. A., Louka, D. A., Nianias, N. P., Siafaka, P., Tsagkalias, I., Tsintzou, G., Achilias, D. S., Andriotis, L., Koutsidis, I. A., Louka, D. A., Nianias, N. P., Siafaka, P., Tsagkalias, I. & Tsintzou, G. (2012). Recent advances in the chemical recycling of polymers (PP, PS, LDPE, HDPE, PVC, PC, Nylon, PMMA). En Dimitris S. Achilias (Ed.), *Material Recycling – Trends and Perspectives* (pp. 3–64). IntechOpen. https://doi.org/10.5772/33457

Ahvenainen, R. (Ed.). (2003). Related titles from Woodhead's food science, technology and nutrition list. En *Novel Food Packaging Techniques* (p. ii). Woodhead Publishing. https://doi.org/10.1016/B978-1-85573-675-7.50001-8

Almeida, D. & Marques, M. de F. (2016). Thermal and catalytic pyrolysis of plastic waste. *Polímeros, 26*, 44–51. https://doi.org/10.1590/0104-1428.2100

Anukiruthika, T., Sethupathy, P., Wilson, A., Kashampur, K., Moses, J. A. & Anandharamakrishnan, C. (2020). Multilayer packaging: Advances in preparation techniques and emerging food applications. *Comprehensive Reviews in Food Science and Food Safety*, *19*(3), 1156–1186. https://doi.org/10.1111/1541-4337.12556

Arenas-Vivo, A., Beltrán, F. R., Alcázar, V., de la Orden, M. U. & Martinez Urreaga, J. (2017). Fluorescence labeling of high density polyethylene for identification and separation of selected containers in plastics waste streams. Comparison of thermal and photochemical stability of different fluorescent tracers. *Materials Today Communications, 12*, 125–132. https://doi.org/10.1016/j.mtcomm.2017.07.008

Avérous, L. & Pollet, E. (Eds.). (2012). *Environmental Silicate Nano-Biocomposites*. Springer. https://doi.org/10.1007/978-1-4471-4108-2

Balla, E., Daniilidis, V., Karlioti, G., Kalamas, T., Stefanidou, M., Bikiaris, N. D., Vlachopoulos, A., Koumentakou, I. & Bikiaris, D. N. (2021). Poly(lactic Acid): A versatile biobased polymer for the future with multifunctional properties – From monomer synthesis, polymerization techniques and molecular weight increase to PLA applications. *Polymers, 13*(11), 1822. https://doi.org/10.3390/polym13111822

Barceló, D. & Kostianoy, A. (2022). *The Handbook of Environmental Chemistry*. www.springer.com/series/698

Bassyouni, M., Zoromba, M. S., Abdel-Aziz, M. H. & Mosly, I. (2022). Extraction of nanocellulose for eco-friendly biocomposite adsorbent for wastewater treatment. *Polymers*, *14*(9), 1852. https://doi.org/10.3390/polym14091852

Bertin, S. & Robin, J.-J. (2002). Study and characterization of virgin and recycled LDPE/PP blends. *European Polymer Journal*, *38*(11), 2255–2264. https://doi.org/10.1016/S0014-3057(02)00111-8

Biblox. (2021, 18 octubre). Los envases multicapa y el dilema de su reciclabilidad. *Biblox*. https://biblox.es/los-envases-multicapa-y-el-dilema-de-su-reciclabilidad/

Brouwer, M. T., Thoden van Velzen, E. U., Augustinus, A., Soethoudt, H., De Meester, S. & Ragaert, K. (2018). Predictive model for the Dutch post-consumer plastic packaging recycling system and implications for the circular economy. *Waste Management, 71*, 62–85. https://doi.org/10.1016/j.wasman.2017.10.034

Cabrera, G., Touil, I., Masghouni, E., Maazouz, A. & Lamnawar, K. (2021). Multi-micro/nanolayer films based on polyolefins: New approaches from eco-design to recycling. *Polymers*, *13*(3), 413. https://doi.org/10.3390/polym13030413

Campos, G., Costantino, A., Pettarin, V. & Morales, F. (2022, octubre). *Influence of a novel drying technique on the mechanical performance of recycled PET*. 14th World Congress and Expo on Recycling, Barcelona, España.

CEFLEX | A circular economy for flexible packaging. (s. f.). CEFLEX. Recuperado 1 de enero de 2023, a partir de https://ceflex.eu/

CETEC. (2017). *Mejora de propiedades de films de polietileno mediante tecnología capa a capa*. Centro Tecnologico del Calzado y del Plastico.

Circular economy: definition, importance and benefits | News | European Parliament. (2015, 2. diciembre). www.europarl.europa.eu/news/en/headlines/economy/20151201STO05603/circular-economy-definition-importance-and-benefits

Costantino, M. A., Pettarin, V., Pontes, A. & Frontini, P. (2015). Mechanical performance of double gated injected metallic looking polypropylene parts. *Express Polymer Letters*, *9*(11), 1040–1051. https://doi.org/10.3144/expresspolymlett.2015.93

Cristalizador / secador infrarrojo: Plastics Technology México. (s. f.). Recuperado 22 de diciembre de 2022, a partir de www.pt-mexico.com/banco-de-conocimiento/secado-de-plasticos/tipos-de-secadores/cristalizador-secador-infrarrojo

Dai, L., Karakas, O., Cheng, Y., Cobb, K., Chen, P. & Ruan, R. (2023). A review on carbon materials production from plastic wastes. *Chemical Engineering Journal, 453*, 139725. https://doi.org/10.1016/j.cej.2022.139725

Datos y cifras de Tetra Pak, incluidos los resultados del desempeño financiero. (s. f.). Recuperado 20 de diciembre de 2022, a partir de www.tetrapak.com/es-ar/about-tetra-pak/the-company/facts-figures

Davis, R. (s. f.). *Los Retos del PET Reciclado | Textiles Panamericanos.* Recuperado 22 de diciembre de 2022, a partir de https://textilespanamericanos.com/textiles-panamericanos/2019/07/los-retos-del-pet-reciclado/

Echeverria, T., Rosales, C., Palazzo, G. & Pettarin, V. (2022). Viability of milk pouch recycling into monolayer blown films. *Sustainable Materials and Technologies, 32*, e00412. https://doi.org/10.1016/j.susmat.2022.e00412

Engineer, C., Parikh, J. & Raval, A. (2011). Review on hydrolytic degradation behavior of biodegradable polymers from controlled drug delivery system. *Trends in Biomaterials and Artificial Organs*, *25*(2), 79–86.

EUBIO_Admin. (s. f.). Market. *European Bioplastics e.V.* Recuperado 20 de diciembre de 2022, a partir de www.european-bioplastics.org/market/

Faraca, G. & Astrup, T. (2019). Plastic waste from recycling centres: Characterisation and evaluation of plastic recyclability. *Waste Management, 95*, 388–398. https://doi.org/10.1016/j.wasman.2019.06.038

Fávaro, S. L., Freitas, A. R., Ganzerli, T. A., Pereira, A. G. B., Cardozo, A. L., Baron, O., Muniz, E. C., Girotto, E. M. & Radovanovic, E. (2013). PET and aluminum recycling from multilayer food packaging using supercritical ethanol. *Journal of Supercritical Fluids, 75*, 138–143. https://doi.org/10.1016/j.supflu.2012.12.015

FBR BP Biorefinery & Sustainable Value Chains, FBR Sustainable Chemistry & Technology, Biobased Products, van den Oever, M., Molenveld, K., van der Zee, M. & Bos, H. (2017). *Bio-based and Biodegradable Plastics: Facts and Figures: Focus on Food Packaging in the Netherlands.* Wageningen Food & Biobased Research. https://doi.org/10.18174/408350

Food Packaging Market Size, Share & Growth Report, 2030. (s. f.). Recuperado 1 de enero de 2023, a partir de www.grandviewresearch.com/industry-analysis/food-packaging-market

Foolmaun, R. K. & Ramjeeawon, T. (2012). Disposal of post-consumer polyethylene terephthalate (PET) bottles: comparison of five disposal alternatives in the small island state of Mauritius using a life cycle assessment tool. *Environmental Technology*, *33*(5), 563–572. https://doi.org/10.1080/09593330.2011.586055

Fortelný, I., Kruliš, Z., Michálková, D. & Horák, Z. (1996). Effect of EPDM admixture and mixing conditions on the morphology and mechanical properties of LDPE/PP blends. *Die Angewandte Makromolekulare Chemie*, *238*(1), 97–104. https://doi.org/10.1002/apmc.1996.052380109

Gan, L., Qiu, F., Yue, X., Chen, Y., Xu, J. & Zhang, T. (2021). Aramid nanofiber aerogel membrane extract from waste plastic for efficient separation of surfactant-stabilized oil-in-water emulsions. *Journal of Environmental Chemical Engineering*, *9*(5), 106137. https://doi.org/10.1016/j.jece.2021.106137

García, J. M. (2016). Catalyst: design challenges for the future of plastics recycling. *Chem, 1*(6), 813–815. https://doi.org/10.1016/j.chempr.2016.11.003

Ghanbarzadeh, B., Almasi, H., Ghanbarzadeh, B. & Almasi, H. (2013). Biodegradable polymers. En Rolando Chamy and Francisca Rosenkranz (Eds.), *Biodegradation – Life of Science* (pp. 141–185). IntechOpen. https://doi.org/10.5772/56230

Ghosal, K. & Nayak, C. (2022). Recent advances in chemical recycling of polyethylene terephthalate waste into value added products for sustainable coating solutions – Hope vs. hype. *Materials Advances, 3*(4), 1974–1992. https://doi.org/10.1039/D1MA01112J

Hidalgo Molina, A. (2013). *Diseño de un proceso para la elaboración de tableros aglomerados a partir de envases tetra pak*. Escuela Superior Politécnicxa de Chimborazo.

Horodytska, O., Kiritsis, D. & Fullana, A. (2020). Upcycling of printed plastic films: LCA analysis and effects on the circular economy. *Journal of Cleaner Production, 268*, 122138. https://doi.org/10.1016/j.jclepro.2020.122138

Horodytska, O., Valdés, F. J. & Fullana, A. (2018). Plastic flexible films waste management – A state of art review. *Waste Management, 77*, 413–425. https://doi.org/10.1016/j.wasman.2018.04.023

Iglesias Montes, M. L., Cyras, V. P., Manfredi, L. B., Pettarín, V. & Fasce, L. A. (2020). Fracture evaluation of plasticized polylactic acid / poly (3-HYDROXYBUTYRATE) blends for commodities replacement in packaging applications. *Polymer Testing, 84*, 106375. https://doi.org/10.1016/j.polymertesting.2020.106375

Jiang, L. & Zhang, J. (2017). 7–Biodegradable and biobased polymers. En M. Kutz (Ed.), *Applied Plastics Engineering Handbook (Second Edition)* (pp. 127–143). William Andrew Publishing. https://doi.org/10.1016/B978-0-323-39040-8.00007-9

Jose, S., Aprem, A. S., Francis, B., Chandy, M. C., Werner, P., Alstaedt, V. & Thomas, S. (2004). Phase morphology, crystallisation behaviour and mechanical properties of isotactic polypropylene/high density polyethylene blends. *European Polymer Journal, 40*(9), 2105–2115. https://doi.org/10.1016/j.eurpolymj.2004.02.026

Kaiser, K., Schmid, M. & Schlummer, M. (2018). Recycling of polymer-based multilayer packaging: A review. *Recycling, 3*(1), 1. https://doi.org/10.3390/recycling3010001

Kale, G., Auras, R. & Singh, S. P. (2006). Degradation of commercial biodegradable packages under real composting and ambient exposure conditions. *Journal of Polymers and the Environment, 14*(3), 317–334. https://doi.org/10.1007/s10924-006-0015-6

Karthika, M., Shaji, N., Johnson, A., Neelakandan, M. S., A. Gopakumar, D. & Thomas, S. (2019). Biodegradation of green polymeric composites materials. En P.M. Visakh, Oguz Bayraktar, and Gopalakrishnan Menon (Eds.), *Bio Monomers for Green Polymeric Composite Materials* (pp. 141–159). John Wiley & Sons, Ltd. https://doi.org/10.1002/9781119301714.ch7

Katiyar, V., Tripathi, N., Patwa, R. & Kotecha, P. (2014). Environment friendly packaging plastics. In Sajid Alavi, Sabu Thomas, K. P. Sandeep, Nandakumar Kalarikkal, Jini Varghese, and Srinivasarao Yaragalla (Eds.), *Polymers for Packaging Applications*. CRC Press.

Kindsvater, R., Munarriz, E. & Nudelman, N. S. (2020). Desafios que presentan algunos aditivos químicos para el reciclado de residuos plásticos. Industria plástica argentina y situación internacional. En Academia Nacional de Ciencias Exactas, Físicas y Naturales (Ed.), *Residuos plásticos en Argentina* (pp. 208–224). Academia Nacional de Ciencias Exactas, Físicas y Naturales.

Kirwan, M. J., Plant, S. & Strawbridge, J. W. (2011). Plastics in Food Packaging. En Richard Coles and Mark Kirwan (Eds.), *Food and Beverage Packaging Technology* (pp. 157–212). John Wiley & Sons, Ltd. https://doi.org/10.1002/9781444392180.ch7

Klemm, D., Shmauder, H. & Heinze, T. (2002). Cellulose. En E.J. Vandamme, S. De Baets, A., Steinbüchel (Eds.), *Polysaccharides II, Biopolymers* (Vol. 6, pp. 275–319). Wiley VCH.

Kol, R., Roosen, M., Ügdüler, S., Geem, K. M. V., Ragaert, K., Achilias, D. S., Meester, S. D., Kol, R., Roosen, M., Ügdüler, S., Geem, K. M. V., Ragaert, K., Achilias, D. S. & Meester, S. D. (2021). Recent advances in pre-treatment of plastic packaging waste.

En Dimitris S. Achilias (Ed.), *Waste Material Recycling in the Circular Economy – Challenges and Developments*. IntechOpen. https://doi.org/10.5772/intechopen.99385

Korkmaz, A., Yanik, J., Brebu, M. & Vasile, C. (2009). Pyrolysis of the tetra pak. *Waste Management*, *29*(11), 2836–2841. https://doi.org/10.1016/j.wasman.2009.07.008

Kulkarni, A. K., Daneshvarhosseini, S., & Yoshida, H. (2011). Effective recovery of pure aluminum from waste composite laminates by sub- and super-critical water. *Journal of Supercritical Fluids*, *55*(3), 992–997. https://doi.org/10.1016/j.supflu.2010.09.007

La Mantia, F. P. & Vinci, M. (1994). Recycling poly(ethyleneterephthalate). *Polymer Degradation and Stability*, *45*(1), 121–125. https://doi.org/10.1016/0141-3910(94)90187-2

Larrain, M., Van Passel, S., Thomassen, G., Van Gorp, B., Nhu, T. T., Huysveld, S., Van Geem, K. M., De Meester, S. & Billen, P. (2021). Techno-economic assessment of mechanical recycling of challenging post-consumer plastic packaging waste. *Resources, Conservation and Recycling*, *170*, 105607. https://doi.org/10.1016/j.resconrec.2021.105607

Lase, I. S., Bashirgonbadi, A., van Rhijn, F., Dewulf, J., Ragaert, K., Delva, L., Roosen, M., Brandsma, M., Langen, M. & De Meester, S. (2022). Material flow analysis and recycling performance of an improved mechanical recycling process for post-consumer flexible plastics. *Waste Management*, *153*, 249–263. https://doi.org/10.1016/j.wasman.2022.09.002

Leja, K. & Lewandowicz, G. (2010). Polymer biodegradation and biodegradable polymers – A review. *Polish Journal of Environmental Studies*, *19*(2), 255–266.

Luchese, C. L., Abdalla, V. F., Spada, J. C. & Tessaro, I. C. (2018). Evaluation of blueberry residue incorporated cassava starch film as pH indicator in different simulants and foodstuffs. *Food Hydrocolloids*, *82*, 209–218. https://doi.org/10.1016/j.foodhyd.2018.04.010

Luckachan, G. E. & Pillai, C. K. S. (2011). Biodegradable polymers – A review on recent trends and emerging perspectives. *Journal of Polymers and the Environment*, *19*(3), 637–676. https://doi.org/10.1007/s10924-011-0317-1

Mangaraj, S., Yadav, A., Bal, L. M., Dash, S. K. & Mahanti, N. K. (2019). Application of biodegradable polymers in food packaging industry: A comprehensive review. *Journal of Packaging Technology and Research*, *3*(1), 77–96. https://doi.org/10.1007/s41783-018-0049-y

Maris, J., Bourdon, S., Brossard, J.-M., Cauret, L., Fontaine, L. & Montembault, V. (2018). Mechanical recycling: Compatibilization of mixed thermoplastic wastes. *Polymer Degradation and Stability*, *147*, 245–266. https://doi.org/10.1016/j.polymdegradstab.2017.11.001

McKeen, L. (2012). 12–Renewable resource and biodegradable polymers. En L. McKeen (Ed.), *The Effect of Sterilization on Plastics and Elastomers (Third Edition)* (pp. 305–317). William Andrew Publishing. https://doi.org/10.1016/B978-1-4557-2598-4.00012-5

Mendes, A. C. & Pedersen, G. A. (2021). Perspectives on sustainable food packaging: Is bio-based plastics a solution? *Trends in Food Science & Technology*, *112*, 839–846. https://doi.org/10.1016/j.tifs.2021.03.049

Morris, B. A. (2017). 4–Commonly used resins and substrates in flexible packaging. En B. A. Morris (Ed.), *The Science and Technology of Flexible Packaging* (pp. 69–119). William Andrew Publishing. https://doi.org/10.1016/B978-0-323-24273-8.00004-6

Mourad, A.-H. I., Akkad, R. O., Soliman, A. A. & Madkour, T. M. (2009). Characterisation of thermally treated and untreated polyethylene–polypropylene blends using DSC, TGA and IR techniques. *Plastics, Rubber and Composites*, *38*(7), 265–278. https://doi.org/10.1179/146580109X12473409436625

Nasri, Y., Benaniba, M. T. & Bouquey, M. (2022). Elaboration and characterization of polymers used in flexible multilayer food packaging. *Materials Today: Proceedings, 53*, 91–95. https://doi.org/10.1016/j.matpr.2021.12.390

Ncube, L. K., Ude, A. U., Ogunmuyiwa, E. N., Zulkifli, R. & Beas, I. N. (2020). Environmental impact of food packaging materials: A review of contemporary development from conventional plastics to polylactic acid based materials. *Materials, 13*(21), 4994. https://doi.org/10.3390/ma13214994

OECD. (s. f.). *Global Plastic Outlook* [Data set]. Organisation for Economic Co-operation and Development. https://doi.org/10.1787/c0821f81-en

OSP. (2022, 19. julio). Envase monomaterial ¿más sostenible que el multicapa? *Osona Seal Pack*. www.osonasealpack.com/envase-monomaterial/

Penava, N. V., Rek, V. & Houra, I. F. (2013). Effect of EPDM as a compatibilizer on mechanical properties and morphology of PP/LDPE blends. *Journal of Elastomers & Plastics, 45*(4), 391–403. https://doi.org/10.1177/0095244312457162

Picuno, C., Alassali, A., Chong, Z. K. & Kuchta, K. (2021). Flows of post-consumer plastic packaging in Germany: An MFA-aided case study. *Resources, Conservation and Recycling, 169*, 105515. https://doi.org/10.1016/j.resconrec.2021.105515

Plastics Europe • Enabling a sustainable future. (s. f.). Plastics Europe. Recuperado 21 de diciembre de 2022, a partir de https://plasticseurope.org/

Pooresmaeil, M., Behzadi Nia, S. & Namazi, H. (2019). Green encapsulation of LDH(Zn/Al)-5-Fu with carboxymethyl cellulose biopolymer; new nanovehicle for oral colorectal cancer treatment. *International Journal of Biological Macromolecules, 139*, 994–1001. https://doi.org/10.1016/j.ijbiomac.2019.08.060

Procesabilidad y sustentabilidad del polipropileno: reciclado mecánico. (s. f.). Recuperado 22 de diciembre de 2022, a partir de www.pt-mexico.com/articulos/procesabilidad-y-sustentabilidad-del-polipropileno-reciclado-mecanico

Punia Bangar, S., Chaudhary, V., Thakur, N., Kajla, P., Kumar, M. & Trif, M. (2021). Natural antimicrobials as additives for edible food packaging applications: A review. *Foods, 10*(10), 2282. https://doi.org/10.3390/foods10102282

Ragaert, K., Delva, L. & Van Geem, K. (2017). Mechanical and chemical recycling of solid plastic waste. *Waste Management, 69*, 24–58. https://doi.org/10.1016/j.wasman.2017.07.044

Ragaert, K., Hubo, S., Delva, L., Veelaert, L. & Du Bois, E. (2018). Upcycling of contaminated post-industrial polypropylene waste: A design from recycling case study. *Polymer Engineering & Science, 58*(4), 528–534. https://doi.org/10.1002/pen.24764

Ratnayake, W. S., Hoover, R., Shahidi, F., Perera, C. & Jane, J. (2001). Composition, molecular structure, and physicochemical properties of starches from four field pea (Pisum sativum L.) cultivars. *Food Chemistry, 74*(2), 189–202. https://doi.org/10.1016/S0308-8146(01)00124-8

Research, Z. M. (2018, 22. junio). *Global Cosmetic Products Market Will Reach USD 863 Billion by 2024: Zion Market Research*. GlobeNewswire News Room. www.globenewswire.com/news-release/2018/06/22/1528369/0/en/Global-Cosmetic-Products-Market-Will-Reach-USD-863-Billion-by-2024-Zion-Market-Research.html

Rosales, C., Aranburu, N., Otaegi, I., Pettarin, V., Bernal, C., Müller, A. J. & Guerrica-Echevarría, G. (2022). Improving the mechanical performance of LDPE/PP blends through microfibrillation. *ACS Applied Polymer Materials, 4*(5), 3369–3379. https://doi.org/10.1021/acsapm.1c01932

Rosales, C., Bernal, C. & Pettarin, V. (2020). Effect of blend composition and related morphology on the quasi-static fracture performance of LLDPE/PP blends. *Polymer Testing, 90*, 106598. https://doi.org/10.1016/j.polymertesting.2020.106598

Rosales, C., Brendstrup, D., Bernal, C. & Pettarin, V. (2020). Morphology/tensile performance relationship for LLDPE/PP double gated injected blends. *Advanced Materials Letters, 11*(2), 1–6. https://doi.org/10.5185/amlett.2020.021472

Rosales, C., Costantino, A., Palazzo, G., Bernal, C., Defacio Dutra, R. & Pettarin, V. (2020). Influence of different copolymer based compatibilizers on performance of pristine and recycled PP/PE blends. *Advanced Materials Letters, 11*(11), 20111572–20111572. https://doi.org/10.5185/amlett.2020.111572

Şahin, G. G. & Karaboyacı, M. (2021). Process and machinery design for the recycling of tetra pak components. *Journal of Cleaner Production, 323*, 129186. https://doi.org/10.1016/j.jclepro.2021.129186

Sahota, A. (2013). Front Matter. En Amarjit Sahota (Ed.), *Sustainability: How the Cosmetics Industry is Greening Up* (pp. i–xxviii). John Wiley & Sons, Ltd. https://doi.org/10.1002/9781118676516.fmatter

Saini, R. (2017). Biodegradable polymers. *International Journal of Applied Chemistry, 13*(2), 179–196.

Salehi Morgani, M., Jalali Dil, E. & Ajji, A. (2021). Effect of processing condition and antioxidants on visual properties of multilayer post-consumer recycled high density polyethylene films. *Waste Management, 126*, 239–246. https://doi.org/10.1016/j.wasman.2021.03.005

Sarasam, A. R., Krishnaswamy, R. K. & Madihally, S. V. (2006). blending chitosan with polycaprolactone: Effects on physicochemical and antibacterial properties. *Biomacromolecules, 7*(4), 1131–1138. https://doi.org/10.1021/bm050935d

Schwarz, A. E., Ligthart, T. N., Godoi Bizarro, D., De Wild, P., Vreugdenhil, B. & van Harmelen, T. (2021). Plastic recycling in a circular economy; determining environmental performance through an LCA matrix model approach. *Waste Management, 121*, 331–342. https://doi.org/10.1016/j.wasman.2020.12.020

Shah, T. V. & Vasava, D. V. (2019). A glimpse of biodegradable polymers and their biomedical applications. *e-Polymers, 19*(1), 385.

Shaikh, S., Yaqoob, M. & Aggarwal, P. (2021). An overview of biodegradable packaging in food industry. *Current Research in Food Science, 4*, 503–520. https://doi.org/10.1016/j.crfs.2021.07.005

Singh, P., Saengerlaub, S., Abas Wani, A. & Langowski, H. (2012). Role of plastics additives for food packaging. *Pigment & Resin Technology, 41*(6), 368–379. https://doi.org/10.1108/03699421211274306

Singla, R. & Mehta, R. (2012). Preparation and characterization of polylactic acid-based biodegradable blends processed under microwave radiation. *Polymer-Plastics Technology and Engineering, 51*(10), 1014–1017. https://doi.org/10.1080/03602559.2012.680562

Siracusa, V., Rocculi, P., Romani, S. & Rosa, M. D. (2008). Biodegradable polymers for food packaging: A review. *Trends in Food Science & Technology, 19*(12), 634–643. https://doi.org/10.1016/j.tifs.2008.07.003

Soares, C. T. de M., Ek, M., Östmark, E., Gällstedt, M. & Karlsson, S. (2022). Recycling of multi-material multilayer plastic packaging: Current trends and future scenarios. *Resources, Conservation and Recycling, 176*, 105905. https://doi.org/10.1016/j.resconrec.2021.105905

Solis, M. & Silveira, S. (2020). Technologies for chemical recycling of household plastics – A technical review and TRL assessment. *Waste Management, 105*, 128–138. https://doi.org/10.1016/j.wasman.2020.01.038

Statistics | Eurostat. (s. f.). Recuperado 26 de diciembre de 2022, a partir de https://ec.europa.eu/eurostat/databrowser/view/ENV_WASPACR__custom_1512763/default/bar?lang=en

Strapasson, R., Amico, S. C., Pereira, M. F. R. & Sydenstricker, T. H. D. (2005). Tensile and impact behavior of polypropylene/low density polyethylene blends. *Polymer Testing*, *24*(4), 468–473. https://doi.org/10.1016/j.polymertesting.2005.01.001

Taherimehr, M., YousefniaPasha, H., Tabatabaeekoloor, R. & Pesaranhajiabbas, E. (2021). Trends and challenges of biopolymer-based nanocomposites in food packaging. *Comprehensive Reviews in Food Science and Food Safety*, *20*(6), 5321–5344. https://doi.org/10.1111/1541-4337.12832

Tai, C. M., Li, R. K. Y. & Ng, C. N. (2000). Impact behaviour of polypropylene/polyethylene blends. *Polymer Testing*, *19*(2), 143–154. https://doi.org/10.1016/S0142-9418(98)00080-4

The Plastic Leak Project Guidelines. (2020). *Quantis*. https://quantis.com/report/the-plastic-leak-project-guidelines/

Tian, K. & Bilal, M. (2020). Chapter 15 – Research progress of biodegradable materials in reducing environmental pollution. En Pardeep Singh, A. Kumar & A. Borthakur (Eds.), *Abatement of Environmental Pollutants* (pp. 313–330). Elsevier. https://doi.org/10.1016/B978-0-12-818095-2.00015-1

Vert, M., Santos, I. D., Ponsart, S., Alauzet, N., Morgat, J.-L., Coudane, J. & Garreau, H. (2002). Degradable polymers in a living environment: where do you end up? *Polymer International*, *51*(10), 840–844. https://doi.org/10.1002/pi.903

Vollmer, I., Jenks, M. J. F., Roelands, M. C. P., White, R. J., van Harmelen, T., de Wild, P., van der Laan, G. P., Meirer, F., Keurentjes, J. T. F. & Weckhuysen, B. M. (2020). Beyond mechanical recycling: Giving new life to plastic waste. *Angewandte Chemie International Edition, 59*(36), 15402–15423. https://doi.org/10.1002/anie.201915651

Wong, A. C.-Y. & Lam, F. (2002). Study of selected thermal characteristics of polypropylene/polyethylene binary blends using DSC and TGA. *Polymer Testing*, *21*(6), 691–696. https://doi.org/10.1016/S0142-9418(01)00144-1

Zallaya, S., El Achkar, J. H., Chacra, A. A., Shatila, S., El Akhdar, J. & Daher, Y. (2023). Steam gasification modeling of polyethylene (PE) and polyethylene terephthalate (PET) wastes: A case study. *Chemical Engineering Science, 267*, 118340. https://doi.org/10.1016/j.ces.2022.118340

Zawadiak, J., Wojciechowski, S., Piotrowski, T. & Krypa, A. (2017). Tetra pak recycling – Current trends and new developments. *American Journal of Chemical Engineering*, *5*(3), 37. https://doi.org/10.11648/j.ajche.20170503.12

Zhang, K., Huang, T.-S., Yan, H., Hu, X. & Ren, T. (2020). Novel pH-sensitive films based on starch/polyvinyl alcohol and food anthocyanins as a visual indicator of shrimp deterioration. *International Journal of Biological Macromolecules, 145*, 768–776. https://doi.org/10.1016/j.ijbiomac.2019.12.159

Zmijková, D., Švédová, B. & Růžičková, J. (2022). Polycyclic aromatic hydrocarbons in biochar originated from pyrolysis of aseptic packages (Tetra Pak®). *Sustainable Chemistry and Pharmacy, 27*, 100682. https://doi.org/10.1016/j.scp.2022.100682

Zúñiga-Muro, N. M., Bonilla-Petriciolet, A., Mendoza-Castillo, D. I., Duran-Valle, C. J., Silvestre-Albero, J., Reynel-Ávila, H. E. & Tapia-Picazo, J. C. (2021). Recycling of Tetra pak wastes via pyrolysis: Characterization of solid products and application of the resulting char in the adsorption of mercury from water. *Journal of Cleaner Production, 291*, 125219. https://doi.org/10.1016/j.jclepro.2020.125219

5 A Comprehensive Techno-commercial Analysis of Biomedical & COVID Waste-related Situation in India

Abetted with a Case Study

Atun Roy Choudhury, Sankar Ganesh Palani, Surajit Chakraborty, Brajesh Dubey, and Sanjay Garg

5.1 INTRODUCTION

Transitioning from ancient Ayurveda naturopathy to modernized diverse allopathic medicinal services, the healthcare industry has witnessed a metamorphic boon in treating ailments. However, the rampant adaptation of allopathy in parallel magnified biomedical waste generation due to the increased use of synthetic compounds, heavy metals, and other cytotoxic elements. Biochemically synthesized medicines have limited self-life, beside treatment and surgeries yielding highly infectious anatomical wastes. The patient count also took an uncanny diversion due to multiple reasons enlisted as population boom, poor sanitation, new terminal ailments etc. Undoubtedly, improper handling and management of medical waste shall further trigger the risk of secondary infection spearing through pathogenic virulence. An estimate publicized by the Centers for Disease Control and Prevention, USA, revealed that the mismanagement of biomedical waste (BMW) could lead to Acquired Immune Deficiency Syndrome (AIDS) proliferation among sanitation workers at a ratio of 1 in 250. Also the probability of contracting Hepatitis B can be as high as 30% (Chand et al., 2021).

Domestically, the All India Institutes of Medical Sciences (AIIMS) has raised the utmost concern over the non-segregation and dereliction of BMW (Kumar et al., 2017). In this regard, the ramifications are not restricted only to the healthcare

DOI: 10.1201/9781003364467-5

establishments and their active dwellers but rigorously to the surrounding habitations and environment.

As per BMW rules 2016, the waste which gets generated during the treatment, diagnosis and immunization of human beings/animals in the healthcare units like hospitals, primary health centers and also in health camps are known as Biomedical waste. Biomedical waste is often case-specific and primarily comprises non-infectious (75–88%), infectious (10–15%), and hazardous (5–10%) matter (CPCB, 2018a). Due to the rise in the highly infectious diseases and mismanagement of BMW, in 1995 the government of India started drafting a rule for the management of BMW. Later in 1998, the rule entitled "Biomedical Waste (Management and Handling) Rules, 1998" was published and implemented. According to the 2017 annual report on BMW by Central Pollution Control Board (CPCB), about 559 tons of BMW was generated in India on a regular basis. Approx. 3000 hospitals, 21000 quarantine facilities, 1500 sample collection outlets, and 274 testing labs are the primary contributors to BMW in India. The regular comprehensive waste generation in India is standing about 600 tons of regular BMW and 100 tons of COVID-related BMW (CBMW) (Ilyas et al., 2020). The sudden surge in BMW generation has put a question on the preparedness of the country in terms of cumulative incineration-based BMW handling capacity of 850 tons per day. The majority of the states have already reached the threshold value of their incineration capacities and are forced to look for alternatives in terms of the disposal of BMW in treatment storage and disposal facilities (TSDFs). Quite a few northeastern states and relatively smaller states with no existing incinerator facilities were forced to opt for disposal through deep burial/landfilling or transport of CBMW to neighbor states (CPCB, 2019a). The sudden escalation in the waste quantity was the outcome of zero source segregation from quarantine facilities and households. Anyhow, the practice was mended to a certain extent during the second wave leading to stipulated waste generation against the gigantic addition in daily caseload (Bhatt et al., 2021; Kar et al., 2021).

With COVID discoveries happening regularly, the repository of literature is getting upgraded almost every day. Numerous literature has so far tried to comprehend multiple dimensions of COVID-related information sharing. But with consistent mutations, COVID-19 never failed to bring new challenges in front of the scientific fraternity. The existing review works have mostly summarized various waste management-related aspects and discussed the comparative potential of available technologies (Chand et al., 2021; Kothari et al., 2021; Joshi and Mehendale, 2021; Bhatt et al., 2021; Kar et al., 2021). Community participation, role of administrative and statutory bodies, ground reality of CBMW management, sustainable business model for waste management facilities etc. received little to no attention. Thus, the present review tried to exercise a holistic approach of emphasizing mostly the unexplored aspects of COVID-19 assorted with the case study of Ghaziabad common biomedical waste treatment facility (CBMWTF).

5.2 GENESIS OF BMW LEGISLATION IN INDIA AND OTHER RELATED LEGISLATION

Visualizing the catastrophic impact of BMW mismanagement, India framed its Biomedical Waste Management Rule as early as 1998, exercising the power of

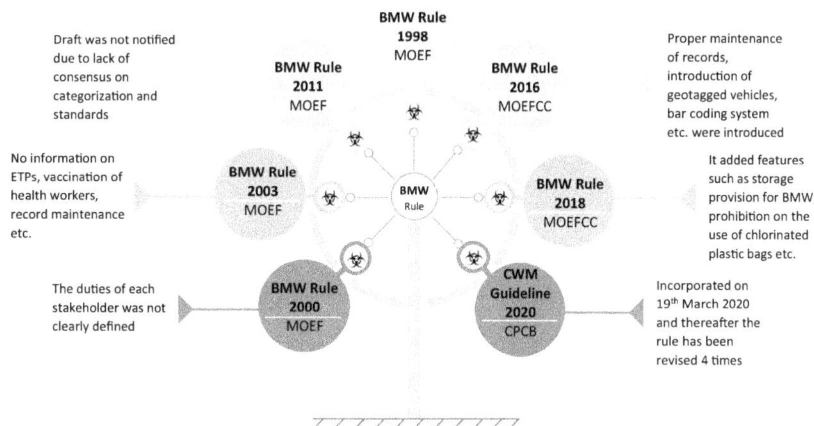

FIGURE 5.1 Genesis of BMW legislations.

rulemaking under the umbrella of the act coined as Environment (Protection) Act, 1986, which was then successively amended in 2000, 2003, 2011, 2016, and 2018 (Figure 5.1).

Under the new Biomedical Waste Management rules 2016, State Pollution Control Board is considered the prescribed authority for implementing the state's provisions. Healthcare facilities (HCFs) are ensured to segregate the waste according to the biomedical waste classification as per the regulation framed.

New laws have enunciated the duties of the corresponding authorities in Schedule III of Biomedical Waste (Management and Handling) Rules 2016. The rule has specified the responsibilities and duties of the occupier and service provider. The newly introduced responsibilities of the occupier include: the waste from yellow and blue colored bags are pretreated, while the wastes from the red and green colored bags could be directly handed over to the CBWTF, source segregation of liquid chemical waste and ensure it is either pretreated or neutralized before mixing it with the effluent. While the specified operator duties include: collected waste is treated as per schedule II of BMW rules, 2016. Moreover, the detailed amendments in BMW rule 2016 are delineated in Table 5.1.

It has also framed standards for treatments and disposal techniques based on the classification of wastes in Schedule I and II of Biomedical Waste (Management and Handling) Rules 2016 (Kashyap, 2021).

5.3 GENERATION OF BMW FROM DIFFERENT CITIES AND STATES

The amount of BMW produced in India is estimated to be between 0.3kg and 1kg per bed per day. The comprehensive BMW generation has anyhow risen up from 517 tons/day to 618 tons/day between 2016 and 2018 (Figure 5.2). The alarming increment of almost 100 tons a day even elevated multiple manifolds during 2020 due to the introduction of the global pandemic. Though the overall BMW generation

TABLE 5.1
Amendments between BMW Rule 1998 and 2016

No.	Area of amendment	Particular
1	Amendment in the inclusion of new healthcare programs	Various healthcare-oriented events such as immunization, enucleation, and blood donation camps etc. have been declared as the potential source of biomedical waste and were taken into provision.
2	Amendment in the duties of the occupier	New occupier responsibilities are as follows: pretreatment of pathological lab waste, abolishing the usage of chlorinated plastics, separate handling and management of liquid waste in ETP, extending necessary training and immunizing the staff, enabling barcoding system for the BMW bags to minimize the mismanagement, and all the major incidents need to be reported to the pollution control board.
3	Amendment in the duties of the CBMWTF	New service provider responsibilities are as follows: enable GPS tracking and barcoding for BMW carriers, minimum 5 years of record maintenance with complete details of individual cycles like duration, time, and date etc. for both disinfection and destruction treatment units
4	Amendment in the establishment of decentralized BMWTFs	In case of the availability of CBMWTFs within a circle of 75 km, HCEs are not permitted to establish individual BMW treatment facilities. While in the reverse scenario, it can be established with prior permission from the regulatory body.
5	Amendment in treatment and disposal	• Chemical disinfection dosage of 1% of hypochlorite is amended to 10% with minimum availability of 30% residual chlorine for 20 min. • Deep burial is only restricted to remote and rural places with no disposal facilities. • Disposal method of Cytotoxic drugs and waste has been amended from secured landfill to incineration. • All types of drugs should be disposed of only using incineration or else should not be accepted. Source segregation of different recyclables in color-coded bags and the same can be handed over to the authorized recycler with valid registration after proper disinfection.
6	Amendment in emission standard	• Suspended particulate matter permissible limit of discharge is reduced from 150 to 50 mg/Nm^3. • Retention time in the secondary incineration chamber is extended from 1 to 2 seconds. • New standards for dioxin and furans were introduced.

Source: Bhalla et al., 2019.

followed the path of gradual increment during the above-mentioned tenure, the figure for the maximum generation was inconsistent. Maharashtra, Karnataka, and Kerala topped the chart with the highest regular BMW generation of 71.5, 67.3, and 71.9 TPD respectively between 2016 to 2018. While Lakshadweep and Andaman Nicobar were the lowermost BMW generators with values as low as 80, 187, and 199 KgPD for the same period. Further, the average regular BMW generation followed a linear progression. The values for the individual waste calendars were 14.4, 15.5, and 16.9 TPD with constant increments of 7.6% and 9% respectively (Figure 5.2). Additionally, the generation of BMW has risen by around 15 times since the COVID-19 outbreak. A single COVID-19 patient in India is estimated to generate between 4.5 and 15 kg of BMW every day.

As per the data published by MoEFCC, India has a total of 2,69,823 HCFs with the largest and lowest count of 60,410 and 25 in Maharashtra and Lakshadweep respectively. Contrarily, the cumulative count of CBMWTFs is only limited to 2000. Maharashtra is enlisted at the top of the count with 31 facilities. A large number of states such as Andaman Nicobar, Arunachal Pradesh, Goa, Lakshadweep, Mizoram, Nagaland, Sikkim, Tripura, Puducherry, Manipur, Meghalaya, Kerala, DD and DNH, Chandigarh, Assam etc. either have 0 or 1 CBMWTFs. Based on the lacuna between the regular quantity of BMW generation (608.4 TPD) and treatment (492.6 TPD) the discrepancy was as high as 19%. Citing the hazard component, the gap of 19% may prove to be fatal. Kerala and Andaman Nicobar were the highest and lowest BMW generators during 2018 with values of 71.97 TPD and 0.2 TPD respectively. While Karnataka and Lakshadweep were topmost and lowermost on the list in terms of effective handling and management. With 26 CBMWTFs Karnataka had addressed

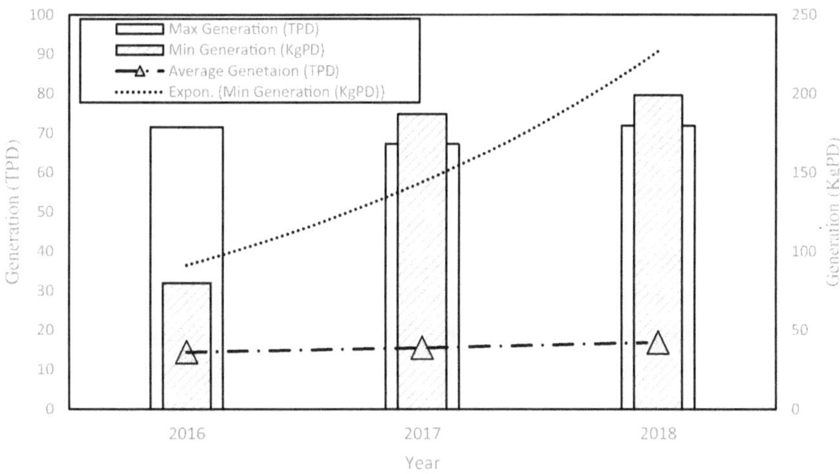

FIGURE 5.2 Variation in BMW generation in India between 2016 and 2018.

Source: CPCB, 2016a; CPCB, 2017a; CPCB, 2018a; CPCB, 2019a; CPCB, 2019b, Banerjee, 2020; MOHFW, 2020.

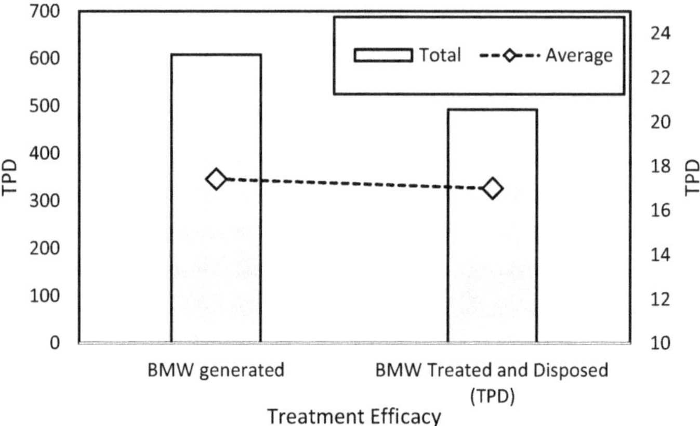

FIGURE 5.3 Comprehensive BMW management scenario in India.

Source: CPCB, 2018a.

the entire BMW generated while Lakshadweep with no existing CBMWTFs only managed to address 0.11 TPD against 0.53 TPD generation through in-house captivity and micro incineration techniques (Figure 5.3) (CPCB, 2018a).

For the purpose of managing biomedical waste, both bedded and non-bedded healthcare facilities need to obtain consent from the concerned State Pollution Control Board/Pollution Control Committee (SPCB/PCC).

5.4 GENERATION OF COVID WASTE FROM DIFFERENT CITIES AND STATES

Since the COVID-19 epidemic, the processing, collection, control, and disposal of waste have been major issues for all nations (MoEFCC, 2020a; MoEFCC 2020b). The global generation of BMW has increased by 40% as a result of the pandemic. While detailed data is yet to be released, biomedical waste appears to be on the rise in India. According to a leading newspaper, regular biomedical waste produced by medical institutions in Delhi has increased from 500 grammes per bed to an average of 2.5 kg to 4 kg per bed in recent months. Many HCFs faced issues such as proper waste disposal during COVID-19. Sanitation staff and waste pickers were the ones who were most at risk as a result of this. However, COVID-19 has boosted the generation of plastic and biomedical waste (Swain and Kumar, 2021).

In comparison with data delineated for 2018, the average generation of BMW has sharply escalated in the COVID era. The variation in quantity was as high as 10.34 times. The average generation of COVID-19 BMW has linearly decreased between August to December 2020 and dipped to a value of 129.4 tons per month. The generation of the above waste was consecutively minimum for Lakshadweep except for the month of August 2020 with an average value of 0.325 tons per month. Whereas,

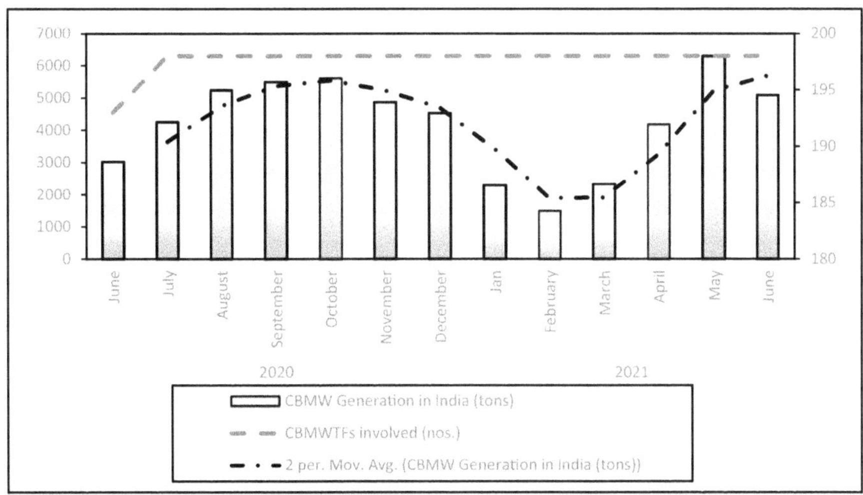

FIGURE 5.4 Variation in CBMW generation in India during first and second wave.

Source: CPCB, 2020f; CPCB, 2020g; CPCB, 2020h; CPCB, 2020i; CPCB, 2020j; CPCB, 2021a; CPCB, 2021b.

Maharashtra had hit the peak of generation thrice during August, November, and December with an average quantity of 865.8 tons per month. Gujrat and Kerala were the other two states with the topmost waste generation of 623 and 641.98 tons per month for September and October respectively. It's needless to mention that waste generation has a direct correlation with the number of active cases. Citing the dip in the regular addition of new cases and speedy recovery of the affected patients, the cumulative waste generation had dipped from approx. 5241 to 4528 tons per month between August and December 2020. The waste generation successively followed a decremental path until February 2021. During the second wave waste generation consecutively escalated to the threshold value of 6293 TPD in the month of May 2021 before witnessing the dip in June. The complete trend of waste generation against the availability of CBMTFs is portrayed in Figure 5.4.

5.5 QUALITATIVE MONITORING BY THE STATUTORY BODIES AND COMPLIANCE STATUS OF RULES

Regulatory compliance of CBMWTFs is governed by the BMW Rule, 2016 and the guidelines issued by CPCB. The state pollution control board (SPCB) is enabled with the authority of pertinent execution and stringent periodic monitoring to ensure a wholesome compliance state.

Irrespective of period scrutiny and rigorous monitoring, quite a few cases of open dumping and burning were reported lately, mostly in rural areas. Citing such issues, National Green Tribunal (NGT) has lately adopted a Suo-moto cognizance approach to evict improper disposal of BMW by CBMWTFs (App. No. 110, 2020). Further, NGT issued an order (dated: August 20, 2020) directing CPCB for a standalone guideline for the monitoring of CBMWTFs and an exclusive drive to be conducted by SPCBs to scrutinize and identify illegal disposal of BMW by CBMWTFs if any.

The generic monitoring of CBMTFs begins by verifying the availability of the consent and authorization of the facility under BMW rule, 2016. Detailed monitoring involves components such as adequacy of infrastructure, operational compliance, data sharing etc.

Operational compliance is of primary criticality from the monitoring perspective. This involves scrutinization of each unit operation in BMW handling and management such as collection and transportation, vehicle tracking, pertinent and complaint processing and disposal of waste, appropriate usage of PPEs etc. The onsite scrutiny incorporates physical inspection and inspection cum monitoring. The frequency for inspection is once a month and this includes physical verification of logbooks, continuous emission monitoring system, site condition etc. Ultimately, a detailed field visit report is documented in tables A–C of Annexure I–III. While monitoring is performed on a quarterly basis and the condition of the shredder, autoclave, incinerator stack, effluent treatment plant (ETP) etc. are evaluated as per the template of Annexure-IV.

Further, to ensure zero illegal disposal, CPCB has recently mandated the implementation of a barcoding system applicable to all CBMWTFs. Later Part C of Annexure IV has been introduced in the guideline to enable COVID-19-related data collection. An interpreted, simplified, and summarized form of the data collection template is delineated in Figure 5.5 (Rajak et al., 2021; Kulkarni, 2020; Bhattacharjee, 2014).

The pollution control committee also has the authority to evaluate the availability of mandatory infrastructure such as geotagged vehicles, mechanical waste feeding systems, upgraded primary and secondary incineration chambers, adequate air pollution control systems and effluent treatment plants, adequacy of green belt etc.

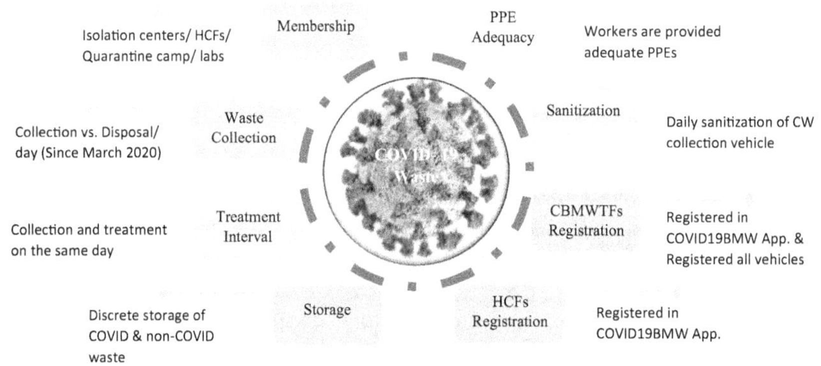

FIGURE 5.5 Model template for COVID-related data collection.

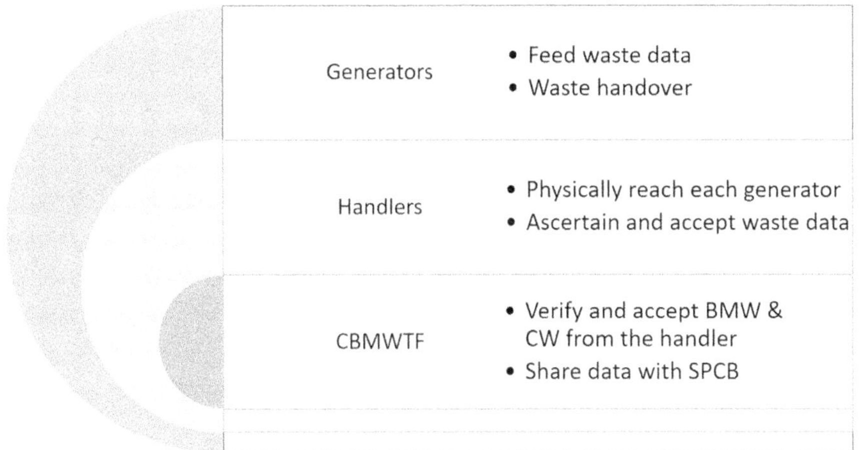

FIGURE 5.6 COVID19BMW application waste tracking system.

Further, the secondary monitoring by SPCBs involves data assessment and record maintenance. It is practiced as per the checklist provided in part C of Annexure III. The collated data is often shared with the local healthcare department to ascertain the lacuna between the actual waste generation and waste scientifically tackled.

As a recent advancement in view of the COVID outbreak, CPCB introduced COVID-19 BMW mobile application as early as May 2020. The platform was established to bring all the HCF and CBMTF stakeholders under a single umbrella and ensure zero tolerance against waste mismanagement. The users were classified under three categories, namely generators, waste handlers, and CBMTFs, and each of the users are provided with a discrete login interface. The interface accepts data (no. of bags and weight of each color category) exclusively as per the color-coded segregation format and this helped in identifying the change in the composition of the waste during the transition phase and both the waves. The model of well-defined responsibility distribution was followed to ensure zero confusion and negligence. The duties of each of the stakeholders are summarized below (CPCB, 2020k) (Figure 5.6).

In this context, the availability of CBMTFs against waste generation and understanding the major accessible BMW management technologies are of primary concern. Figure 5.7 depicts a broad image of the state-wise availability of CBMWTFs and the variability in their numbers between 2008 and 2020.

Figure 5.8 reveals mostly northeastern states have an acute scarcity of incineration-based CBMTFs and primarily rely on the risky method of deep burial. Also, most of the islands under Indian territory such as Andaman and Nicobar, Lakshadweep etc. are also dependent on neighboring states for BMW disposal. Maharashtra has yet to abolish the obsolete burial techniques, irrespective of possessing the highest number of operating CBMTFs. A detailed technology-oriented BMW management geographical map is portrayed below with ancillary information on the present state-wise availability of CBMTFs (Kothari et al., 2021).

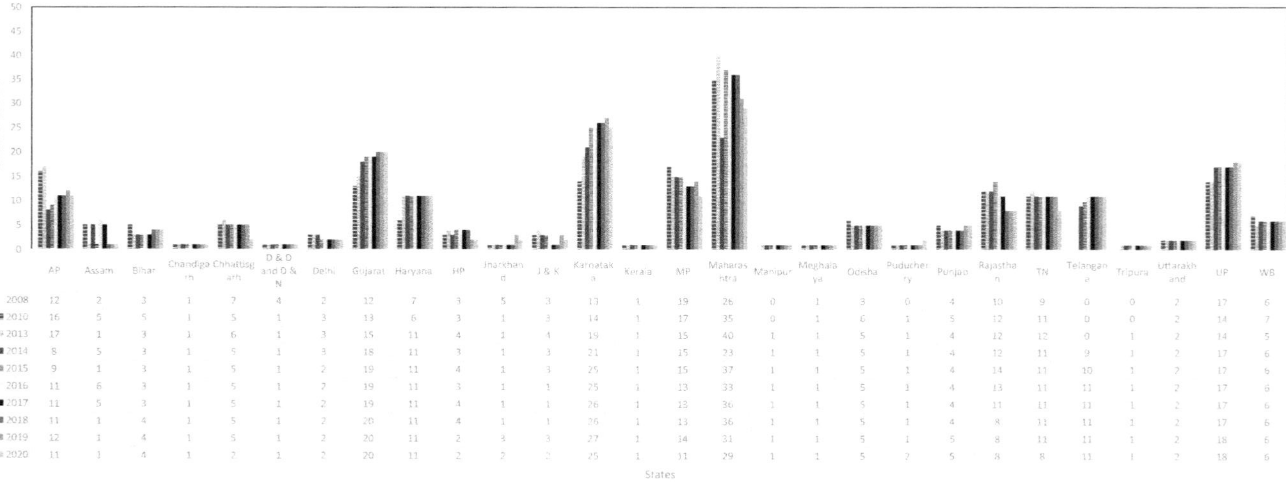

FIGURE 5.7 State-wise variation in number of CBMWTF availability between 2008 and 2020.

Source: CPCB, 2008; CPCB, 2010b; CPCB, 2013; CPCB, 2014; CPCB, 2015; CPCB, 2016b; CPCB, 2017b; CPCB, 2018b; CPCB, 2019c; CPCB, 2021c.

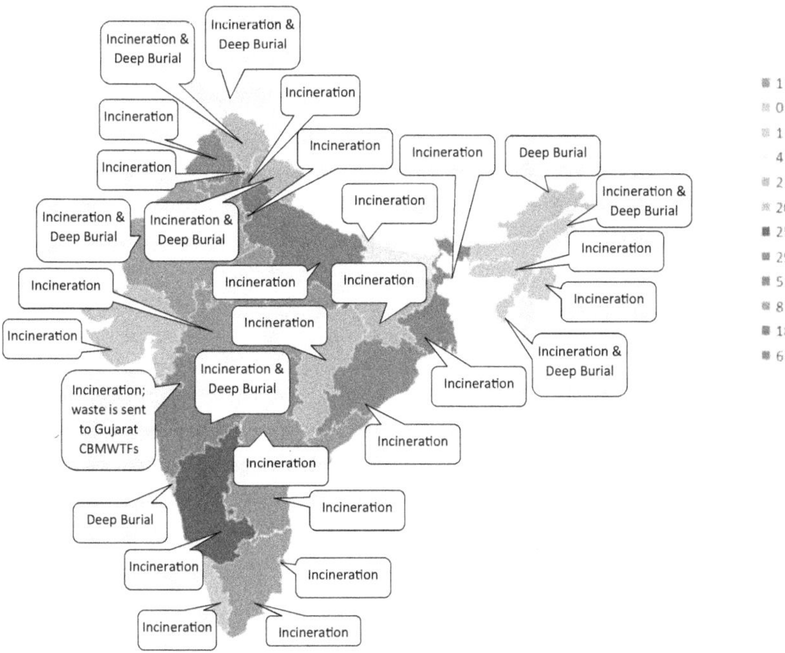

FIGURE 5.8 State-wise identified major BMW disposal technologies and practices.

5.6 FORCEFUL/ILLEGAL DISPOSAL OF BMW IN MUNICIPAL SOLID WASTE MANAGEMENT FACILITY

The disposal of biomedical waste is a critical issue as it can pose significant risks to public health and the environment. The forceful or illegal disposal of BMW in municipal solid waste (MSW) management facilities has become a significant concern in recent years. This article explores the impact of this practice on public health and the environment, and discusses possible solutions to address the issue (Chamberlain, 2019; Jaseem et al., 2017; Joshi and Ahmed, 2016).

The forceful or illegal disposal of biomedical waste in MSW management facilities is a prevalent practice in many countries, particularly in developing nations where regulations are less stringent. This practice has significant adverse impacts on public health and the environment. Biomedical waste contains pathogens, such as bacteria, viruses, and parasites, which can cause serious illnesses, including respiratory and gastrointestinal infections. The disposal of biomedical waste in MSW management facilities can lead to the spread of these pathogens through contaminated air, water, and soil.

The disposal of biomedical waste in MSW management facilities also poses risks to the environment. Biomedical waste contains hazardous chemicals and pharmaceuticals, which can contaminate soil and water sources. The improper disposal of these chemicals can lead to pollution of groundwater and surface water, resulting in long-term environmental damage. The accumulation of biomedical waste in MSW management facilities can also result in the emission of toxic gases and odors, which can have adverse effects on air quality and public health (Datta et al., 2018; Zodpey and Farooqui, 2018).

Various studies have shown the prevalence of forceful or illegal disposal of biomedical waste in MSW management facilities. For instance, a study conducted in India reported that nearly 80% of biomedical waste generated in the country was disposed of improperly, with a significant proportion ending up in MSW management facilities (Kumar et al., 2019). Another study conducted in Nigeria reported that biomedical waste generated from healthcare facilities was often mixed with MSW, and the majority of waste handlers in the facilities were not aware of the dangers of this practice (Oguntoke et al., 2019). The gap between BMW generation and successfully collected and treated waste in India from 2017 to 2019 is depicted in Table 5.2. Constantly a gap of 6.8% to 7.8% was significant during this tenure. Mostly, the unaddressed waste either ended up in MSW facilities alongside the other mixed waste or got dumped illegally. However, this gap was successfully diminished in 2020 and later due to the digitalization of waste collection and treatment systems with the help of the COVID-19 BMW mobile application and website.

To address the issue of forceful or illegal disposal of biomedical waste in MSW management facilities, there is a need for effective regulations and enforcement mechanisms. Governments must establish and enforce stringent regulations to govern the handling, transportation, and disposal of biomedical waste. Regulations should include guidelines on the segregation of biomedical waste from MSW, the use of proper packaging and labeling, and the use of appropriate disinfectants during the disposal process.

In addition to regulations, awareness campaigns should be conducted to educate healthcare workers, waste handlers, and the public on the dangers of improper disposal of biomedical waste. Training programs should also be developed to equip healthcare workers and waste handlers with the necessary skills to handle biomedical waste safely.

Another possible solution is the use of alternative disposal methods, such as incineration, autoclaving, and microwave treatment. These methods have been shown to be effective in the safe disposal of biomedical waste and can significantly reduce the risks associated with forceful or illegal disposal in MSW management facilities.

TABLE 5.2
Gap between BMW Generation and Treatment in India, 2017 to 2019

Year	Biomedical waste generated (MT)	Biomedical waste treated (MT)	Percentage of biomedical waste treated (%)	Reference
2017	6,09,000	5,61,300	92.2	MoEFCC, 2017
2018	6,31,000	5,88,000	93.2	MoEFCC, 2018
2019	6,60,000	6,15,000	93.2	MoEFCC, 2019

However, the implementation of these methods requires significant investments in infrastructure and equipment, which may not be feasible in developing countries.

The forceful or illegal disposal of biomedical waste in MSW management facilities is a significant concern that poses significant risks to public health and the environment. There is a need for effective regulations and enforcement mechanisms, as well as awareness campaigns and training programs to educate healthcare workers, waste handlers, and the public on the dangers of this practice. Alternative disposal methods, such as incineration, autoclaving, and microwave treatment, should also be considered to address the issue. By taking these steps, we can ensure the safe and responsible management of biomedical waste, and protect public health and the environment.

5.7 ILLEGAL REUSE OF CONTAMINATED BMW IN THE HEALTHCARE UNITS AND RECYCLING

In 1992, the government recognized the importance of biomedical waste management and handling and considered it hazardous in nature and included it in the "Basel Convention on the Control of Transboundary Movements of Hazardous Wastes and their Disposal". Due to its acute toxicity and hazardous nature, the Government of India framed and initiated the Biomedical Waste (Handling and Management) rules in 1998 which has been improved again in 2011. Despite stringent rules, mismanagement had been a prominent issue associated with BMW. Presently, BMW is disposed of through three major ways, namely deep burial/secured landfill, disposal after disinfection (autoclaving, microwaving, hydrolyzing etc.), and destruction. Materials of commercial interest such as plastics and metals can be recycled and reused by authorized recyclers post-sterilization. But the involvement of middlemen and unwillingness to pay waste management fees leads to the interim movement of non-sterilized waste to unauthorized recyclers. Additionally, if the quality control and assurance of disinfected waste are not in practice in CBMWTFs then the presence of microbial spores may incur a threat to biosafety. Further, unauthorized disposal of pharmaceutical waste into the CBMWTFs is another illegal instance that is rampantly spreading across the nation due to the non-availability of the TSDFs. The above practice not only incurs an ancillary burden on the CBMWTFs but also diminishes the possibility of alternate fuel and raw material recovery (Sarkodie and Owusu, 2021).

5.8 ISSUES WITH WASTE COLLECTION

Safe and scientific collection of CBMW is a vital part of the handling of this highly infectious waste. CPCB has stipulated the guideline for the pertinent collection and management of CBMW in April 2020. Further, CPCB introduced a mobile application named COVID19BMW in May 2020 to track waste generation and movement-related data. A multilevel data entry and verification assured zero illegal disposal and also extended ease of data management. The observations related to onsite waste collection scenarios and practices are delineated below (Figure 5.9).

FIGURE 5.9 Source segregation and segregated collection of waste as per BMW Rule 2016.

- Primarily all HCFs in the region of Delhi National Capital Region (NCR) used double-layered non-chlorinated plastic bags for the collection of CBMW and foot-operated large bins were used for the temporary in-house storage.
- Primarily all HCFs in the region of Delhi National Capital Region (NCR) used double-layered non-chlorinated plastic bags for the collection of CBMW and foot-operated large bins were used for the temporary in-house storage.
- Non-segregation of regular MSW and CBMW remained a constant problem during the first wave. It incurred a tremendous burden on the incinerator-based CBMTFs. Often the additional loading of incinerators resulted in reduced efficacy of the air pollution control system (APCS). The discharge of black smoke is a prominent indicator of APCS malfunction.
- Better segregation at the source resulted in the reduction of parallel CBMW generation against the number of active cases. By the end of October 2020, more than 80% of the HCFs and CBMWTFs concluded the registration on CPCB mobile application and initiated data sharing using the common platform.
- Public felicitation of waste collectors was witnessed in different parts of the country for their phenomenal service.

5.9 TREATMENT TECHNOLOGIES

Post-secured collection and transportation, processing, and disposal of the waste in CBMWTFs is the most prominent component of BMW management. The selection of the pertinent treatment method is driven by factors such as availability of land, capital and operational expenditure, volume of waste, administrative requirement, anticipated efficacy, regulatory compliance conditions etc. Anyhow, secondary pollution from the wastewater and ash generated is one of the prominent existing issues attributed to the current practices of BMW management. The details of commonly practiced mechanisms are delineated below (Ilyas et al., 2020; Capoor and Bhowmik, 2017; Voudrias, 2016).

5.9.1 Mechanical Processing

The unit operations of mechanical processing involve shredding, granulating, pulverizing, mixing, and crushing. The method is adopted for volume reduction (~ 60%). It is an interim tackling mechanism in case of high volume. The limitations of the process are as follows: no disinfection offered, the process does not facilitate any resource recovery, particulate pollution, likely risk of secondary contamination, the requirement of ancillary treatment for ultimate disposal etc.

5.9.2 Incineration

Incineration of BMW is performed at 1100^0 C to ensure complete destruction. Besides thermal degradation (volume reduction ~ 90%), elevated temperature oxidizes harmful gases such as dioxin and furan in the secondary combustion chamber. Incineration is a mandatory treatment for BMW categories 1, 2, 3, and 6 primarily including anatomical wastes and body fluids. Incineration is the most pertinent mechanism to ensure complete pasteurization. Anyhow, disposal of bottom ash and fly ash is an additional burden attributed to this process. The system also requires a robust air pollution control system in-line to diminish the higher concentration of gasses such as SOx, NOx, PM_{10}, $PM_{2.5}$ etc. Other thermal destruction mechanisms include pyrolysis and gasification (Kulkarni, 2020).

5.9.3 Chemical Treatment

Chlorine compounds are commonly used for disinfecting BMW as they effectively oxidize hazardous chemical compounds and nullify pathogens. Ethylene oxide is also considered for similar applications. However, it is considered costlier. Chemical treatment is most appropriate for treating liquid medical waste.

5.9.4 Microwave Radiation Method

It is an advanced disinfection mechanism for BMW performed with the help of high-frequency (~ 2450 Hz) microwave radiation. The higher operating frequency leads to an increment in operating temperature (~ 100 ^{0}C) and thereby the formation of vapor to ensure sterilization. The typical operating period is of 20 minutes per batch and the efficacy of the process is directly correlated to the influent particle size and surface area. It makes shredding a mandatory pre-treatment for microwave disinfection. The process is primarily suitable for addressing recyclables. Biological (spore) and virological tests are often performed to evaluate the efficacy of disinfection on a periodic basis. The other comparative disinfection technologies include autoclaving (Operating condition: 121 ^{0}C, 15 Psi, 1 h), hydrolysis, ozonation etc.

5.9.5 Irradiation Method

The process pasteurizes the infectious inert fraction of BMW with the help of gamma rays and an electron beam. It is similar to the radiation therapy used for cancer treatment. Similarly, like microwave disinfection, prior shredding enhances the process's efficacy. Presently, the on-field implementation of the process is subjected to hindrances such as a large operational footprint, high OPEX, compromised worker safety etc.

5.9.6 Vitrification

Vitrification offers a partial solution in terms of handling BMW, precisely the glass fraction. Glass cullet produced during the shredding operation is transformed into recycled glass products in this process. A higher furnace temperature of 1500 ^{0}C assures absolute pasteurization. This method also helps in minimizing the metal leachability-related concern intrigued by the disposal of banned drugs.

5.9.7 Inertization and Landfilling

It is a passive mechanism of dilution and encapsulation. The process involves the stabilization of waste with the help of fly ash, lime, and cement. This conventional method is flexible in terms of waste acceptability. Certain fractions of BMW with relatively less moisture content such as discarded drugs and medicines, non-sharp inert, incineration ash etc. can be more effectively handled in this method. The process ensures zero metal leachability with a higher percentage of cement dosage (up to 40%) (Table 5.3).

5.10 DISPOSAL OF CBMW IN TSDFS

India was one of the worst COVID-hit nations across the globe. In parallel, waste generation reflected a similar trend. With about 200 operational CBMWTFs and approx. 4200 tons of waste to tackle every month, CPCB came up with an approach to handle the excess COVID-19-related yellow bag wastes in hazardous waste management facilities. TSDFs originally designed to handle hazardous waste from industries were granted special permission to dispose of yellow bags with the existing incinerators. Typically, the twin-chambered incinerators commissioned in TSDFs are comparable in terms of emission characteristics with the CBMTF incinerators. Anyhow in absence of other ancillary infrastructures such as a separate mechanical feeding system, disposal of BMW is not encouraged in TSDFs. Though it is permitted in unavoidable cases such as prolonged regulatory non-compliance of CBMWTF, provisional breakdown of CBMTF, and excessive waste generation in

TABLE 5.3
Comparative Techno-commercial Assessment of Various BMW Management Technologies

City or Town	CAPEX	OPEX	Coverage	Efficacy	Acceptability
Mechanical processing	Moderate	Moderate	Except liquid BMW & Anatomical waste	Low	Least Preferred
Incineration	High	High	Complete BMW	High	Most Preferred
Chemical treatment	Low	Moderate	Liquid BMW	High	Least Preferred
Microwave disinfection	Moderate	Moderate	Except liquid BMW & Anatomical waste	High	Preferred
Autoclaving	Low	Low	Except liquid BMW & Anatomical waste	Moderate	Preferred
Ozonation	Moderate	Low	Except liquid BMW & Anatomical waste	High	Preferred
Irradiation	High	High	Except liquid BMW & Anatomical waste	High	Least Preferred
Vitrification	Low	Moderate	Only glass fraction	Moderate	Least Preferred
Inertization and landfilling	Low	Low	Complete BMW	Moderate	Preferred

critical situations such as COVID-19. In all of the prior mentioned situations, special consent is required from the SPCB with validity for the TSDFs. Further, the guideline specifies if the TSDF intends to continue with the disposal of BMW, a separate authorization would be required under BMWM rule 2016 alongside the HWM rule, 2016. Also as per the regulatory protocol, the installation of continuous emission monitoring system and its connectivity with the CPCB and SPCB server is one of the primary eligibility criteria for TSDFs to take up the BMW disposal task. Additionally, post-COVID, registration and data sharing through the COVID19BMW application have also made a mandate for TSDFs handling BMW. The specified guideline for TSDFs involved in the handling and management of BMW is summarized below (CPCB, 2021c) (Figure 5.10).

5.11 TECHNOLOGIES WITH PROVISIONAL APPROVAL

Being infectious in nature, BMW always carries a potential threat to environmental and biotic safety. So, the availability of conclusive technical evidence and practical validation is utterly mandated for the approval of any new technologies in this gamut. Lack of innovation has always remained a stigma in waste management. That forces the end-users to opt for conventional systems mostly in third-world countries. Moreover, advanced machinery and tools are still beyond the reach of such nations due to their expensive nature. Anyhow, this leads to the encouragement of indigenously developed technologies across developing nations. In India, CPCB has been provided with the authority to review, scrutinize, and approve promising technologies in the waste management domain. Lately, technologies such as plasma

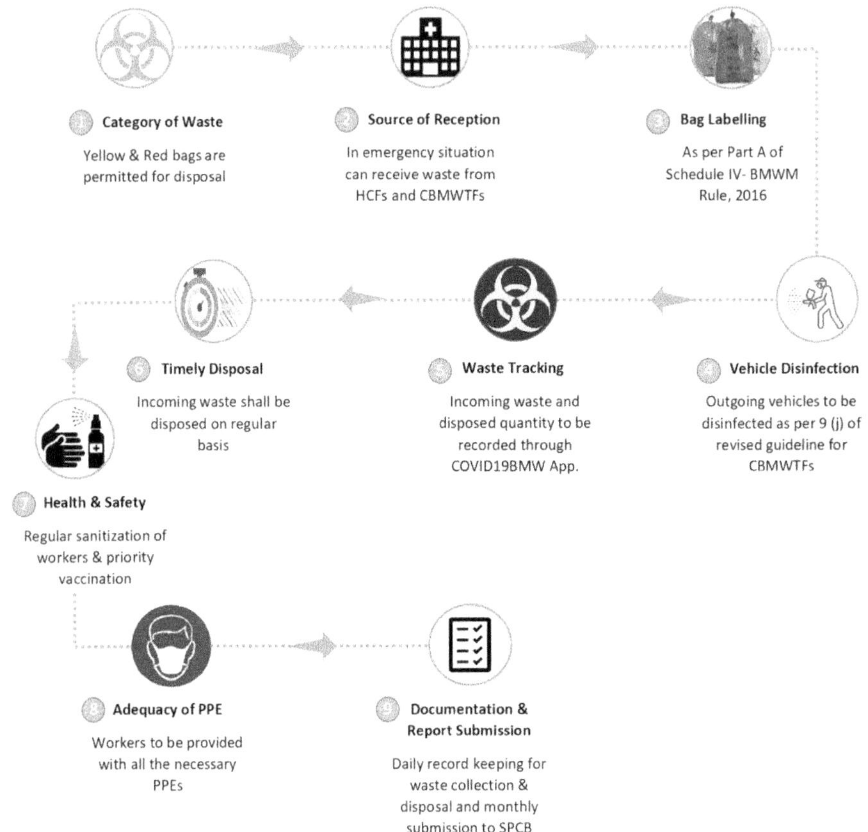

FIGURE 5.10 Summarized guideline for TSDFs to be involved in BMW Management.

pyrolysis, PIWS-3000, and Sharp blaster have obtained provisional authorization from CPCB in the BMWM domain. Since the approval in 2019, the technologies are yet to make a mark possibly due to the outbreak of the pandemic and the no-risk theory of CPCB.

5.11.1 Plasma Pyrolysis Technology

This is one of the lately introduced technologies developed by the Facilitation Centre for Industrial Plasma Technologies, Gujarat for handling the category 1 and 2 of BMW as mentioned in Schedule I by destruction mechanism. Any HCFs or CBMTFs with existing incinerators are eligible to switch over to the above technology with prior approval from the SPCB as per the stipulated guideline mentioned in Umbrella Act, 1986. The possible benefits and difficulties of adapting plasma pyrolysis are summarized below.

5.11.2 Sharp Blaster Technology

The aforesaid innovation is meant for handling category 4 of BMW (sharp wastes except for glass) as mentioned in Schedule I of BMWM Rule, 2016 through dry heat sterilization technology. It is an automated batch process with an operational temperature of 185 °C for 1.5 hours of sterilization cycle. The system is designed to ensure 100% pasteurization with both vials/spore (with at least 1 x 10^6 spores per milliliter) and strip test against the highest design capacity. The process consumables such as compressed canisters and disinfected used carbon filters are permitted to be disposed of in sanitary landfills by the regulatory authority.

5.11.3 Positive Impact Waste Solution (PIWS- 3000) Technology

PIWS-3000 yields a conjugated effect of fragmentation and disinfection. The operation involves shredding followed by dry chemical disinfection. The technology is validated to be effective against categories 1, 2, 3, 4, 6, and 7 of BMW as mentioned in Schedule I of BMWM Rule, 2016. The waste is shredded to the particle size of 1–1.5 inches and a disinfectant named Cold-ster (composed of a little more than 85% of CaO) is added to ensure 100% pasteurization within 30 min at 90–95 °C temperature and 11–12.5 pH against the full design capacity. The maximum dosage of Cold-ster is restricted to 7.5% against the incoming waste load. The treated waste is approved to be disposed of in sanitary landfills alongside municipal solid waste.

Detailed application, benefit, and disadvantage-related analytical observations for each of the above-mentioned technologies are delineated in Table 5.4.

5.12 INDIAN APPROACH AGAINST COVID-19

The government of India has ensured rigorous measures since the outbreak of COVID-19 to contain its spread and mitigate its detrimental impact on community health. India started screening and isolation of international passengers from the affected countries as early as January 2020. Further, a nationwide lockdown was imposed when the total tally of active cases was even lower than 500. Fear-mongering amid the public remained a constant stigma of the first wave. It resulted in a huge volume of CBMW generation which incurred a significant burden on the CBMWTFs. Better preparedness, a tactical approach, and source segregation helped minimize waste generation during the second wave. The primary differences identified between the first and second waves in terms of waste generation and management are detailed below.

- Source segregation at HCFs was better and non-mixing of MSW resulted in a decrement in BMW generation.
- The composition of the waste was largely varied between both waves. Components such as MSW, bedding materials, cloths etc. substantially decreased in the overall CBMW consortium.
- Major escalation in the quantity of PPE waste was observed, exceeding the previously logged peak value by nearly 30%.

TABLE 5.4
Comparative Technical Assessment of Various BMW Management Technologies with Provisional Approval

No.	Technology	Applicable category	Advantage	Disadvantage
1	Plasma Pyrolysis	Human Anatomical Waste, Animal Waste	No toxic gas emission	Higher CAPEX & OPEX than the incineration process
			Process firm slag can be used in construction	Shortage of skilled manpower
			Combustion gas can be used in the WtE concept	Wet feedstock results in higher energy consumption and low syngas generation
2	Sharp Blaster	Waste sharps (Needles, scalpels, syringes, and blades)	Noise and odor-free operation	Consumables such as canisters and carbon filters are expensive
			Little administrative control requirement	The treated waste requires further disposal in landfills.
3	PIWS- 3000	Human Anatomical Waste, Animal Waste, Microbiology and Biotechnology Wastes, Waste sharps, Soiled Waste, Solid Waste	Offers flexibility in terms of mobility	Landfill disposal required
			Inhibits odor generation by the use of Cold-ster	Skilled manpower required
			A continuous process with cost-effective operation	Emission from 150 KW capacity DG set attached
			Does not produce any wastewater or harmful emission	It cannot treat radioactive and chemotherapy-related wastes.

Source: CPCB, 2010a; CPCB, 2011; CPCB, 2012; Cai and Du, 2021; Datta et al., 2018.

Further, community participation and constant innovation pioneered the path to minimization of COVID cases in India. The country rolled out indigenously developed and manufactured vaccines, namely Covaxin® and Covishield™ in the January 2021. Massive vaccination resulted in a constant decrement in active cases and, in parallel, waste generation.

Frontline workers such as health professionals, waste managers etc. were sincerely acknowledged for their service and safeguarded with early vaccination.

The overwhelming response in respect of these frontline workers was recorded in different parts of the nation. Moreover, effective vaccination helped reduce the risk of secondary COVID-related infections among the waste managers and no such instances were recorded (Table 5.5).

TABLE 5.5
COVID Waste Management-related Observations

Year	Month	Major response
2020	January	• COVID-related passenger screening got stringent at major airports. • COVID-19 was declared the "Public Health Emergency of International Concern" by World Health Organization.
	February	• Indians were evacuated from the COVID-affected countries. • Government of India (GOI) initiated the complete suspension of visas of foreign travelers.
	March	• India invokes the "Epidemic Disease Act" citing the declaration of COVID-19 as a pandemic by WHO. • GOI mandated 14 days of self-quarantine of travelers from affected countries. • Sentinel surveillance started, movement restrictions intensified and an advisory on social distancing was issued by 16 March 2020. • CPCB issued the first version of the COVID waste management guideline on 19 March 2020. • GOI imposed 21 days nationwide lockdown on 25 March 2021.
	April	• GOI launched the Arogya Setu application to make individuals aware of the exposure and risk. • GOI declared advisory and strategy on the usage of rapid antibody testing kits. • GOI extended the lockdown (14 April 2020) but extends certain relaxations. • CPCB revised the COVID waste management guideline on 18 April 2020. • NGT ordered (20 and 24 April 2020) CPCB to submit an action taken report for effective management of COVID BMW.
	May	• GOI launched the "Vande Bharat Mission" to evacuate stranded Indians from abroad. • GOI announced the "Atmanirbhar Package" as financial support to vulnerable groups and MSMEs. • CPCB developed and introduced the COVID19BMW mobile application and a web interface for the collection and maintenance of wholesome COVID waste-related data. • A High-Level Task Team (HLTT) was created under the governance of the CPCB Chairman to monitor the COVID waste management situation in individual states and UTs. • A draft template of the model plan for COVID waste management in village Panchayats/ Sub- Divisions/Tehsils was prepared by CPCB. • CPCB advised non-compliant 15 SPCB to facilitate waste tracking through the COVID19BMW app. Rather than data submission through email.

(*continued*)

TABLE 5.5 (Continued)
COVID Waste Management-related Observations

Year	Month	Major response
	June	• CPCB published the third revision of the COVID waste management guideline in consultation with the subject area experts which includes guidance on segregation and training and awareness of waste handlers.
	July	• India initiated Phase-1 clinical trials of its indigenous COVID-19 vaccine (Covaxin) on 15 July 2020. • Final revised version of the COVID waste management guideline (Fourth version) was issued. • Total of 106 CBMTFs were issued show cause notice by CPCB for not using the COVID19BMW app.
	August	• Indian Council of Medical Research reports more than one million COVID-19-related diagnostic tests. • National Expert Group on Vaccine Administration was formed by GOI.
	September	• Registration of total of 150 CBMWTFs (out of 198 operational) and 5000 HFCs in COVID19BMW App (8 September 2020). • Country's first Clustered Regularly Interspaced Short Palindromic (Feluda) COVID-19 test was approved by the Drug Controller General of India for launch on 19 September 2020. • CPCB issued a letter to 33 non-registered CBMWTFs imposing an "Environmental Compensation Charge" (25 September 2020). • The count of registered HFCs went up to 6800 by the end of the month.
	October	• Number of registered CBMTFs increased to 181 out of 198.
	November	• Number of registered CBMTFs increased to 184 out of 198.
	December	• Number of registered HCFs increased to 8000.
2021	January	• Empowered Group on Vaccine Administration was performed by GOI. • GOI launched a free vaccination drive with 3,006 operating vaccination centers PAN India basis on 16 January 2021.
	February	• A fresh travel advisory was issued by MOHFW to restrict non-essential travel to countries such as Italy, Singapore, Iran, and Korea citing the emergence of fresh Corona cases. • Proper source segregation and dip active cases resulted in the lowest ever COVID waste generation rate of 53 TPD.
	March	• The second phase of vaccination began on 1 March 2021 for people above 45 years of age with comorbidities.
	April	• GOI up-scaled local manufacturing of COVID-related drugs and eased the import. • Oxygen produced for industrial applications was diverted as medical supply. • GOI temporarily banned the export of Remdesivir and its manufacturing ingredients to cope with the demand.

May	• Localized lockdown was imposed by the GOI to flatten the curve in worse-affected areas. • Indian Railways introduced an emergency oxygen supply train service to facilitate the movement of medical oxygen to the scarcer areas. • GOI exempted import duty on multiple COVID-related relief items until 31 July 2021. • All adult Indians were granted eligibility for vaccination and doses were administrated through the CoWin app. • Peak COVID waste generation reached the figure of 250 TPD exceeding the former peak value of 200 TPD during the first wave in 2020. • No proportional increment in waste generation with the number of active cases, due to proper segregation at HCFs and non-mixing of regular MSW and food waste.
June	• COVID BMW generation reduced by 39 TPD (~17%) when compared to the month of May. • GOI eased second dose-related minimum tenure restrictions for Covishield recipients travelling abroad for educational purposes.
July	• GOI launched "India COVID-19 Emergency Response & Health System Preparedness Package: Phase-II" worth US$3,000 million for FY 2021–22.
August	• India reached the 500 million vaccination mark covering approx. 36% of its population. • The rate of vaccination was increased to more than 4 million doses a day by the sheer efforts of GOI, NGOs, actors, and healthcare stakeholders.
September	• GOI burst the myth of the correlation between COVID vaccination and miscarriage in pregnant women.
October	• Expert panel of GOI provided Covaxin with emergency approval for 2–18-year-olds. • India reached the 1 billion vaccination mark covering approx. 72% of its population.

Source: Ratna et al., 2021; Bhatt et al., 2021; Kar et al., 2021; GRID COVID-19 Study Group, 2020; CPCB 2020a; CPCB 2020b; CPCB 2020c; CPCB 2020d; CPCB 2020e; Joshi and Mehendale, 2021.

5.13 CASE STUDY OF MEDICARE ENVIRONMENTAL MANAGEMENT PRIVATE LIMITED, RAMKY

Medicare Environmental Management Private Limited, Ghaziabad (MEMPLG) is a state-of-art CBWTF that has a legacy of treating over 30,000 tons of BMW since its establishment in 2003. The facility is operating at an approved industrial outset of the UPSIDC industrial area of Ghaziabad, Uttar Pradesh (UP) with a land footprint of approx. 1 acre. Presently, it serves more than 2500 healthcare establishments spanning a radius of up to 125 km. The plant was designed to handle an utmost capacity of 4.5 tons per day (TPD) with shredding, autoclaving, and incineration being the primary operations. Further, individual capacities of the incinerator, autoclave, and shredder include 300 kg/h, 430 litters/shift, and 50 kg/h respectively. The functional period for the above unit operations involves 12 hours for incineration and two cycles of the autoclave. Since the outbreak of SAARC-COVID-19, the facility has stretched its capacity to its fullest extent and operates in multiple shifts to cater to the requirement. The record for regular operation during the peak pandemic hours during August 2020 had even surpassed the plant capacity more than two times. The facility was kept functional round the clock with prior intimation and permission from the state pollution control board (SPCB) to meet the end requirement as a token of commitment from the Ramky group. A typical process flow of the above CBMWTF is portrayed in Figure 5.11.

5.13.1 SERVICE AREA

With only 17 functional CBMWTFs, UP is one of the largest states in India that generated more than 43 tons of BMW on a regular basis in 2018 against the handling capacity of 52 TPD (UPPCB, 2017; MOHFW, 2020). The marginal niche has exceeded multiple manifolds since March 2020. All the CBMWTFs had to cater round the clock and cover a distant service area and MEMPLG is no exception. Until 2017, about 905 HCFs were under the service radar of MEMPLG and the count had gone past 2500 in March 2021 citing reasons such as the outbreak of the novel coronavirus, installation of a parallel incinerator etc. The prominent districts under the serviceability of the facility include Ghaziabad, Noida and Greater Noida, Meerut, Amroha, Sambhal, Chandausi, Moradabad etc. The major clients of the facility include some distinguished names such as Apollo hospital, Nutema hospital, Anand hospital, Bright Star hospital, and Cosmos hospital. Presently, the facility possesses 11 global positioning system (GPS) enabled collection vehicles that cover distances as far as 130 km to ensure safe handling and collection and strengthened biosafety. Since March 2020 two vehicles were exclusively dedicated to the collection of solely COVID waste.

5.13.2 EXISTING COLLECTION METHOD, FREQUENCY, AND FEEDSTOCK STORAGE

BMW is highly infectious and contains a wide variety of potential pathogens. The handling and management (H&M) of BMW requires well-trained workers and the utmost care and precaution. As per the biomedical waste management rule 2016,

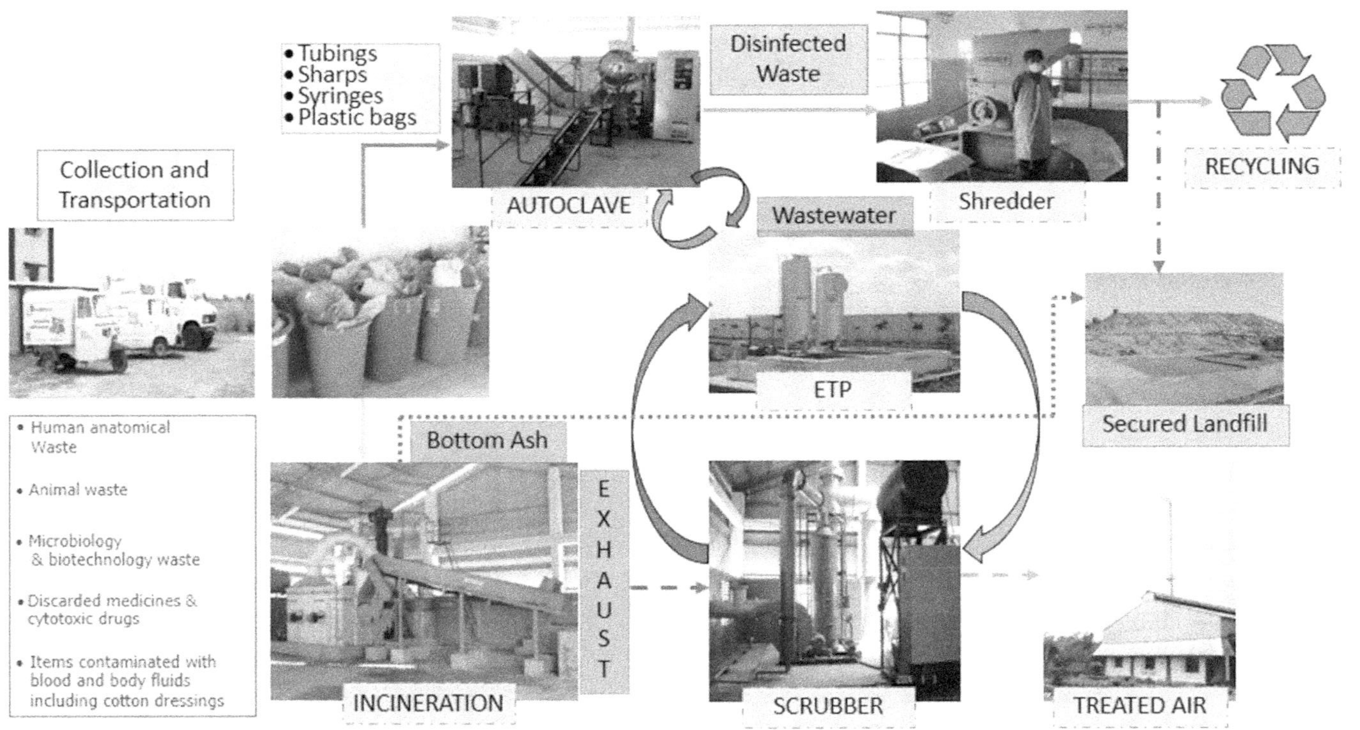

FIGURE 5.11 Process Flow Diagram of MEMPL Ghaziabad.

different colour coding is advised for the source segregation of BMW with proper barcoding for each bag. MEMPLG evidently follows the depicted guidelines both for BMW and CW. Further details are detailed below.

5.13.3 Practice for BMW

The color-coded non-chlorinated plastic bags are handed over to the HCFs for ward-to-ward collection. Anatomical wastes, cotton, beddings, plasters, dressings, and linens are stored in yellow bags. The blue or red bag is prearranged for disposables, blood bags, catheters, syringes, gloves, aprons etc. Last but not least white puncture-proof containers are provided to store the sharps such as needles, blades, scalpels, nails, and lancets. The segregated waste bags are picked up from a common pick-up point by MEMPLG and further safely transported to the facility in an aesthetically pleasing closed container vehicle (Figure 5.12).

The frequency of waste collection is determined based on the generation rate.. The usual frequency for the established HCFs is once a day. However, homoeopathic and Ayurveda centers with a limited number of occupants yield significantly smaller quantities of waste and the same gets collected on an on-call basis.

The collected BMW materials are addressed on a regular basis. Larger quantities (i.e. exceeding the plant capacity) of pharma waste loads incorporating expired and discarded drugs, syrups, injection vials etc. are temporarily stored in a designated 20'x20' closed containment with a discrete record.

5.13.4 Practice for CW

Separate yellow and red color bags are handed over to HCFs and testing and immunization centers for the collection of COVID and vaccination-related waste. The major

FIGURE 5.12 Collection of BMW from HCF.

chunk of these wastes includes empty and broken vials, vacutainers, expired vaccines, empty and broken diluent ampoules, sharps, Eppendorf tubes, blood, and body fluid contaminated cotton, used gloves, and face masks, and all other used personal protective equipment (PPE).

Specific PPEs such as aprons, face shields, hazmat suits, safety goggles, plastic coveralls, and nitrile gloves are collected in red bags. The rest of the PPEs such as used masks, shoe covers, head caps, linen gowns, and non-plastic coveralls are in yellow bags.

Wastes is being collected only in double-layered bags to assure zero leakage and all the bags carry a tag biohazard with "COVID-19" mentioned on the top of it. For biosafety, bags are not filled more than a quarter of their capacity and are sanitized with a 1% sodium hypochlorite (NaOCl) solution before the pickup (Figure 5.13).

COVID waste gets collected on a regular basis from all the HCFs with specific vehicles dedicated to only the prior-mentioned waste and all the vehicles are geotagged and monitored closely.

Each vehicle is intensively sanitized after each trip using 1% NaOCl solution, along with the workers. A separate record is maintained for the COVID-related wastes and the same gets immediate disposal in incineration upon arrival.

5.13.5 Treatment and Disposal Practices

Similar to the collection, handling and management of BMW and CW are quite discrete. Regular BMW is considered less infectious than CW. This leads to the possibility of recovery and reuse through pertinent sterilization. However, at present, the entire chunk of COVID waste is handled by destruction only. A detailed treatment and disposal practice is portrayed below.

FIGURE 5.13 Collection of CW from HCF.

(a) (b)

FIGURE 5.14 (a) Incineration of category 1, 2, 3, & 6 type of BMW (b) Autoclave to handle category 4 & 7 type of BMW.

5.13.5.1 Practices for General BMW

In the case of regular BMW, waste is segregated based on color upon arrival. Yellow bags are mechanically fed to the incinerator and burnt off. The combustion temperature at primary and secondary chambers is maintained at 850^{0}C and 1050^{0}C respectively to ensure effective combustion and oxidation of carcinogenic pollutants such as dioxins and furans. The laden gas is passed through a whole set of air pollution control devices (APCD) to ensure the quality of the flue gas is on par with the CPCB discharge limit (Figure 5.14a). The chimney is attributed to a continuous emission monitoring system (CEMS) to depict the real-time data and the same gets logged within an internal database as well as a CPCB system. Further, white and red/ blue bags are manually opened, and autoclaved at 121^{0}C, 15 Psi pressure for 1 hour to attain effective pasteurization (Figure 5.14b). The sterilized materials such as plastics and metals are shredded and sent to authorized recyclers and metal foundries for recycling and recovery.

5.13.5.2 Practices for CW

CW boxes/bags are sterilized once again with 1% NaOCl after their arrival at the facility. The same is weighed and recorded in a separate logbook and shared with the pollution control board using a mobile app entitled “COVID19BWM”. A special nodal officer verifies all the recorded values on a day-to-day basis. The waste is mechanically fed to the incinerator and the bottom ash is securely collected and disposed of in the secured landfill. The laden gas is treated through a range of APCDs such as a set of scrubbers, activated carbon column, bag house filter etc.

5.13.6 Personnel Handling BMW and CW

The facility strictly follows the COVID waste management guidelines and all further amendments. MEMPLG doesn't involve workers of age group 50 or more in H&M of CW. All the working personnel are immunized with vaccinations against diseases such as Hepatitis B, Tetanus, and COVID-19. Various safety-related training on hand hygiene, respiratory etiquette, social distancing, and infection prevention measures are provided on a mandatory basis to all the workers including group D staff during the induction and repeated once a year. All the workers handling CW are sanitized after every trip. Further, anybody showcasing the symptoms of infection is provided adequate leave with salary protection and incentives. Also, a separate book is maintained to record all sorts of injuries (i.e. needle pricks, cuts etc.), diseases, and measures adopted (CPCB, 2020a).

5.13.7 PPE Used by the Operators and Precautionary Measures

All the workers associated with the H&M and CW are provided with an adequate quantity of necessary PPEs. The list of supplied PPEs includes triple-layered masks, splashproof aprons, nitrile gloves, gum boots, and safety goggles (Figure 5.15). All the PPEs are collected safely in a designated box at the end of the day and incinerated to avoid cross-contamination.

FIGURE 5.15 CBMW collection workers upon arrival at the facility wearing full PPE kit.

5.14 OPERATION ECONOMY AND SUSTAINABILITY

The quality and accessibility of healthcare services in India need to improve as it is the largest sector in India in terms of employment generation and revenue earning. The recent trend reveals that the healthcare sector in India is one of the fastest-growing sectors and was worth approx. US$193.83 billion by 2020 (MoCI, 2021). This increasing trend has also increased the generation of BMW in the country which poses serious challenges in terms of environmental pollution.

BMW is an area of concern which has a direct impact on health and human safety (Datta et al., 2018). Management of BMW may be an ethical, social obligation of all supporting and funding healthcare activities. All the healthcare establishments (HCEs) are required by law to put in place the mechanisms for proper segregation and scientific disposal of BMW to minimize adverse impacts on healthcare workers and on the environment. However, the installation of individual treatment facilities by healthcare establishments requires significant capital investment and trained manpower for the proper operation and maintenance of treatment systems.

India has the second-highest population in the world which accounts for a total of 614 tons per day of BMW generation (CPCB, 2018a). Approximately 534 tons of BMW per are treated in the existing 200 in CBWTFs and captive treatment facilities (CPCB, 2018a). CBWTFs offer huge advantages to healthcare establishments through more efficient treatment and disposal of BMW and through 'Economies of Scale' (significant decrease in the cost of treatment per kilogram). The present COVID-19 scenario has worsened the present BMW generation scenario. In India, 33264.4 tons COVID-related BMW were generated between June to December 2020 (MoEFCC, 2020b). COVID-related BMW includes goggles, face-shield, splashproof apron, Plastic Coverall, Hazmat suit, nitrile gloves etc. According to Kumar et al. (2020), PPE suits (pant, gown, head and shoe cover) made of Polypropylene fabric accounts for 84% COVID-related BMW in terms of weight.

The available disposal system for these COVID-related BMW includes centralized and decentralized incineration and disposal in secured landfills. Any incineration process includes emissions to the air and bottom ash generation which will ultimately result in greenhouse gas (GHG) emissions and environmental pollution. Disposal in secured and lined landfill sites will not result in air emission and it also has the advantage of energy recovery in terms of biogas capture and utilization. Disposal in landfill sites requires long-distance transportation which will indirectly cause GHG emissions. From a sustainable point of view, it can be suggested that proper planning for the collection of waste, developing a small collection center with pressurized air and steam sterilization and shredding arrangements and transportation planning to the landfill site with minimum GHG emission could be the way forward for secured disposal of COVID-related BMW until the availability of new technology to cope with this problem.

5.14.1 Business Model Dealing with BMW and CW

Continuous degradation of the environment is posing several health-related issues for all the developed and developing countries in the world. The healthcare market is globally evaluated to be valued at US$8,452 billion and is likely to grow at a compound rate of 8.9% annually from 2020 to 2027. The Indian healthcare market is one of the fastest-growing markets and is expected to reach US$372 billion by 2022. According to the Department for Promotion of Industry and Internal Trade (DPIIT), the total FDI in India during the COVID-19 situation (from April 2000 to September 2020) was US$16.87 billion in the drugs and pharmaceuticals sector.

With the growing population in the country, the number of patients is also increasing; thereby it will result in more investment in the healthcare sector. According to the Union budget 20–21, US$9.87 billion will be invested in the healthcare sector and another US$5.09 billion will be invested in nutrition-related programs. The growing healthcare sector will also increase the BMW generation in the country. This continuous increase in BMW generation creates business opportunities in the waste management sector, particularly the BMW management sector.

The process of BMW management includes BMW storage, collection, waste transportation, treatment, and finally disposal through authorized vendors and these whole processes have been accompanied by proper documentation. The continuous increase in the volume of BMW along with its existing management practices and stringent regulatory framework applicable to BMW for its safe management and eco-friendly disposal of waste are some of the key drivers of industries involved in BMW management. It may not be possible for small healthcare units to carry out treatment and disposal of biomedical waste generated as per the methods prescribed in the Biomedical Waste Management Rules 2016. Considering the investment associated with the captive treatment plant it is not reasonable for bigger hospitals to install a captive treatment plant for BMW.

Further, various business/revenue models framed and evaluated citing different criteria for CBMWTFs are detailed below (Figure 5.16). Models were compared by formulating a decision matrix and pairwise weightage is summarized.

Eight different business and revenue models were compared for CBMWTFs citing control over inflow, process efficacy, revenue, and social and environmental impact as criteria. Based on the offered flexibility and limitations, alternates were subjected to pairwise comparison using Analytic Hierarchy Process. The resulting weight-based rankings of different alternates obtained through pairwise comparison are delineated in Table 5.6. Further, the decision matrix was formulated based on the resulting weight from the principal eigenvalue of 8.52 against a consistency ratio of 5.3% (Table 5.7). The comprehensive finding for the sequential ranking is summarized in Figure 5.17.

The model predicted the pay-as-you-go business structure as the most preferred while the leasing model was the least preferred alternative. The former model gives the liberty to the waste generator to compensate the service charge against the amount of waste generated. This structure involves no primary financial commitments, rather it is generation-specific. In the case of BMW and CW generation of waste is ascertained against each bed. Contrarily, the later business structure incorporates the establishment of individual waste management facilities by the healthcare units and outsourcing the operation and management to a third party. With minimum resource availability and little to no supervision, the reliability of the reported efficacy is always questionable. Additionally, micromanagement often leads to higher operational expenditure and a greater negative environmental footprint.

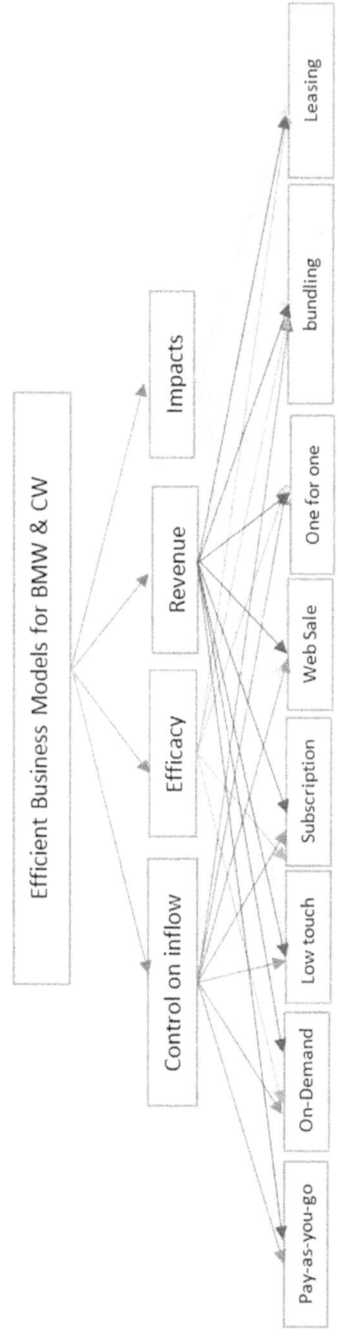

FIGURE 5.16 AHP Hierarchies for prioritizing the efficient treat process for MSW.

TABLE 5.6
Pairwise Comparison Using AHP Analysis

Category		Priority	Rank	(+)	(-)
1	Pay-as-you-go	37.8%	1	16.9%	16.9%
2	On-demand	20.2%	2	7.2%	7.2%
3	Low-touch	16.7%	3	6.8%	6.8%
4	Subscription model	10.6%	4	4.1%	4.1%
5	Web sales	5.8%	5	1.8%	1.8%
6	One-for-one model	4.0%	6	1.2%	1.2%
7	Bundling model	2.8%	7	1.0%	1.0%
8	Leasing model	2.1%	8	0.9%	0.9%

TABLE 5.7
Decision Matrix Using AHP Analysis

	1	2	3	4	5	6	7	8
1	1	3.00	4.00	5.00	6.00	7.00	8.00	9.00
2	0.33	1	2.00	3.00	4.00	5.00	6.00	7.00
3	0.25	0.50	1	3.00	4.00	5.00	6.00	7.00
4	0.20	0.33	0.33	1	3.00	4.00	5.00	6.00
5	0.17	0.25	0.25	0.33	1	2.00	3.00	4.00
6	0.14	0.20	0.20	0.25	0.50	1	2.00	3.00
7	0.12	0.17	0.17	0.20	0.33	0.50	1	2.00
8	0.11	0.14	0.14	0.17	0.25	0.33	0.50	1

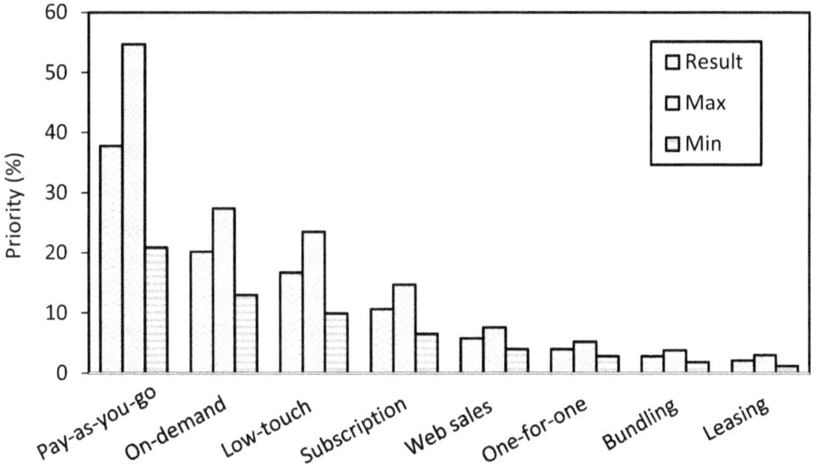

FIGURE 5.17 Consolidated AHP evaluation report for BMW & CW business model.

Further, the pros and cons attributed to each of the business/revenue structures are detailed below.

The on-demand model offers prompt service driven by customer inquiry.

Pros: The generator gets the flexibility of zero capital investment and immediate service against the reequipment.

Cons: No control over the inflow.

The low-touch model is case-specific where certain crucial services are offered at a very competitive rate while advanced services are quite expensive.

Pros: Enables relatively smaller establishments to pay less for non-complex waste.

Cons: Not suitable for large healthcare establishments with a variety of wastes.

The subscription model for BMW and CW management is quite similar to online subscriptions. HCEs need to obtain membership in CBMWTF by paying a certain sum at the beginning of every financial year.

Pros: It is certainly profitable for those gigantic HCEs which generate a huge chunk of waste.

Cons: The service provider does not have any control over the inflow and it may lead to loss due to the havoc amount of waste intake.

The web sales business model is a web-based revenue escalation model which is adopted when CBMWTF fails to meet the desired business due to reasons such as inappropriate location, poor and obsolete handling techniques etc.

Pros: Enhances the drying revenue of certain CBMWTFs.

Cons: Lowers the service rate due to peer competition.

One-for-one model is a way of attracting the attention of customers and local non-governmental organizations by catering to the demand of under-privileged people against the sale.

Pros: Incurs a sense of social service and helps create a positive image.

Cons: Often a considerable amount of revenue is compromised to cater to the ever-rising demand.

The Budling model offers certain services cheaper in clusters than in individuals.

Pros: Helps enhance the revenue of CBMWTFs.

Cons: Packaged service forces the HCEs to opt for unwanted combinations and leads to greater expenditure.

5.15 OBSERVATIONS AND SUGGESTIONS

A profound understanding of the situation helped the authors identify the major challenges and a strong logical interpretation with the on-field waste management expertise helped recognize the most promising solutions. The detailed observations are summarized below.

5.15.1 Major Shortcomings

- Out of 28 states and 8 union territories (UTs) in India, only 8 states have HCFs authorization under BMW rule, 2016.
- Pre-COVID about 70% of the BMW generated is scientifically handled and treated (Singh and Saha, 2020).
- Non-availability of BMW treatment facilities in small towns and villages.
- Most of the northeastern states and other states with micro footprint such as Goa, Andaman &and Nicobar Island, and Lakshadweep do not have separate CBMTF and primarily depend on deep burial or the neighboring states.
- Deep burial is still in practice either partially or fully in 23 states and UTs irrespective of government restriction.
- So far only 12 states have upgraded the existing air pollution control system to comply with the new emission standard.
- A robust barcoding system is yet to be established across the country. It is highly crucial to tackle the illegal disposal of BMW if any.

5.15.2 Suggestions

5.15.2.1 Related to Virus Containment

Maximize the use of robots to ensure no-contact distribution of sanitizers, food, and other necessary medical packages to the COVID wards.

Digital stethoscopes such as AyuSynk can be used to ensure social distancing between the doctor and the patients.

Necessary and periodic disinfection of frequently touched surfaces in public areas and office buildings with verified and eco-friendly products such as COVIKILL.™

5.15.2.2 Related to Waste Minimization

Biodegradable and low-cost materials such as banana tree fiber, tea bags etc. should be used for manufacturing face masks. These masks are equally durable and can be degraded within 60 days by the co-composting method.

All forms of PPEs can be disinfected and reused using innovations such as Vajra. This will ensure the minimization of PPE-related high-volume waste generation.

Used polypropylene PPEs can be recycled as products such as “Brick 2.0” which could be helpful to reduce the burden on existing limited incinerators (MOHFW, 2021).

5.15.2.3 Related to Waste Management

Existing 198 CBMTFs are inadequate to handle the ever-rising volume of BMW. The majority of the states with the existing number of CBMTFs have reached their threshold capacity. So, new CBMWTFs need to be established, especially in small towns and their connectivity can be improved with the HCFs to abolish the practice of deep burial.

Ancillary attention needs to be provided to the seven sister states of India. The states have a comprehensive count of five CBMTFs which is highly inadequate even to tackle the present rate of BMW generation.

Pertinent and frequent monitoring of the HCFs and CBMTFs by the district-level monitoring committee would ensure better operational compliance.

5.15.2.4 Related to Effective PPE Waste Handling

Ozone-based infectious waste sterilization technology can open a new niche in the PPE waste handling gamut. The proposed closed vessel equipment shall consist of two major compartments. The primary compartment would consist of a waste disposal hopper, followed by a shredder, a disinfection channel with an expeller, and an ozone reducer (Figure 5.18). The secondary compartment shall incorporate an ozone generation assembly and a product collection bin. The frequency of emptying the bin can be controlled by artificial intelligence technology. Ozone would be injected both in the shedder and sterilization channel to ensure complete decontamination. The proposed retention period is 30 min. Post-treatment, excess ozone would be reduced to pure oxygen using a reducer and liberated to the environment. This unique feature makes the above product a carbon-negative stand-alone waste treatment unit.

The major constituents of the PPE disinfected mass include polypropylene (PP), polyethene (LDPE/HDPE), latex, nitrile, vinyl (PVC) etc. Further, it can be materialized into granules in a plastic recycling facility by agglomeration, extrusion, and palettization processes.

5.16 CONCLUSION AND FUTURE RESEARCH

Due to the pandemic, there is a tremendous increase in the use of protective gear. It is not only medical professionals who use such PPE but also other professionals, including those involved in waste management, sanitation, and public health. Almost all of the PPEs used are disposed of after single use. The highest chance of infection is during the COVID BMW waste collected from the generator by the sanitary worker. Hence, proper care and protection should be taken at this step and continued to other COVID BMW handling and treatment stages. COVID BMW must be collected in suitable colored bags labelled as "hazardous," and immediately appropriate disinfectants should be applied to the BMW after it is collected. The disinfectant should be effective during the latent period and until BMW reaches the treatment facility. Studies have clearly shown that the COVID-19 virus can persist on many surfaces from a few hours to several days, especially in those materials that are used in PPEs. Hence, sufficient latent/resting time should be provided to the COVID BMW waste before any processing can be done. The latent step is imperative for both eliminating and breaking the transmission of the virus through BMW. Wherever possible COVID BMW has to be either electronically tagged so that its movement can be traced based on GPS surveillance. At the BMW waste treatment facility, appropriate technologies, as discussed earlier in this chapter, have to be used for treating/disposing of various types of waste. COVID BMW management and handling strategies are well-defined for hospitals and healthcare centers that majorly generate this waste.

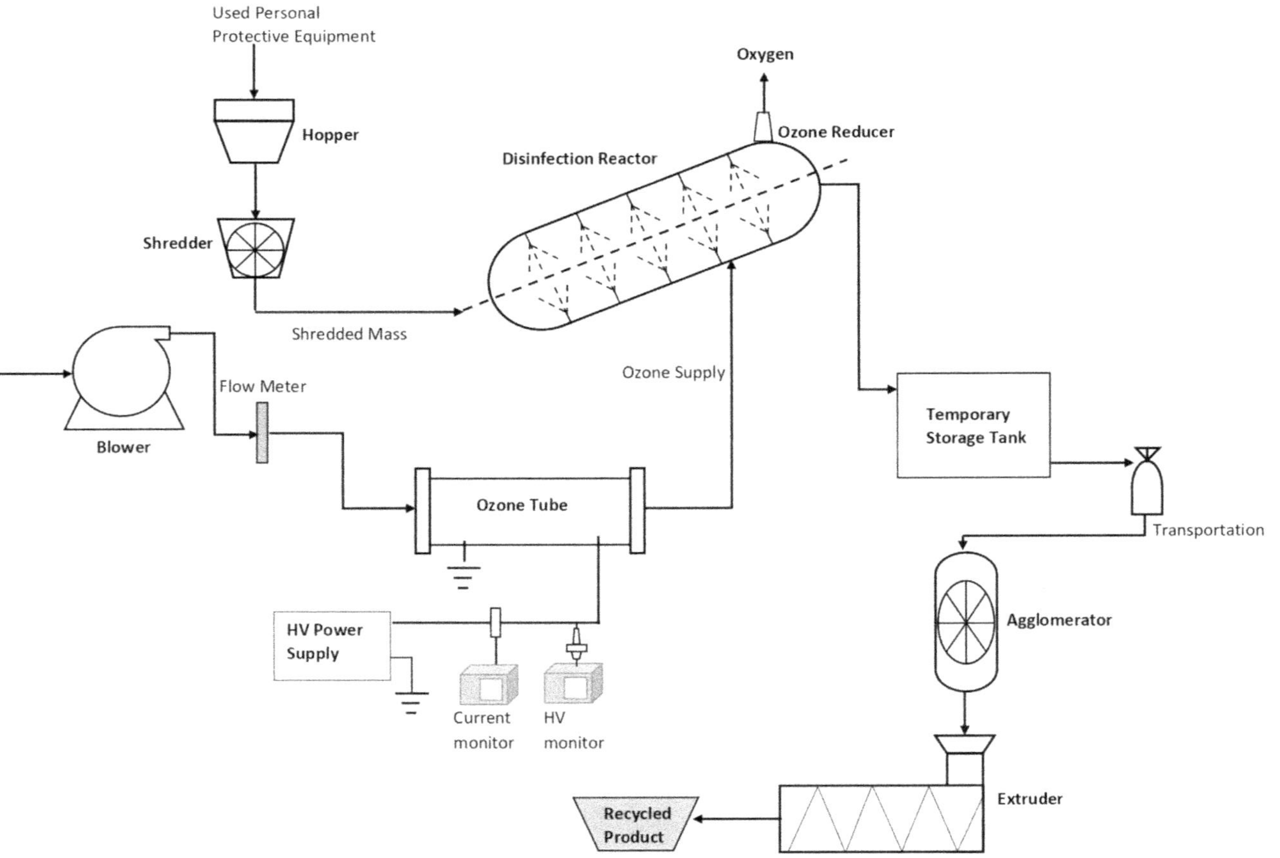

FIGURE 5.18 Proposed engineering diagram of effective PPE waste recycling technology.

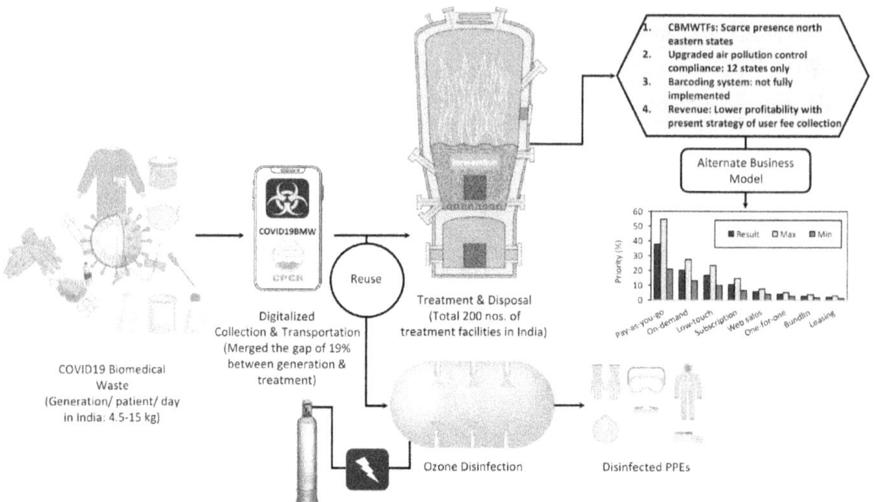

FIGURE 5.19 Comprehensive pathway of handling and valorizing BMW.

However, COVID BMW, especially face masks and hand gloves generated from individual households where patients are getting self-isolated and asymptomatic, serve as potential virus transmission sources because these wastes get mixed with municipal solid wastes that end up in dump yards or landfills. The comprehensive findings of the study are summarized in Figure 5.19.

Further research should focus on recovering energy from the PPEs during the treatment process. Companies should explore the possibilities of using biodegradable/ compostable materials for PPEs, which would reduce the pollution load caused by conventional PPEs, which are essentially made up of petrochemical-based plastics. It is expected that the pandemic situation will continue for a few more years across the globe. Hence it is guaranteed that the COVID BMW generation will continue. Mismanagement of COVID BMW and reusing disposable PPEs serve as one of the primary means of virus transmission. Therefore, at the local government level, proper policies and sufficient budget allocation have to be provided for the safe management and handling of COVID BMW.

REFERENCES

Banerjee, T., 2020. A waste-ful enterprise: COVID-19 and India's waste management story. *Degrees of Change.* www.degreesofchange.in/articles/a-waste-ful-enterprise-Covid-19-india-waste (accessed February 7 2021).

Bhalla, G. S., Bandyopadhyay, K., Sahai, K., 2019. Keeping in pace with the new Biomedical Waste Management Rules: What we need to know!. *Med J Armed Forces India.* 75 (3), 240–245. https://doi.org/10.1016/j.mjafi.2018.12.003

Bhatt, M., Srivastava, S., Schmidt-Sane, M., Mehta, L., 2021. Key considerations: India's deadly COVID-19 second wave: Addressing impacts and building preparedness against future waves. *Brighton: Social Science in Humanitarian Action (SSHAP).* https://doi.org/10.19088/SSHAP.2021.031

Bhattacharjee, S., 2014. Bio-medical waste: Rampant mismanagement, lackadaisical implementation of legislations, flouting of constitutional rights, solutions to tackle effects on ecology and health–The Indian scenario. http://docs.manupatra.in/newsline/articles/Upload/D3D46A20-F27B-4868-BF59-DF2B971C096E.Bio-Medical%20Waste_Subornadeep%20Bhattacharjee_p33-44.pdf

Cai, X., Du, C., 2021. Thermal plasma treatment of medical waste. *Plasma Chemistry and Plasma Processing* 41, 1–46. https://doi.org/10.1007/s11090-020-10119-6

Capoor, M. R., Bhowmik, K. T., 2017. Current perspectives on biomedical waste management: Rules, conventions and treatment technologies. *Indian J Med Microbiol.* 35 (2), 157–164. https://doi.org/10.4103/ijmm.IJMM_17_138. PMID: 28681801.

Central Pollution Control Board (CPCB), 2008. State-wise status of Common Bio-medical Waste Treatment Facilities. https://cpcb.nic.in/uploads/Projects/Bio-Medical-Waste/CBWTF_Status_2008.pdf

Central Pollution Control Board (CPCB), 2010a. Provisional Approval to Plasma Pyrolysis Technology. https://cpcb.nic.in/uploads/Projects/Bio-Medical-Waste/ProvisionalApprovalPlasmaPyrolysisTechnology.pdf

Central Pollution Control Board (CPCB), 2010b. State-wise status of Common Bio-medical Waste Treatment Facilities. https://cpcb.nic.in/uploads/Projects/Bio-Medical-Waste/CBWTF_Status_2010.pdf

Central Pollution Control Board (CPCB), 2011. Provisional Approval to Shrp Blaster Technology. https://cpcb.nic.in/uploads/Projects/Bio-Medical-Waste/ProvisionalApprovalShrpBlasterTechnology.pdf

Central Pollution Control Board (CPCB), 2012 Provisional Approval to PIWS-3000 Technology. https://cpcb.nic.in/uploads/Projects/Bio-Medical-Waste/ProvisionalApprovalPIWS-3000Technology.pdf

Central Pollution Control Board (CPCB), 2013. State-wise status of Common Bio-medical Waste Treatment Facilities. https://cpcb.nic.in/uploads/Projects/Bio-Medical-Waste/CBWTF_Status_2013.pdf

Central Pollution Control Board (CPCB), 2014. State-wise status of Common Bio-medical Waste Treatment Facilities. https://cpcb.nic.in/uploads/Projects/Bio-Medical-Waste/CBWTF_Status_2014.pdf

Central Pollution Control Board (CPCB), 2015. State-wise status of Common Bio-medical Waste Treatment Facilities. https://cpcb.nic.in/uploads/Projects/Bio-Medical-Waste/CBWTF_Status_2015.pdf

Central Pollution Control Board (CPCB), 2016a. Status on Bio-medical Waste Management Scenario and recommendations for ensuring compliance to the Biomedical Waste Management Rules, 2016. https://cpcb.nic.in/uploads/Projects/Bio-Medical-Waste/AR_BMWM_2016.pdf

Central Pollution Control Board (CPCB), 2016b. State-wise status of Common Bio-medical Waste Treatment Facilities. https://cpcb.nic.in/uploads/Projects/Bio-Medical-Waste/CBWTF_Status_2016.pdf

Central Pollution Control Board (CPCB), 2017a. Annual Report on Biomedical Waste Management as per Biomedical Waste Management Rules, 2016. https://cpcb.nic.in/uploads/Projects/Bio-Medical-Waste/AR_BMWM_2017.pdf

Central Pollution Control Board (CPCB), 2017b. State-wise status of Common Bio-medical Waste Treatment Facilities. https://cpcb.nic.in/uploads/Projects/Bio-Medical-Waste/CBWTF_Status_2017.pdf

Central Pollution Control Board (CPCB), 2018a. Annual Report on Biomedical Waste Management as per Biomedical Waste Management Rules, 2016. https://cpcb.nic.in/uploads/Projects/Bio-Medical-Waste/AR_BMWM_2018.pdf

Central Pollution Control Board (CPCB), 2018b. State-wise status of Common Bio-medical Waste Treatment Facilities. https://cpcb.nic.in/uploads/Projects/Bio-Medical-Waste/CBWTF_Status_2018.pdf

Central Pollution Control Board (CPCB), 2019a. Annual Report on Biomedical Waste Management as per Biomedical Waste Management Rules, 2016. https://cpcb.nic.in/uploads/Projects/Bio-Medical-Waste/AR_BMWM_2019.pdf

Central Pollution Control Board (CPCB), 2019b. Gap analysis of the compliance reports submitted by States/ Union Territories. http://www.indiaenvironmentportal.org.in/files/file/gap%20analysis%20of%20the%20compliance%20reports%20submitted%20by%20states.pdf

Central Pollution Control Board (CPCB), 2019c. State-wise status of Common Bio-medical Waste Treatment Facilities. https://cpcb.nic.in/uploads/Projects/Bio-Medical-Waste/CBWTF_Status_2019.pdf

Central Pollution Control Board (CPCB), 2020a. Scientific Disposal of Bio-Medical Waste arising out of COVID-19 treatment–Compliance of BMWM Rules, 2016 before Hon'ble Rajak et al. 11 National Green Tribunal, Principle Bench, New Delhi. https://greentribunal.gov.in/sites/default/files/news_updates/Status%20Report%20in%20O.A%20No.%2072%20 of%202020.pdf

Central Pollution Control Board (CPCB), 2020b. Revision 4: Guidelines for handling, treatment and disposal of waste generated during treatment/diagnosis/quarantine of COVID-19 patients. http://bspcb.bih.nic.in/Rev.4%20COVID%20Guidelines%20CPCB.pdf

Central Pollution Control Board (CPCB), 2020c. ENVIS centre on control of pollution water, air and noise, ministry of environment, forests, and climate change. Government of India. http://cpcbenvis.nic.in/Bio_Medical_waste.html# (accessed January 25 2021).

Central Pollution Control Board (CPCB), 2020d. COVID-19 waste management. Central Pollution Control Board. https://cpcb.nic.in/Covid-waste-management/ (accessed 15 December 2020).

Central Pollution Control Board (CPCB), 2020e. Pictorial guide on biomedical waste management rules 2016 (Amended in 2018 & 2019). https://cpcb.nic.in/uploads/Projects/Bio-Medical-Waste/Pictorial_ guide_Covid.pdf

Central Pollution Control Board (CPCB), 2020f. Generation of COVID19 related Biomedical Waste in States/UTs (June–August 2020). https://cpcb.nic.in/uploads/Projects/Bio-Medical-Waste/COVID19_Waste_Management_status_August2020.pdf

Central Pollution Control Board (CPCB), 2020g. Generation of COVID19 related Biomedical Waste in States/UTs (September 2020). https://cpcb.nic.in/uploads/Projects/Bio-Medical-Waste/COVID19_Waste_Management_status_September2020.pdf

Central Pollution Control Board (CPCB), 2020h. Generation of COVID19 related Biomedical Waste in States/UTs (October 2020). https://cpcb.nic.in/uploads/Projects/Bio-Medical-Waste/COVID19_Waste_Management_status_October2020.pdf

Central Pollution Control Board (CPCB), 2020i. Generation of COVID19 related Biomedical Waste in States/UTs (November 2020). https://cpcb.nic.in/uploads/Projects/Bio-Medical-Waste/COVID19_Waste_Management_status_November2020.pdf

Central Pollution Control Board (CPCB), 2020j. Generation of COVID19 related Biomedical Waste in States/UTs (December 2020). https://cpcb.nic.in/uploads/Projects/Bio-Medical-Waste/COVID19_Waste_Management_status_December2020.pdf

Central Pollution Control Board (CPCB), 2020k. USER MANUAL Android Mobile Application & Web Application. https://cpcb.nic.in/uploads/Projects/Bio-Medical-Waste/V1_COVID-19_BMW_Tracking_App.pdf

Central Pollution Control Board (CPCB), 2021a. Generation of COVID19 related Biomedical Waste in States/UTs (January-May 2021). https://cpcb.nic.in/uploads/Projects/Bio-Medical-Waste/COVID19_Waste_Management_status_Jan_May_2021.pdf

Central Pollution Control Board (CPCB), 2021b. Generation of COVID19 related Biomedical Waste in States/UTs (June 2021). https://cpcb.nic.in/uploads/Projects/Bio-Medical-Waste/COVID19_Waste_Management_status_June_2021.pdf

Central Pollution Control Board (CPCB), 2021c. SOP for disposal of BMW including pandemic medical waste through incineration in common hazardous waste treatment, storage and disposal facility. https://cpcb.nic.in/uploads/Projects/Bio-Medical-Waste/SOP_Covid_1.pdf

Chamberlain, M., 2019. Consequences of improperly disposed pharmaceuticals. *Daniels Health.* www.danielshealth.com/knowledge-center/pharmaceutical-improper-disposal-consequences (accessed January 15 2021).

Chand, S., Shastry, C. S., Hiremath, S., Joel, J. J., Krishnabhat, C. H., Mateti, U. V., 2021. Updates on biomedical waste management during COVID-19: The Indian scenario. *Clin Epidemiology Glob Health* 11, 100715. https://doi.org/10.1016/j.cegh.2021.100715

Datta, P., Mohi, G. K., Chander, J., 2018. Biomedical waste management in India: Critical appraisal. *J Lab Physicians* 10, 6–14. https://doi.org/10.4103/JLP.JLP_89_17

GRID COVID-19 Study Group, 2020. Combating the COVID-19 pandemic in a resource-constrained setting: insights from initial response in India. *BMJ Glob Health* 5 (11), e003416. https://doi.org/10.1136/bmjgh-2020-003416

Ilyas, S., Srivastava, R. R., Kim. H., 2020. Disinfection technology and strategies for COVID-19 hospital and bio-medical waste management. *Sci Total Environ.* 749, 141652. https://doi.org/10.1016/j.scitotenv.2020.141652

Jaseem, M., Kumar, P., John, R. M., 2017. An overview of waste management in pharmaceutical industry. *Pharma Innov* 6, 158–161.

Joshi, R., Ahmed, S., 2016. Status and challenges of municipal solid waste management in India: A review. *Cogent Environ. Sci.* 2, 1–18. https://doi.org/10.1080/23311843.2016.1139434

Joshi, R. K., Mehendale, S. M., 2021. Prevention and control of COVID-19 in India: Strategies and options. *Med J Armed Forces India.* 77, S237–S241. https://doi.org/10.1016/j.mjafi.2021.05.009

Kar, S. K., Ransing, R., Arafat, S. M. Y., Menon, V., 2021. Second wave of COVID-19 pandemic in India: Barriers to effective governmental response. *EClinicalMedicine* 36, 100915. https://doi.org/10.1016/j.eclinm.2021.100915

Kashyap, A., 2021. Bio-medical waste. Legal Service India. www.legalserviceindia.com/legal/article-3209-bio-medical-waste.html (accessed January 17 2021).

Kothari, R., Sahab, S., Singh, H. M., Singh, R. P., Singh, B., Pathania, D., Singh, A., Yadav, S., Allen, T., Singh, S., Tyagi, V. V., 2021. COVID-19 and waste management in Indian scenario: challenges and possible solutions. *Environ Sci Pollut Res* 28, 52702–52723. https://doi.org/10.1007/s11356-021-15028-5

Kulkarni, S. J., 2020. Biomedical waste scenario in India–regulations, initiatives and awareness. *Biomed Eng Int.* 4, 86–92. https://doi.org/10.33263/BioMed22.086092

Kumar, H., Azad, A., Gupta, A., Sharma, J., Bherwani, H., Labhsetwar, N. K., Kumar, R., 2020. COVID-19 Creating another problem? Sustainable solution for PPE disposal through LCA approach. *Environ Dev Sustain.* 23, 9418–9432. https://doi.org/10.1007/s10668-020-01033-0

Kumar, P., Reddy, K. V. K., Mishra S., 2017. AIIMS Bio-medical Waste Management Manual. www.aiims.edu/images/pdf/BMW-Book%20final.pdf

Kumar, S. R., Abinaya, N. V., Venkatesan, A., & Natrajan, M., 2019. Bio-medical waste disposal in India: From paper to practice, what has been effected. *Indian J Health Sci Biomed Res (KLEU)*, 12(3), 202. https://doi.org/10.4103/kleuhsj.kleuhsj_112_19

Ministry of Commerce & Industry (MoCI), 2021. Healthcare Industry in India. India Brand Equity Foundation. www.ibef.org/industry/healthcare-india.aspx#:~:text=Healthcare%20has%20become%20one%20of,health%20insurance%20and%20medical%20equipment (accessed February 22 2021).

Ministry of Environment, Forest and Climate Change (MoEFCC), 2017. Annual Report 2016–17. Central Pollution Control Board. http://cpcb.nic.in/uploads/Projects/BMW/annual-report-2016-17.pdf

Ministry of Environment, Forest and Climate Change (MoEFCC), 2018. Annual Report 2017-18. Central Pollution Control Board. http://cpcb.nic.in/uploads/Projects/BMW/annual-report-2017-18.pdf

Ministry of Environment, Forest and Climate Change (MoEFCC), 2019. Annual Report 2018–19. Central Pollution Control Board. http://cpcb.nic.in/uploads/Projects/BMW/annual-report-2018-19.pdf

Ministry of Environment Forest & Climate Change (MoEFCC), 2020a. Guidelines for Handling, Treatment and Disposal of Waste Generated during Treatment/Diagnosis/Quarantine of COVID-19 Patients. www.mppcb.mp.gov.in/proc/BMW/18-4-2020-CPCB-Gidelines-for-disposalofCOVID-19.pdf

Ministry of Environment Forest & Climate Change (MoEFCC), 2020b. Guidelines for Handling, Treatment and Disposal of Waste Generated during Treatment/Diagnosis/Quarantine of COVID-19 Patients (Revision 4). https://cpcb.nic.in/uploads/Projects/Bio-Medical-Waste/BMW-GUIDELINES-COVID_1.pdf

Ministry of Health and Family Welfare (MOHFW), 2020. Bio-Medical Waste. https://pib.gov.in/PressReleseDetail.aspx?PRID=1602353 (accessed December 22 2020).

Ministry of Health and Family Welfare (MOHFW), 2021. Environment-friendly solution to disinfect Personal Protective Equipment (PPE): reduces biomedical waste generation Environment-friendly solution to disinfect Personal Protective Equipment (PPE): reduces biomedical waste generation. www.pib.gov.in/PressReleasePage.aspx?PRID=1723073 (accessed 12 September 2021).

Oguntoke, O., Emoruwa, F. O., & Taiwo, M. A., 2019. Assessment of air pollution and health hazard associated with sawmill and municipal waste burning in Abeokuta Metropolis, Nigeria. *Environ Sci Pollut Res*. 26(32), 32708–32722. https://doi.org/10.1007/s11356-019-04310-2

Rajak, R., Mahto, R. K., Prasad, J., Chattopadhyay, A., 2021. Assessment of bio-medical waste before and during the emergency of novel Coronavirus disease pandemic in India: A gap analysis. *Waste Manag Res*. 27, 734242X211021473. https://doi.org/10.1177/0734242X211021473

Ratna, R. S., George, J., Aliani, A. H., 2021. Steps taken by India to Contain the Second Wave of COVID-191. www.unescap.org/sites/default/d8files/knowledge-products/15.9.2021%20India%20Paper%20Covid_final.pdf

Sarkodie, S. A., Owusu, P. A., 2021. Impact of COVID-19 pandemic on waste management. *Environ Dev Sustain* 23, 7951–7960. https://doi.org/10.1007/s10668-020-00956-y

Singh, A., Saha, K., 2020. COVID-19 and biomedical waste management. www.sprf.in/post/Covid-19-and-biomedical-wastemanagement (accessed November 20 2020).

Swain, S. K., Kumar, S., 2021. Infection control measures during COVID-19 pandemic–An otorhinolaryngological and head-and-neck perspective. *Indian J Health Sci Biomed Res*. 14, 3–11. https://doi.org/10.4103/kleuhsj.kleuhsj_252_20

Uttar Pradesh Pollution Control Board (UPPCB), 2017. Installed capacity of common biomedical waste treatment facility. www.uppcb.com/pdf/common_061017.pdf

Voudria, E. A., 2016. Technology selection for infectious medical waste treatment using the analytic hierarchy process. *J Air Waste Manag Assoc.* 66, 663–672. https://doi.org/10.1080/10962247.2016.1162226

Zodpey, S., Farooqui, H. H., 2018. Universal health coverage in India: Progress achieved & the way forward. *Indian J Med Res* 147 (4), 327–329. https://doi.org/10.4103/ijmr.IJMR_616_18

6 Selective Collection for Optimized Recycling of Waste

Case Study: The City of Constantine (Algeria)

Omar Redjal, Oualid Meddour, and Slimane Merouani

6.1 INTRODUCTION

At the urban level, it is conceivable that the environment must be pleasant and attractive. However, many cities around the world appear to be undergoing recent changes and upheavals due to demographic shifts in the urban population, resulting in sprawling growth. In fact, excessive urbanization, changes in lifestyles, consumption, and production are factors, among many others, that have contributed to the unbridled production of waste of all kinds, responsible for nuisances and pollution with multiple and varied harmful consequences (Popkin, 1999; Santos et al., 2022). Thus, the question of waste is now more acute. Moreover, waste and environmental problems seem to lead to the implementation of new policies for sustainable development (Luís Padilha and Luiz Amarante Mesquita, 2022). Clearly, cities have always, and still, today, sought to manage their waste production, which is growing alarmingly by the day. To remedy this, it is likely that management strategies will have to operate in a new, more balanced and sustainable way (Sharholy et al., 2008).

In the field of waste, relating to the transition of societies toward sustainable development, it seems appropriate to develop, in this respect, proactive policies with all the necessary means (financial means, control, awareness, education, etc.) in favor of waste reduction, in the first instance, and treatment, in the second (Hoornweg and Bhada-Tata, 2012). Instead, measures should be taken to reduce waste flows at the source, to develop treatment, recycling of raw materials, and recovery. Certainly, these actions will bring undeniable benefits to limit the use of disposal but also to control the health and environmental impacts (Ferronato et al., 2021). Overall, an efficient environmental policy seems to have to be pursued by consolidating these initiatives and references for sustainable and integrated waste management (Gupta et al., 1998). Theoretically, sustainable waste management must first involve the definition of the policy, the organization of the planning, and finally the execution of

DOI: 10.1201/9781003364467-6

the predefined policy through management and handling (Singh and Basak, 2018). Indeed, the most important principles of sustainable management are: environmental sustainability, the polluter pays, precaution, the responsibility of actors, the reduced cost of provisions, source reduction, and the use of adequate and available technology (Cucchiella, D'Adamo and Gastaldi, 2014; Ibrahim and Mohamed, 2016; Iqbal, Liu and Chen, 2020).

In addition, solid waste management must promote, in concrete terms, responsibility, coherence, and citizenship through the appropriate and eligible participation and governance of all social, economic, and political actors (populations, traders, and governments) (Awasthi et al., 2019; Study, 2022). The production of waste, especially household waste, often causes complex problems especially in terms of its management on several scales: the unit (home), the neighborhood, the city, or even the region (Pandey, Sharma and Nathawat, 2012; Besen and Fracalanza, 2016).

At present, most of the agglomerations in developing countries are unable to take full responsibility for the collection of household waste and wild dumps, which are highly detrimental to the health of the neighborhoods. The generalization of selective collection of household waste and the development of valorization equipment rightly require, first, a prior and in-depth knowledge of the needs of the population, the urban composition and the technical and financial means to be employed (Das et al., 2019). In fact, everything must begin with an analysis of the urban fabric of neighborhoods, and careful observation of the rhythms of life of urban dwellers in order to understand the diversity of areas and needs in the territories to be served. Moreover, in waste management, knowledge of the composition of the waste is as essential as the determination of its quantity. It enables us to determine management methods and to promote processing treatment and valorization systems.

The present work is then intended to be an awakening for the inhabitants and the local authorities of Constantine city (Algeria) in order to operate for a healthy and clean city worthy of attention. The overall objective is to study the strategy for innovation conducted locally, the main problems posed by the management, organization and use of the selective collection of household waste in the collective habitat, and finally to contribute to a real integration of the latter. In order to do this, it was suggested that the evolution of the quantity and composition of household waste generated in the city through a study area (in this case DAKSI ABDESLEM) should be highlighted and that the quality of the proposed service should be assessed in a correlative manner, the degree of compatibility between the new measures adopted and current urban social contexts.

6.2 WASTE MANAGEMENT IN URBAN AREAS (ELEMENTS OF APPRAISAL)

By referring to previous studies, our comparisons between cities and countries have revealed that the generation and composition of waste are associated with the standard of living of the people who generate it (Table 6.1) (Charnay, 2005). From Table 6.1, we affirm the heterogeneity of household waste. However, it is observed and thus found that the fermentescible fragment is important in developing countries, because

TABLE 6.1
Composition of Household Waste in Developing Countries (In % by Weight on Dry)

Country	Fermentable and vegetable products	Glass	Plastics	Paper and board	Metals
Benign	45	(?)	3–4	(?)	2
Burkina Faso	39	3	10	9	4
Egypt	**60**	2.5	1.5	13	3
Guinea	**69**	0.3	**22.8**	4,1	1,4
Mauritius	**68**	1	13	12	1
India	38.6	1	6.03	5.57	0.23
Lebanon	**62.4**	5.6	11	11.3	2.9
Malaysia (Petaling Jaya)	36.5	3.2	18.4	27	3.9
Malaysia (Seberang Perai)	30.1	1.5	12	**30.8**	3.2
Morocco	**65–70**	0.5–1	2–3	18–20	**5,6**
Mexico	55	4	4	15	6
Mauritania	4.8	3.8	20	3.6	4.2
Peru	34.7	**7.1**	7.2	6	2.8
Tunisia	**68**	2	7	11	4
Turkey	36.1	1.2	3.1	11.2	4.6

Source: Charnay, 2005.

it represents the highest percentages reaching 70% (for Morocco and Guinea). The percentages of this fraction will indeed be indicators of the mode of consumption of the population. The other parts are of low importance, at least, are as revealing for some countries: Peru (glasses 7.1%); Guinea (plastic 7.1%); Malaysia – Seberang Perai (paper and board 30.8%); and Morocco (metals 5.6%). What is more, the lifestyles of large cities in developing countries now seem to be gradually moving closer to cities in developed countries showing a similar level of consumption. For these agglomerations, the management operations to be implemented will also be similar to those undertaken in industrialized countries. It is, in any case, a question of being able to manage on a daily basis but above all to benefit from this important deposit of waste, especially household waste (Table 6.2). This effectively requires the promotion of a just transition from the linear model "the whole at the landfill," practiced to date in Algeria, to a circular model, more environmentally friendly and economically profitable. In addition, that must necessarily require a better knowledge of the waste that society produces.

In the Maghreb countries (Table 6.2), the fermentable fraction of waste is still dominant, with an average percentage between 50% and 80% for Algeria, Tunisia, and Morocco. For the remaining components (paper, chalk, and plastic), although they show much lower figures, they are also significant reflecting food consumption patterns.

TABLE 6.2
Composition of Household Waste in Maghreb Countries

Country	Organic fraction Putrescible	Paperboard	Plastics	Inert fraction fine
Algeria	**67–89**	**7–9**	**2–3**	**0,2–23**
Libya	42–48	16–19	2	3
Morocco	50–70	5–20	2–8	5–20
Mauritania	4,6	3	17	44,5
Tunisia	37–81	1–23	1–16	0–2

Source: Ammar, 2006; Mezouari-Sandjakdine, 2011.

Relatively, in developed and modern countries, even if reducing the amount of waste they produce is still the essential objective, the priorities are indeed oriented toward other objectives. The management of their waste already forms a highly coveted economic activity that responds to current economic and environmental challenges. On the other hand, for developing countries, these objectives are still being pursued. Difficulties seem to be encountered which can be attributed to multiple and significant reasons, but above all, linked to the specificities of the countries and the needs of the populations concerned.

These comparisons between cities and countries reveal that the production and composition of waste is associated with the standard of living of the populations that generate it. Moreover, the lifestyles of large urban centers in developing countries now seem to be moving toward cities in developed countries with similar levels of consumption (Thonart and Diabate, 2005). Thus, for these agglomerations, the management operations to be implemented will also be similar to those undertaken in the industrialized countries. It is really a question of being able to manage large quantities of especially household waste every day. It is thus a choice of management that is most integrated, viable, and sustainable, but more inspired and adapted to each policy pursued. It is with this in mind that we have highlighted the evolution of the quantity and composition of domestic waste generated at the national level.

In this study, it was found that household waste in Algeria has changed considerably: the organic fraction, which was around 80% in 1983, has undergone a decrease and shows a percentage of 62.12% in 2010. What amounts to the profit of other materials as well as plastics, paperboards that records a remarkable increase, enabling incentive treatment tools and actions to minimize significant amounts of waste generated (ANGed, 2014) (Figure 6.1).

In practice, several projects have been carried out at the national level, but it has been found that they are still insufficient. The figures showed a low rate of development of valorization chains in relation to the number of technical landfill sites completed. With a rate of collection of municipal solid waste, ranging from 85% to 90% in urban areas, only 7% of waste is recycled and 1% is composted. The city of Constantine (Figure 6.2), for its part, is witnessing a societal conversion to

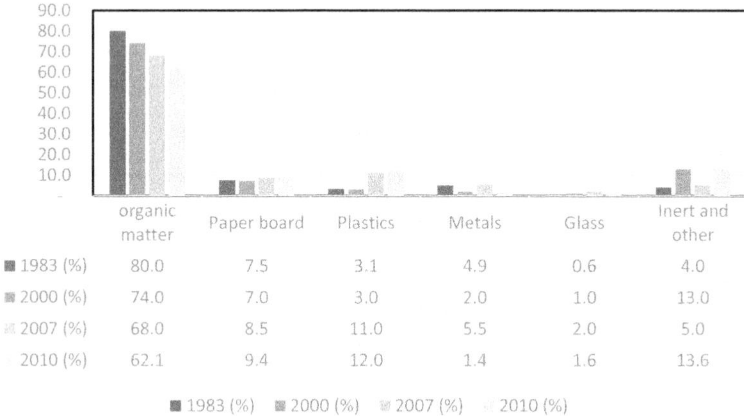

	organic matter	Paper board	Plastics	Metals	Glass	Inert and other
1983 (%)	80.0	7.5	3.1	4.9	0.6	4.0
2000 (%)	74.0	7.0	3.0	2.0	1.0	13.0
2007 (%)	68.0	8.5	11.0	5.5	2.0	5.0
2010 (%)	62.1	9.4	12.0	1.4	1.6	13.6

FIGURE 6.1 Changes in the composition of household waste in Algeria.

FIGURE 6.2 Satellite view of the city of Constantine.

Source: Google Earth version 6.0.3.21.

FIGURE 6.3 Status of household waste collection points.

a disproportionate production of waste that undermines its environment and marks several of its sites and landscapes, as provided in the example of Figure 6.3. Some 600 tons of solid waste are produced daily through the wilaya (EDC, 2016). In order to cope with these quantities, the local authorities have made extensive efforts to implement the guidelines of the national strategy for the preservation of the environment, and those of the National Plan of Environmental Action and Sustainable Development.

The strategy to adopt cleaner technologies and sustainable management methodologies has now enabled innovative projects to be launched in the sector. In fact, the city's neighborhoods have adopted an ecological system of selective collection through the air and buried containers. However, the methods undertaken have given rise to enormous difficulties in implementation combined with multiple factors: financial, technical, functional, and social.

6.3 ISSUES AND OBJECTIVES

This state of affairs fully expressed our main concern. Moreover, our study was based on the following question: What are the strategic means and tools to better integrate the selective collection of household waste (as a crucial step for valorization) in the districts of the city of Constantine? The stated objectives were then to study mainly the strategy for innovation carried out locally, the main problems posed by management, the organization and use of the selective collection of household waste in the collective habitat, and finally to contribute to real integration of it.

6.4 THE CITY OF CONSTANTINE: URBAN CONTEXT AND HOUSEHOLD WASTE MANAGEMENT STRATEGIES

In our specific case, based on an analysis of the urban context, the city of Constantine was built over the years on a difficult, irregular site with a very rugged topography. In fact, these physical constraints have not retained their evolution. Urban sprawl was carried out rapidly, with little control and with less control, creating a heterogeneous urban setting and very varied and differentiated forms of organization. Because of its original site, and its particular characteristics, Constantine had to be a city of exemplary quality in terms of environment, services, and living environment. However, in its development, the city has exposed significant deficiencies and difficulties in the organization and operation of certain services of general interest, especially in waste management.

In this regard, in the context of the implementation of the local strategy, significant progress has been made, as well as several initiating projects undertaken in Constantine such as the technical landfill site of BOUGHAREB, the waste disposal of the 13th km, and the establishment in some parts of the city of a pre-collection system for sorting and selective collection. The methods undertaken have unfortunately raised enormous problems in the implementation. This state of affairs has led us to characterize the actions carried out, identify responsibilities, and search for the main causes of the difficulties encountered. That is why our study focused on more comprehensive fieldwork in one of the neighborhoods in the city. This is the "DAKSI ABDESLEM" pilot district.

6.4.1 Case Study: DAKSI ABDESLEM

This district is situated on the southwest side of the SIDI MABROUK Urban Group, in the context of the evolution of the city towards the east (Figure 6.4). At present, this fraction of this grouping seems to be a very important urban area with the function of multi-services because of its location along the axis linking the city to the other central districts. In short, the neighborhood forms a clearly defined urban complex with a discontinuous built area (collective buildings for residential use) that differentiates it from its immediate environment (DPCC, 2011).

Nevertheless, despite this urban profile, recurring waste management problems have been noted. At neighborhood level, collection points are constantly overflowing with waste (Figure 6.5). The same is still true for newly installed buried bins, which cannot contain all the waste that is piling up and literally littering the ground in the alleys.

6.4.2 Study of the District: Diagnostic Elements (Area, Densities, and Spatial Organization)

Our target area (pilot district) comprises 646 buildings gathering 3780 dwellings characterized by a strong predominance of collective housing. The gross density of the district is of the order of 63 housing/ha. A density that is significantly high

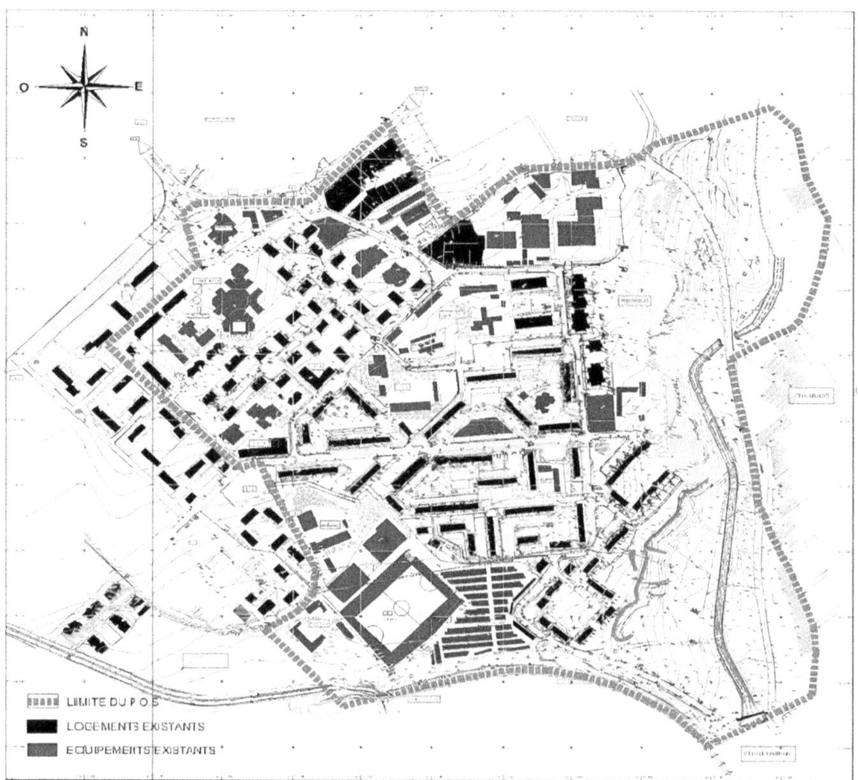

FIGURE 6.4 Land use plan of the DAKSI ABDESLEM district.

Source: DPCC, 2011.

compared to the rest of the neighborhoods of the SIDI MABROUK urban area. As for the population of the district, it is estimated at 17,767 residents, representing nearly 4.34% of the population of the city of Constantine estimated at 409,450 residents. This number allowed us to highlight a large density of 296 residents/ha. In fact, this intensity of residential space occupancy explicitly reflects the amount of household waste generated in the study area.

Our analysis of the urban context has shown that the current composition forms a ventilated fabric with low-space-consuming apartment buildings, which is a favorable aspect to make forward-looking adjustments for often-neglected interstitial spaces, poorly maintained and without cleanliness equipment. Indeed, the composition of the whole is without an apparent scheme. However, the structuring of the district was carried out on the basis of structural elements, which are the ways that obey directions imposed by constitutive and structuring barriers, thus giving rise to unequal zones (Figure 6.6).

FIGURE 6.5 Stockpiling of household waste at collection points.

6.4.3 The Built Environment (Housing Park)

There are now 3780 **housing** units in the neighborhood (NSO, 2020), of which almost 10% are uninhabited, and 3.3% are for professional use. The current occupancy rate, on the basis of dwellings only is 5.42 persons per dwelling, indicating the importance of large families in this neighborhood. In addition, it should be noted that they are not served by an internal storage room. The interior of these buildings has not been fitted with equipment and has no facilities for pre-collection of waste. These are evacuated outside to existing collection points.

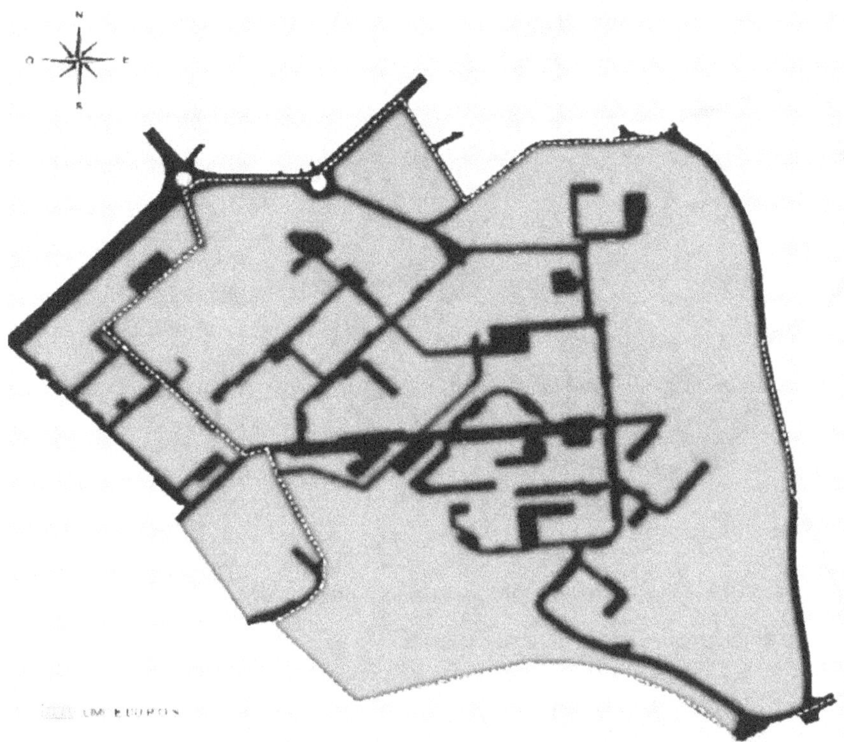

FIGURE 6.6 Spatial organization of the DAKSI ABDESLEM district.

Source: Authors treatment.

6.4.4 Study of Household Waste Management (Survey with Users via the Questionnaire)

To determine "social acceptability" on the issue of the promotion of selective collection in the pilot district DAKSI ABDESLEM, and to know the conditions that could make compatible the integration of this new equipment, we used a survey questionnaire as an instrument (Ghiglione, 2008). For this, a quota sample of 329 households (1/10 of neighborhood households) was chosen (Ghiglione and Matalon, 1978). Thus, the investigation we carried out allowed us:

- (i) To know the practices of the residents of the district regarding the management of their household waste;
- (ii) To survey the satisfaction of these residents with regard to the quality of the service provided, in particular the establishment of sorting and selective collection;
- (iii) To highlight weaknesses (malfunctions) and strengths for effective service improvement.

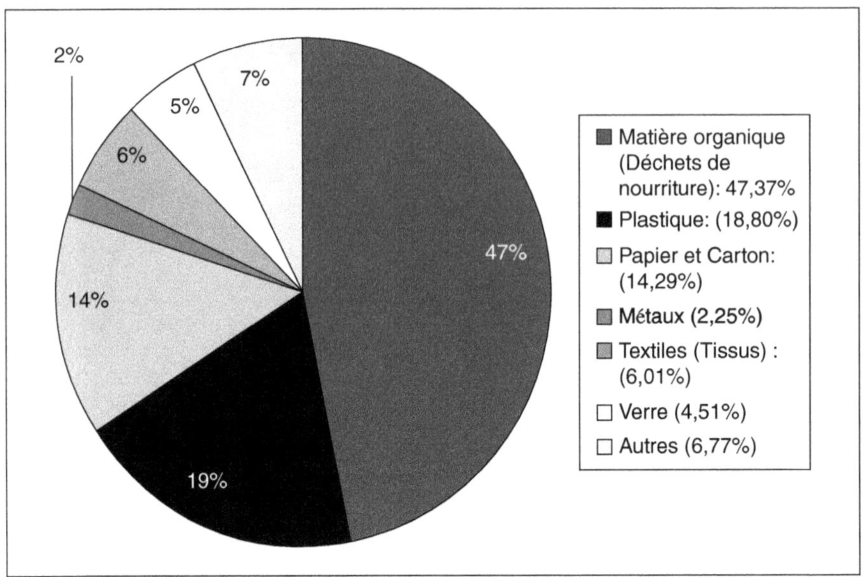

FIGURE 6.7 Composition of household waste in the district.

Source: Authors treatment (Field investigation).

6.5 RESULTS AND DISCUSSION

From a quantitative point of view, the results showed the importance of the amount of household waste generated at the neighborhood level. Considering a daily production of 0.8 kg of waste per capita (national average in urban areas 2014). Residents can produce 14.20 tons of household waste per day, or 5183 tons per year. This estimated quantity, which appears to be significant, is mainly distributed among households by weight between 3 and 5 kg. In addition, the investigation determined the residents' practices. Even if this waste is completely disposed of without any sorting beforehand, it is specified that the sorting operation is carried out, at least at the level of certain households, by limiting itself to a shorter separation of plastic bottles. Sorting information is just as flawed, and rightly so, the vast majority of residents are unaware of it.

From a qualitative point of view, the results also showed an increase in the volume of household waste. This is due to changes in the consumption and production patterns of the residents (Figure 6.7). These resulting facts and indicative parameters can usefully serve as a framework for reuse, recycling, and valorization. However, it should be remembered that selective collection in collective habitat is more difficult or even more complex to implement than in individual habitat areas, for behavioral, technical, and organizational reasons.

In our case, it was found that for 530 residents it is dedicated to a pre-collector/sector agent, and 1 vehicle for 8870 residents/sector (which is clearly far from the

FIGURE 6.8 Land organization and distribution of daily household waste production by island.

Source: Authors treatment (Field investigation).

universal standard which is a vehicle for 4000 residents) (Figure 6.8). In theory, these figures may be convincing, but in reality, the current situation relates to another interpretative logic, shortcomings, and difficulties have been recorded; a single daily rotation inducing the accumulation of large garbage bags overflowing the drop-off points. The result is an image of a dirty neighborhood burdened with garbage.

This state of affairs is reflected in part by the discrepancy between the frequency of the deposit of waste by households and that of collection by the departments concerned and by the absence of cleaning of containers, to which is added the manual and mechanical removal that is missing. Apart from the limited use of mechanized collection, household waste is collected, in any case manually, by agents without adequate personal protective equipment and transported by obsolete and incompatible rolling stock.

FIGURE 6.9 Buried bins (containers): amenities that beautify the neighborhood

Source: SOPT–Constantine.

6.5.1 Investigations and Readings on the Modernization of Household Waste Management Systems

Pre-collection devices (buried bins) are an economical alternative to the traditional collection, especially in collective habitat, a container can replace up to eight rolling bins (Figure 6.9). What is more, the facilities dedicated to it offer a green and beautifying image to the neighborhood. However, due to a lack of organization, preparation, and lack of information, and awareness, piles of household waste are piled up at these collection points on a daily basis.

The provision and operation of this equipment are based, in part, on pre-screening at the household level and, in others, on a predefined quality of the operating and organizational method. Overall, the investigations and readings carried out on the modernization of household waste management systems through the adoption of the selective collection system revealed some significant aspects:

- **POSITIVE ASPECTS**

  (i) The adoption of the action in an ecological approach ensures effective preservation of the environmental quality.
  (ii) The method adopted is more hygienic and makes it possible to remedy the problem of the recurrent pile of household waste in the open air.
  (iii) The beautification of the area is clearly seen through the facilities dedicated to the buried bin installations.
  (iv) The improvements have brought about changes in the quality of the outdoor spaces experiencing real abandonment.

(v) The improvement of the quality of service through these facilities is a support offering residents the possibility of reducing and sorting their residues.
(vi) The evolution and progressive production of recyclable waste to prioritize reuse and valorization.

- **NEGATIVE ASPECTS**

On the other hand, the results obtained, however, highlight a number of negative aspects and reveal many shortcomings. The survey carried out in the DAKSI ABDESLEM pilot district has enabled us, by analyzing the performance of the selective collection system put in place, to highlight the shortcomings or even constraints in management. Although they are multidimensional – technical, organizational, informational, and financial – we present the essential records as follows:

(i) Absence of preliminary and specific studies for best implementation practices;
(ii) Absence of technical and regulatory requirements defining the methods of source sorting, selective collection, and the responsibilities of the various actors involved;
(iii) Lack of accurate data on household waste produced by sector;
(iv) Lack of financial resources reflecting the lack of human and material resources;
(v) Lack of control over various waste management operations, including pre-collection;
(vi) Lack of source sorting at the household level to facilitate the selective collection, and consequently optimization of the recovery of waste materials (valorization, reuse, recycling, etc.);
(vii) Improper sitting and inadequate containers at household waste collection points in a poorly developed neighborhood resulting in the emergence of wilderness drop-off points in interstitial spaces;
(viii) Mismatch between household waste collection frequencies and storage schedules;
(ix) Lack of information and awareness programs, dissemination, and education on collection and support for rational waste management.

6.5.2 Strategy and Outline of Sustainable Solutions for True Ecological Management of Household Waste

To find solutions to this problem is to build appropriate management financing schemes. Above all, consider the question upstream, in all the operations of taking charge of the wastes from their production to their final treatment. In Constantine, through the study district, the participation of residents, which is an essential link in the innovation chain and better waste management, cannot yet be taken for granted. If management policy is to be effective and respond to social needs, it is more than necessary to meaningfully involve the population in the strategic approach. In order to do so, residents must have the means to participate, but first a basic information system that is easily accessible and comprehensible, and then be involved in the organization of management.

6.5.3 Special Provisions for a More Suitable Choice

In order to show the contribution of the selective collection service, it would therefore be necessary to structure its organization and to gather human, material, technical, or even intellectual means to implement correctly the planned activity, it is strongly recommended to (Redjal, 2019):

(i) Promote human resources corresponding to the objectives of the selective collection;
(ii) Adapt the frequency of collection (in the case of a combined collection) to the needs of the inhabitants;
(iii) The maintenance of equipment and the improvement of the performance of the means engaged;
(iv) Provide for landscaping of aggregation points that meet the typology of settlement areas, the requirements of the method, and the needs of the residents.

6.6 CONCLUSION

Integrated waste management remains dependent on the country's development process and reflects the strategies, actions, and methods chosen. Today's companies clearly share the same concern, the common goal of which is to remedy this recurring problem and reduce the quantities of waste produced. However, in developed countries and modern societies, priorities are shifting toward other objectives, management is already forming highly coveted economic activities and the circular economy model thus pays particular attention to waste. In contrast, developing countries are still striving to achieve these goals.

The difficulties encountered in management are largely associated with the specificities of the countries, the needs and the characteristics of the populations concerned. It is not so much a question, in the spirit of sustainability, of unifying the operational model but rather of a choice of management, in the same approach, inspired and adapted to the policies pursued.

This study carried out on the DAKSI ABDESLEM district (Constantine City, Algeria), which is the bearer of the pilot project for the adoption of curbside recycling, served as an image of a proven rupture between action and reality. Overall, the study showed that there is little talk of adjustment to the standards and technical practices of installed equipment, but rather a consideration of a set of fundamental aspects of operation: environmental, urban, economic, but above all social.

Without effective awareness-raising and clear information to users, the technical solutions required in the gradual implementation of the sorting and selective collection of household waste cannot be sufficient and effective, local policies will soon have to take into consideration the new configurations of the city, as well as social and functional diversifications. For this, several studies are necessary, and could be essential, essentially:

(i) A feasibility study and implementation of this new management system must take into account the current urban configuration of the city districts (habitat typology, structural elements of urban fabric, roads, accessibility, etc.);
(ii) A study pointing to future extensions of the city, carrying architectural, landscaping, and urban planning projects (a draft for a coherent realization model);
(iii) From a socioeconomic point of view, it would be wise to rethink the current economic model (Rutqvist, 2016) as well as to examine the origin of products and the destination after use (waste cycle) to improve activities by adapting them to the urban composition of Constantine.

BIBLIOGRAPHY

Ammar, S.B. (2006) 'Les enjeux de la caractérisation des déchets ménagers pour le choix des traitements adaptés dans les pays en développement: résultats de la caractérisation dans le grand Tunis mise au point d'un méthode adaptée'. PhD thesis, Institut National Polytechnique de Lorraine, France.

ANGed (2014) *Report on solid waste management in Algeria*. Tunis Tunisia: SWEEP-Net/GIZ.

Awasthi, M.K. et al. (2019) 'Chapter 6–Sustainable management of solid waste', in M.J. Taherzadeh et al. (eds). *Sustainable Resource Recovery and Zero Waste Approaches*. Elsevier, pp. 79–99. Available at: https://doi.org/https://doi.org/10.1016/B978-0-444-64200-4.00006-2

Besen, G.R. and Fracalanza, A.P. (2016) 'Challenges for the sustainable management of municipal solid waste in Brazil', *disP–The Planning Review*, 52(2), pp. 45–52. Available at: https://doi.org/10.1080/02513625.2016.1195583

Charnay, F. (2005) 'Composting of urban waste in developing countries: developing a methodological approach for sustainable compost production'. PhD thesis, University of Limoges, France.

Cucchiella, F., D'Adamo, I. and Gastaldi, M. (2014) 'Sustainable management of waste-to-energy facilities', *Renewable and Sustainable Energy Reviews,* 33, pp. 719–728. Available at: https://doi.org/https://doi.org/10.1016/j.rser.2014.02.015

Das, S. et al. (2019) 'Solid waste management: Scope and the challenge of sustainability', *Journal of Cleaner Production,* 228, pp. 658–678. Available at: https://doi.org/https://doi.org/10.1016/j.jclepro.2019.04.323

DPCC (2011) 'Department of Planning and Construction of Constantine (land use plan of DAKSI 2011'. Constantine, Algeria.

EDC (2016) 'Environnemental Directorat of Constantine'. Constantine, Algeria.

Ferronato, N. et al. (2021) 'Environmental assessment of construction and demolition waste recycling in Bolivia: Focus on transportation distances and selective collection rates', *Waste Management & Research*, 40(6), pp. 793–805. Available at: https://doi.org/10.1177/0734242X211029170

Ghiglione, R. (2008) *Survey techniques in social sciences*. Paris: DUNOD Edition.

Ghiglione, R. and Matalon, B. (1978) *Sociological surveys; Theories and practice*. Paris: Armand Colin.

Gupta, S. et al. (1998) 'Solid waste management in India: Options and opportunities', *Resources, Conservation and Recycling*, 24(2), pp. 137–154. Available at: https://doi.org/https://doi.org/10.1016/S0921-3449(98)00033-0

Hoornweg, D. and Bhada-Tata, P. (2012) 'What a waste: A global review of solid waste management'. Urban development series;knowledge papers no. 15. World Bank, Washington, DC. © World Bank. Available at: https://openknowledge.worldbank.org/handle/10986/17388

Ibrahim, M.I.M. and Mohamed, N.A.E.M. (2016) 'Towards sustainable management of solid waste in Egypt', *Procedia Environmental Sciences,* 34, pp. 336–347. Available at: https://doi.org/https://doi.org/10.1016/j.proenv.2016.04.030

Iqbal, A., Liu, X. and Chen, G.-H. (2020) 'Municipal solid waste: Review of best practices in application of life cycle assessment and sustainable management techniques', *Science of The Total Environment,* 729, p. 138622. Available at: https://doi.org/https://doi.org/10.1016/j.scitotenv.2020.138622

Luís Padilha, J. and Luiz Amarante Mesquita, A. (2022) 'Waste-to-energy effect in municipal solid waste treatment for small cities in Brazil', *Energy Conversion and Management*, 265, p. 115743. Available at: https://doi.org/https://doi.org/10.1016/j.enconman.2022.115743

Mezouari-Sandjakdine, F. (2011) 'Conception et exploitation des centres de stockage des déchets en Algérie et limitation des impacts environnementaux'. Phd doctoral thesis, Ecole polytechnique d'Architecture et d'Urbanisme, Algeria.

NSO (2020) 'National Statistics Office'. Algeria: www.ons.dz/

Pandey, P.C., Sharma, L.K. and Nathawat, M.S. (2012) 'Geospatial strategy for sustainable management of municipal solid waste for growing urban environment', *Environmental Monitoring and Assessment*, 184(4), pp. 2419–2431. Available at: https://doi.org/10.1007/s10661-011-2127-2

Popkin, B.M. (1999) 'Urbanization, lifestyle changes and the nutrition transition', *World Development*, 27(11), pp. 1905–1916. Available at: https://doi.org/https://doi.org/10.1016/S0305-750X(99)00094-7

Redjal, O. (2019) 'Towards new urban waste management strategies for a modern city–case of the city of Constantine'. PhD thesis, University of Constantine 3 Salah Boubnider, Algeria.

Rutqvist, L.P. (2016) *From waste to wealth: The advantages of the circular economy*. Paris: MA editions-ESKA.

Santos, A.A. et al. (2022) 'Recyclable waste collection – Increasing ecopoint filling capacity to reduce energy for transportation', *Energy Reports,* 8, pp. 430–436. Available at: https://doi.org/https://doi.org/10.1016/j.egyr.2022.01.066

Sharholy, M. et al. (2008) 'Municipal solid waste management in Indian cities–A review', *Waste Management*, 28(2), pp. 459–467. Available at: https://doi.org/https://doi.org/10.1016/j.wasman.2007.02.008

Singh, A. and Basak, P. (2018) 'Economic and environmental evaluation of municipal solid waste management system using industrial ecology approach: Evidence from India',

Journal of Cleaner Production, 195, pp. 10–20. Available at: https://doi.org/https://doi.org/10.1016/j.jclepro.2018.05.097

Study, C. (2022) 'Optimization of solid waste collection system in a tourism destination', *Global Journal of Environmental Science and Management*, 8(3), pp. 419–436. Available at: https://doi.org/10.22034/gjesm.2022.03.09

Thonart, P. and Diabate, I. (2005) *Practical Guide to the Management of Household Waste and Technical Landfills in Southern Countries*. Canada: IEPF.

7 Appraising the Natural Bio-Processes over Thermal Treatments to Treat Municipal Solid Waste

A Step Toward a More Sustainable Environment

Arumugam Poornima, Subramani Yashini Vidhya, Kanagaraj Suganya, Balraj Sudha, Sundaravadivelu Sumathi, Sanjeeb Kumar Mandal, and Sumithra Salla

7.1 INTRODUCTION

In the current fastidious and modern world, with increased trends in industrialization and urbanization, the growing population is driving toward a scenario in which there could be two major challenges. One of them is the steady rise in the energy needs which is not expected to completely be met by the conventional non-renewable sources and on the other hand is the generation of high amounts of municipal solid wastes (MSW) that can be a possible threat to mankind and environment if left untreated. Thus, the discipline of "Waste to energy" can serve a promising way to overcome both these challenges efficiently and effectively finally leading to a more sustainable environment (Shah et al., 2019). Energy from waste can possibly be one of the most prominent future alternative sources of energy that are renewable as its exploitation assures sustained and steady power generation. With environment compatibility being one major property of alternative energy source, the choice of MSW as the basic feedstock for energy plants would serve best for both sustainable management of waste and in order to produce renewable energy (Kang and Yuan, 2017). More than 2 billion tonnes of municipal garbage are thought to be produced globally, and the World Bank estimates that number could rise to 3.4 billion tonnes by 2050 (The World Bank, 2020 and Waste Atlas, 2018). Hence the

DOI: 10.1201/9781003364467-7

necessity for sustainable management strategies and systems of municipal solid waste is highly recommended.

MSW are basically those which are collected from household, commercial enterprises and small-scale institutions in a municipality. It typically comprises both degradable and non degradable fractions such as yard waste, kitchen waste, paper wood, plastic, glass, metal, electronic waste, inert materials, rubber, and some other miscellaneous trash which is the most heterogeneous consisting of fabrics, hygiene products, cosmetics, textiles, pharmaceuticals, polymeric residues, and biomedical wastes. The management process might vary with each community, considering the differences in municipal localities, cities, states, and countries but the practice undergoes certain basic stages such as generation, collection, handling, transfer, processing, disposal, and treatments of waste. It is known that well-established management systems exist in developed nations whereas the developing nations are still striving to manage the solid wastes efficiently (Baran et al., 2016). With landfills being one major disposal method, lack of new space for new wastes generated in the developing nations due to urbanization is one key challenge. This drives the management toward various other strategies to treat the municipal solid wastes and recover energy (Nanda and Berruti, 2021a).

Some such energy from waste technologies are incineration, gasification, pyrolysis, plasma gasification, torrefaction, and hydrothermal carbonization falling under thermochemical treatments and composting and anaerobic digestion falling under biological treatments. Thermal treatments obtain bio-oil, char, syngas, and other by-products and biological treatments yield biogas, biomethane, and fertile compost from MSW (Saidi et al., 2020). Though thermal treatments serve better in reducing the waste volume and efficient energy recovery they are not preferable due to their significant harmful effects on the environment (Zhang et al., 2018).

Around 40–70% of the total MSW constitute the degradable organic fraction (Asim et al., 2017; Panigrahi and Dubey, 2019) and thus biological treatments can be a potential option to recover energy and manage waste in a more sustainable and eco-friendly approach without harming the environment. In that concern, this chapter would discuss the various thermal and biological treatments to drive MSW to energy plants and emphasize the significance of biological treatments over thermochemical treatments which would lead to zero harmed environment. On that note, the chapter is aimed at stressing the necessity of biological treatments over other treatments stepping into a greener and more sustainable environment.

7.2 MUNICIPAL SOLID WASTE

MSW is a mixture containing solid wastes that are regularly disposed as trash, garbage, and refuse by both urban and rural communities. It is generated from a collection of solid garbage from homes, workplaces, minor institutions, and commercial entities. Though it differs in its classification and composition between different communities around the world, MSW generally consists of both non-biodegradable and biodegradable fractions (Chen et al., 2020). However, household waste, plastic, rubber, glass, metals, e-waste, yard waste, inert materials, and paper wastes, and various trash constitute MSW in most circumstances. The organic portion of MSW is made up of both yard and kitchen wastes. The miscellaneous garbage, including textiles, fabrics, biological wastes (such as sharps and goggles), personal hygiene goods, cosmetics, drugs, health care items, rubber, leather, and polymeric residues, are the most diverse and commonly seen components of MSW. Public health and the environment will be at risk if MSW are not handled securely through segregation, collection, transfer, treatment, elimination, recycling, and reuse (Malav et al., 2020). The WHO has drawn attention to the risks of improper solid waste disposal in terms of water, air, and soil pollution, as well as the resulting health effects on the communities in the region of the affected areas. It is predicted that 3.40 billion tonnes of MSW will be produced globally by 2050 (Sharma and Jain, 2020).

Around 71% of the municipal solid waste that the municipalities collect is recycled, 19% is dumped in landfills or other places, and 11% is used for energy recovery. Municipal solid waste is a significant renewable and affordable source for energy and resource recovery through either the "Waste-to-Energy" or "Energy-from-Waste" systems (Shah et al., 2021). In order to meet the rising need for energy, it can be transformed into a range of fuels using various waste-to-energy methods, such as thermochemical and biological treatments. While incineration is the most popular waste-to-energy process, it also greatly reduces the volume of trash. It involves burning municipal solid waste to produce steam, energy, and combined power and heat (Yadav and Karmakar, 2020). On the other side, MSW can be transformed into products like char, bio-oil, and syngas via a number of well-known thermochemical processes, including pyrolysis, liquefaction, torrefaction, and gasification. The MSW can be converted into biogas or biomethane and fertile compost using the right biological conversion processes, such as aerobic composting and anaerobic digestion (Nanda and Berruti, 2021a). The treatment strategies of MSW are shown in Figure 7.1.

7.3 MUNICIPAL SOLID WASTE MANAGEMENT

Municipal Solid Waste Management (MSWM) is the process of managing municipal waste according to the principles of Integrated Solid Waste Management (ISWM). This integrated management of the solid waste implies the implementation of effective methods that encompass all types of waste materials to acquire the two primary objectives of waste reduction and effective treatment of waste (Khan et al., 2022). To reduce waste, it is now widely acknowledged that society in general must generate "more with less," or increasing goods and services while using fewer resources globally and with less pollution and garbage. As part of solid waste minimization programs, production, as well as product improvements, has been implemented in several nations employing internal material recycling or on-site energy recovery (Roy et al., 2022).

To provide improved human health and safety, effective solid management systems are required. They must protect public health by limiting the spread of disease while also being safe for workers. They also need to be economically and environmentally sustainable in addition to these requirements. Minimizing both costs and environmental impacts at the same time is challenging (Vyas et al., 2022). The key is to strike a compromise between reducing the waste management system's total environmental

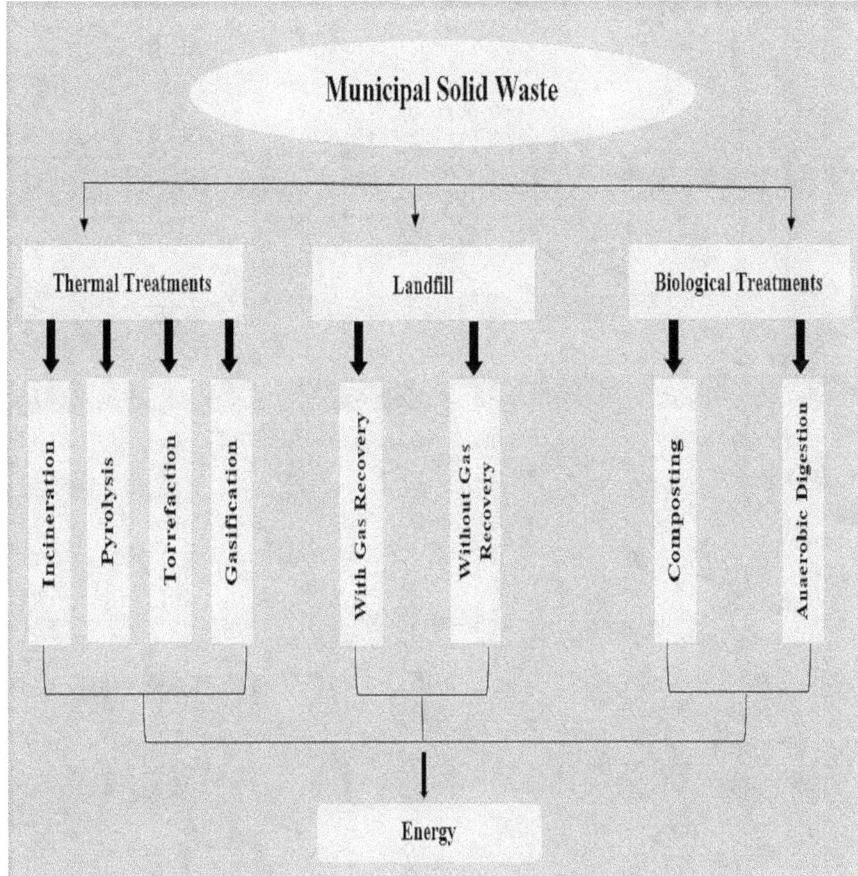

FIGURE 7.1 Municipal solid waste treatment – Waste to energy.

consequences as much as feasible while keeping costs within reasonable bounds. A solid waste management system must adopt an integrated strategy that handles all categories of solid waste and all sources of solid waste in order to be both economically and environmentally sustainable. A multi-material, multi-source management strategy often substitutes a material- and source-specific strategy in terms of the environment and the economics. Such a system should manage specific wastes in various streams (Sliusar et al., 2022). Figure 7.2 shows the overview of management of MSW.

An effective waste management system should include the elements listed below:

- Waste collection and transportation;
- Resource recovery, which entails the recovery of materials and energy through biological, thermal, or other processes;
- Waste transformation which involves lowering the toxicity, volume, or other physical/chemical characteristics of waste to make it appropriate for disposal.

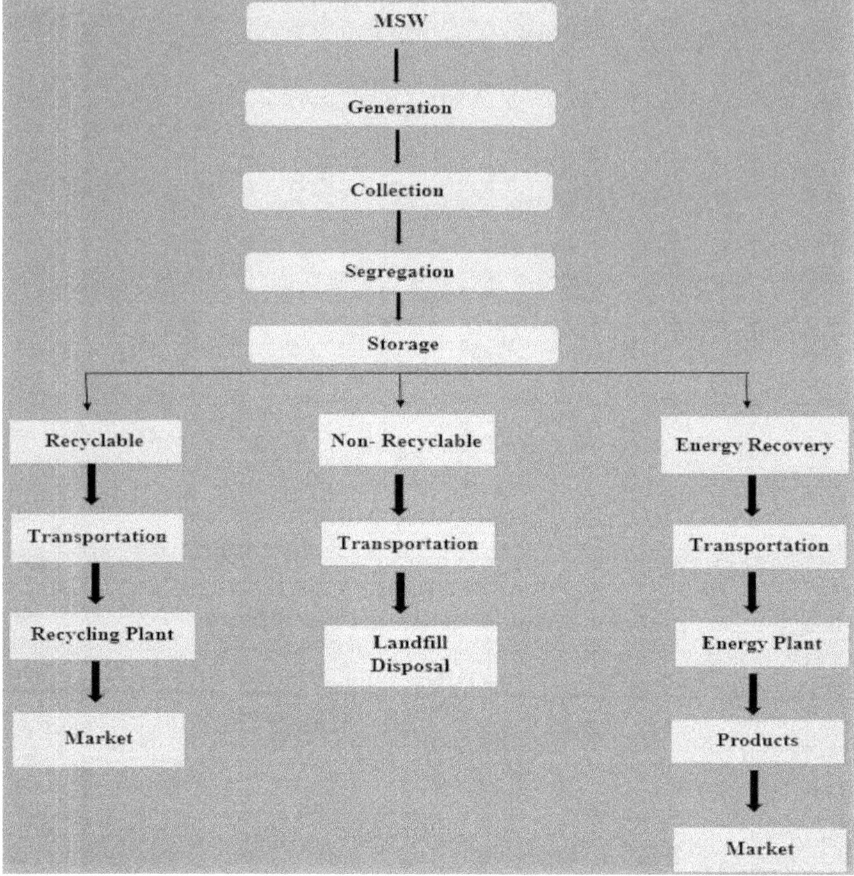

FIGURE 7.2 An overview on management of municipal solid waste.

7.3.1 Elements of MSWM

The six functional aspects that can be used to categorize the actions involved in managing municipal solid wastes from the point of generation to final disposal are waste generation, waste handling and sorting, storage, and processing at the source, collection, sorting, processing, and transformation, transfer, and transport, and disposal (Prajapati et al., 2021).

7.3.2 Waste Generation

Waste generation includes processes that identify materials as being unusable and either discard them or collect them for disposal. According to the data from World Bank, the total volume of global municipal solid waste generated is estimated to reach around 3.4 billion tonnes by the year 2050. Each person on average is said to generate around 0.74 kg of municipal solid waste per day. Though it can differ with various regions, the municipal solid waste generated seems to rise before the rise in population (https://ourworldindata.org/waste-management). The generation of waste is currently an activity that is difficult to control (Miller et al., 2020). However, it is expected that the development of waste will be under more control in the future. Even though it is not under the control of solid waste managers, waste reduction at the source is now taken into account in system evaluations as a way to regulate the amount of waste produced (Wang et al., 2021).

7.3.3 Waste Handling, Sorting, Storage, and Processing at the Source

The methods used to manage trash until they are put in storage containers for collection include garbage handling and sorting. According to the World Bank, 30–40% of the MSW is not properly collected and those collected wastes are sometimes not handled and sorted properly. Also there seems to be more than a million informal waste pickers around the world. Handling includes transporting loaded containers to the collecting destination. A critical stage in the treatment and storage of solid waste at the source is the separation of waste components. For instance, the point of generation is the ideal location to segregate waste materials for recycling and reuse. Households should handle newspaper, cardboard, bottles, glass, kitchen trash, ferrous objects, and non-ferrous objects separately (Zhu et al., 2021). Due to public health issues and aesthetic considerations, on-site storage is crucial. At many residential and commercial sites, it's common to see unsightly improvised containers or even open-ground storage, both of which are unpleasant. The household, in the case of an individual, or the management of commercial and industrial properties, often bears the cost of providing storage for solid wastes at the source. Processing at the source includes activities like composting yard garbage (Taşkın and Demir, 2020).

7.3.4 Collection

According to statistics, the collection rate of MSW seems to be relatively low when compared to that of high-income countries. The gathering of recyclables and solid wastes is just one part of the collection's functional component; another is transporting these commodities from the collection site to where the collection vehicle is emptied. This place could be a facility for processing resources, a transfer point, or a disposal site for waste (Yadav and Karmakar, 2020).

7.3.5 Sorting, Processing, and Transformation of Solid Waste

The fourth functional component is the sorting, processing, and transformation of solid waste. It is very much necessary as almost half the waste that is being generated does not get into the upcoming processes, instead are just being dumped as landfills. The recovery of sorted materials, processing of solid waste, and transformation of solid waste are all included in this functional component, which primarily takes place at locations far from the point of garbage generation (Yin et al., 2020). Mixed trash is typically sorted in composting plants, transfer stations, combustion facilities, and disposal sites. Large items are frequently removed during sorting, industrial wastes are separated by size using screens, waste components are manually separated, and ferrous and non-ferrous metals are separated. To recover conversion products and energy, garbage is processed (Nanda and Berruti, 2021). The municipal solid waste component can be altered using a wide range of biological and thermal approaches. Aerobic composting is the biological transformation method that is most frequently applied. Incineration is the most typical thermal transformation process. Without resource recovery, waste transformation reduces the quantity, weight, size, or toxicity of waste. Many mechanical, thermal, or chemical processes can modify materials (Wang et al., 2021).

7.3.6 Transfer and Transport

The functional aspect of transfer and transport consists of two processes: (i) transferring waste from a relatively small collection vehicle to a larger transport vehicle; and (ii) transporting the waste from the larger transport equipment to a processing or disposal facility, typically over long distances. The transfer typically takes place at a transfer station (Taşkın and Demir, 2020).

7.3.7 Disposal

Nowadays, all solid wastes are disposed of by landfilling or unregulated dumping, whether they are residential wastes that are collected and delivered straight to a landfill site, residual materials from materials recovery facilities (MRFs), residue from the combustion of solid waste, rejects of composting, or other products from various solid waste-processing facilities (Alam and Qiao, 2020). An engineered facility known as a municipal solid waste landfill plant is used to dispose of solid wastes on land or deep into the earth's mantle without endangering the public's health or safety or causing inconveniences like rodent and insect breeding or groundwater contamination (Diaz et al., 2020).

7.4 THERMAL TREATMENTS

Several thermal treatments have been employed to treat the municipal solid wastes. Some of the best known ones are incineration, combustion, and gasification treatments. The other known methods are plasma gasification, pyrolysis, and torrefaction (Table 7.1).

7.4.1 Gasification

Gasification is another destructive thermochemical process lying between pyrolysis and combustion involving partial oxidation of waste carried out using steam, air, carbon dioxide, oxygen, or a mixture of them, in an ambience containing sub-stoichiometric oxygen levels, meaning that the oxygen supplied becomes insufficient for complete combustion. Gasifiers may be co-current, counter-current, or cross flow

TABLE 7.1
Features of Major Thermal Treatments

	Incineration	Gasification	Plasma Gasification
Type of reaction	Oxidizing	Reducing	Reducing
Process temperature	850–1200°C	400–900°C	1500–5000°C
Input energy requirements	None	Auto-thermal (75%-88% of waste)	Very High (1200-1500 MJ/ tonne of waste)
Products	COz, H, O, waste heat that could be recovered	CO, CO_2, H_2, H_2O, CH_4	CO, CO_2, H_2, H_2O, CH_4, waste heat from cooling syngas that could be recovered
Waste	Ash (approx. 30%of original volume)	Char, Ash	Inert slag (6%–15% of original volume)

on the basis of interaction between the agent and the substrate (Table 7.2). Though being a largely exothermic process it is operated at a temperature ranging within 800–1200°C, to initiate and sustain the reaction depending on the composition of the feedstock material and reactor type. As the changes in waste composition is one major factor, this complex multiple reaction process requires pretreatment of MSW involving mechanical preparation, separation of inert material, metals and glass, sorting, shredding and blending, and later undergoes the phases of drying, pyrolysis, reduction, and oxidation. The reactors in practice include entrained packed-bed gasifiers, cyclone gasifiers, flow gasifiers, and fluidized bed gasifiers (Hameed et al., 2021; Hoang et al., 2021).

The major product is syngas comprising a mixture of methane, carbon monoxide, carbon dioxide, hydrogen, and some low-molecular-weight hydrocarbons (Table 7.3). Similarly certain undesired by-products such as alkali metals, sulphide, particulate matter, tar and chlorine are seen (Hoang et al., 2021).

One key challenge in using syngas for energy recovery is tar formation which is a high molecular weight hydrocarbons mixture leading to corrosion, fouling, and blocking. Incomplete oxidation can generate toxic gases which can be hazardous to both environment and human health, hence care should be taken to avoid seepage using an effective gas clean-up system and a proper conditioning system. The application of higher temperature secondary processing phase with filters, wet electrostatic precipitators, scrubbers thermal, and catalytic cracking can be used to 'crack' the tars for more pure syngas. Blending biomass containing more carbon contents with MSW is another promising option to reduce tar with production of other useful contents (Hameed et al., 2021).

Gasification deftly reduces the volume of MSW driven to landfills possibly by 95% and also reduces the methane emissions from it. It lowers the CO_2 emissions and thus can replace fossil-fuel based production of electricity. Steam gasification is recently known to produce syngas associated with much reduced tar formation and higher yields of hydrogen (Shafiq et al., 2021).

TABLE 7.2
Gasification Reactions

Gasification Treatments		Reactions	Energy /Enthalpy
Reactions involving steam	Waster–gas reaction	$C + H_2O \rightarrow CO + H_2$	+131MJ /kmol
	Secondary Water gas reaction	$C + 2H_2O \rightarrow CO_2 + 2H_2$	–90MJ /kmol
	Steam methane reforming	$CH_4 + H_2O \rightarrow CO + 3H_2$	+206MJ /kmol
Oxidation reactions	Water–gas shift reaction	$CO + H_2O \rightarrow CO_2 + H_2$	–41MJ /kmol
	Carbon partial oxidation	$C + 1/2O_2 \rightarrow CO$	–111MJ /kmol
	Carbon oxidation	$C + O_2\ CO_2$	–394MJ /kmol
	Carbon monoxide oxidation	$CO + 2O_2 \rightarrow CO_2$	–283MJ /kmol
	Hydrogen oxidation	$H_2 + 1/2O_2 \rightarrow H_2O$	–242MJ /kmol
Reactions involving hydrogen	Methanation	$11CO + 3H_2 \rightarrow CH4 + H_2O$	–227MJ /kmol
	Hydrogasification	$10C + 2H_2 \rightarrow CH_4$	– 75MJ /kmol
Decomposition reactions	Carbonization	$C_nH_m \rightarrow nC + m/2H_2$	Endothermic
	Dehydrogenation	$pC_xH_y \rightarrow qC_nO_m + rH_2$	Endothermic
Reactions involving carbon dioxide	Boudouard reaction	$C + CO_2 \rightarrow 2CO$	+172MJ /kmol
	Dry reforming	$C_nH_m + nCO_2 \rightarrow 2nCO + m/2H_2$	Endothermic

TABLE 7.3
Calorific Values of Different Components of MSW

Material	Calorific value (BTU /lb)	Ash content (wt %)	Moisture content (wt %)
Soft wood	6330	0.1	19
Damp wood	7600	4.6	7.5
Sludge material (steel mill)	9150	24.5	1.9
Nitrile rubber	15,240	3.4	
Cardboard, granulated	8592	12.3	6.4
Carbon residue	13,681	8.7	0.0
Wood waste, sawdust	7500	0.8	14

High temperature gasification by-product, the inert, glassy slag can be used to make roofing tiles, asphalt filler or as sandblasting grit. Though a large energy input is required compared to incineration, gasification shows relatively higher net energy output (Sudibyo et al., 2017). Gasification seems to be more cost effective compared to incineration (Pham, 2014; Luo, 2018).

7.4.2 Plasma Gasification

Plasma gasification is a thermochemical treatment of MSW that uses high temperature plasma to degrade waste. When a large electric current is fed through a gasifying agent, where the electrons dissociate and produce a plasma steam of ionized gas, thermal plasma is produced. The steam then breaks down the feedstock resulting in significant amounts of highly active radicals, excited molecules, electrons, and ions and in reactor. It is more prominent in waste disposal with improved energy recovery efficiency in relation to its fast reaction times, high heat flux densities and low oxidant amounts. Fast reaction times can lead to high flow rate of material with more hydrogen yield in the syngas with reduced formation of tars and no combustion taking place within the compact reactors. Independent of the fluctuations seen in the feedstock flow rate, the heat source of plasma gasification can be kept in control for temperature regulation within reactor. The difficulty here is that generating the high temperature plasma arc can need a significant amount of energy, yet this technology appears to be effective if certain defects are fixed (Sesotyo et al., 2019; Hoang et al., 2021). This treatment in our country is not economically feasible because of its very high energy requirements. It is estimated that approximately 1.5 kWh/kg of the waste being treated for small plants belong to less than 100 MT capacity and around 1.2 kWh/kg of waste is being processed for plants with greater than 100 MT capacity. Also the rate of electrode consumption is high accounting to ~ 500 mg/kg of waste processed. Additionally, this would result in a rise in recurring costs; as a result, the method is not widely seen as an economically rational choice (Ministry of Science & Technology, 2020)

7.4.3 Incineration

Incineration is the process of treating garbage by igniting any organic materials that are present. Without additional processing, waste material is transformed into heat, fuel gas, and ash product later which is released into the atmosphere. Compared to landfills, incinerating MSW is a promising alternative for waste treatment. It also produces energy and heat that might be used for other reasons, reducing the volume and the solid waste weight by 90% and 70%, respectively (Chen et al., 2022). Currently, more than 10% of the MSW is produced (Joseph et al., 2018).In numerous European nations, including Sweden, France, the Netherlands, and Denmark, the ratio might reach above 50% (Yin et al., 2016). In recent years, incineration plants have quickly increased in many countries, including China. According to statistics, China generated 228 million tonnes of solid garbage in 2018, and 102 million metric tonnes (or 45% of the total) were burned (Zhu et al., 2020).

The vast number of by-products of incineration including bottom ash, fly ash, and air pollution control (APC) residues are the primary causes of worry. According to studies, each ton of MSW burned produces 25–30 kg of fly ash, 250–300 kg of APC residues, and 250–300 kg of bottom ash (Ashraf et al., 2019). Numerous researchers have looked into the recycling and reuse of these leftovers in various industries because they include valuable composites like CaO, Al_2O_3, SiO_2, MgO, etc. Electricity can be generated from a significant fraction of heat. The other contents such as nitrogen, and sulphur dioxide and carbon dioxide have better utilization when they are used optimally. Carbon dioxide is used as fire extinguishers, nitrogen, as fertilizer to enhance crop productivity, and sulphur, separated from sulphur dioxide, can be used in dental care system. Ash is utilized in the building since it can be obtained as solid lumps. The attractive property of incineration is that it reduces organic waste volume and solid mass content by 95–96% and 80–85%, respectively (Pooja, 2017).The cost of transporting waste can be decreased by placing incinerators close to the source of waste generation. In addition to offering a cheap aggregate, using the ash for environment friendly buildings also lessens the demand for landfill space. To maintain an appropriate slag quality, it is essential to prevent burning garbage that contains heavy metals and other contaminants. Before using the slag, it should first be tested for quality. Energy generated thus can be collected and used to produce heat or electricity. Organic compounds eventually break down into simpler carbon molecules like CO_2 and CH_4 in all waste disposal choices. The balance between these two gases and the reaction's timescale differs depending on the option. The introduction of solid waste disposal through incineration still faces significant air pollution control challenges. The most advanced technology for the incineration facility could cost up to 35% of the entire project cost in the United States. However, it will also be influenced by the rules of air pollution in many less developed nations. When a landfill cannot be found owing to a lack of suitable locations or lengthy hauling distances, which might result in exorbitant costs, waste burning may be helpful.

7.4.4 Pyrolysis

At temperatures between 300°C to 650°C, pyrolysis degradation process is carried out in an oxygen-deficient environment which breaks down large chain hydrocarbons into small chain hydrocarbons (Sharma et al., 2014). The major products are char and condensable gases that include CO, CO_2, H_2, and CH_4. Different factors such as type of feedstock, contact time, reactor system, temperature, heating rates, effect of catalysts, presence of hydrogen gas and pressure ranges affect the composition and yield of different pyrolysis products (Kaminsky and Zorriquetta, 2007). MSW is pyrolyzed in reactors, primarily rotary kilns and tubular reactors for larger plants, and fluidized bed and fixed-bed reactors for lab-scale facilities. For the most part, pre-treatments like shredding and drying are still frequently necessary, especially for diverse MSW fractions. The ferrous and nonferrous metals that can be recovered from the pyrolytic residues using sorting and screening procedures are value-added products. The dark-brown liquid formed from the oxidized MSW after it has vaporized and condensed is known as bio-oil (Lu et al., 2020).

Pyrolysis can be used for treatment of defined wastes such as plastics and used tyres (Czajczyńska et al., 2017; Jha and Kannan, 2021). Since plastic serves as one of the major constituents of MSW, when pyrolysed it can produce considerable amounts of fuel. Among all plastics such as polyvinyl chloride, polystyrene, polypropylene, polyethylene, and polyethylene terephthalate are mostly used for pyrolysis. It is seen that high temperature plastic pyrolysis enhanced the yield of aromatic components

TABLE 7.4
Types of Pyrolysis

Technology	Products	Temperature (°C)	Residence Time	Heating Rate
Carbonization	Charcoal	300–500	Hours–Days	Very Low
Pressure Carbonization	Charcoal	450–550	15min–2hours	Medium
Conventional Pyrolysis	Char, oil, syn–gas	400–600	Hours	Low
	Char, syn–gas	700–900	5–30mins	Medium
Vacuum Pyrolysis	Oil	350–450	2–30sec	Medium
Flash /Rapid Pyrolysis	Oil	450–650	0.1–2sec	High
	Oil, syn–gas	650–900	<1sec	High
	Syn–gas	1000–3000	<1sec	Very High

in the liquid product (Miandad et al., 2016). The char can be applied for long-term carbon sequestration and energy production. Also used as a soil amendment for improved soil fertility due to high content of K and P.

Pyrolysis is applied to produce activated carbon; charcoal and methanol in chemical industries and also waste sorting for pyrolysis obtains the materials like glass, ceramics, soil, and stone which can be used for landfill covers and construction. Pyrolysis products can be stored and used in later phases for energy sources. It has high flexibility for the composition of feedstock subjected to the process, minimizes the risk of contamination of water bodies, reduce landfill waste and gas emission, does not require complex system for gas clean up, reduced corrosion and is easier to control than incineration (Zeng et al., 2015; Werle et al., 2016).

Based on the operating temperature of the pyrolysis treatment, it can be classified into slow, fast, and flash pyrolysis (Table 7.4). Slow pyrolysis (300◦C), is a process of carbonization, where different feedstock materials are treated under pyrolysis without condensation of the pyrolysis products (Lu et al., 2020). Fast pyrolysis (400 to 700◦C) carried out to obtain higher bio-oil yields. Flash pyrolysis (700 to 900 ◦C or above) is a type where larger amounts of syngas and better quality bio-oil with much reduced water content can be recovered (Sipra et al., 2018). The environmental impacts of the process is the formation of inorganic HCl and heavy metals such as P, Br, Ca, Cl, Cr, S, Pb, and Sb found in the syngas and bio-oil produced and organic polychlorinated dibenzo-p-dioxins and dibenzofurans (Gao et al., 2016; Yuan et al., 2014; Miskolczi et al., 2013). Fuel gas scrubbing is used for controlling the emission of both organic and inorganic pollutants that are associated with MSW pyrolysis (Chen et al., 2014).

7.4.5 Torrefaction

In order to prevent combustion and oxidation of the feedstock, torrefaction, a moderate and slow pyrolysis, is conducted at ambient pressure and temperature ranges between 200°C and 350°C in an inert environment (Nordin et al., 2013). It is initiated by moisture evaporation, and later partial devolatilization takes place. Some major components used as feedstock materials include food waste, wood residues, PVC plastic and discarded tires (Bilgic et al., 2016). The main product of the process is the char which mainly depends on the crucial factor, the torrefaction temperature wherein increased temperatures lead to higher yields of char. In various applications, such as entrained flow gasification, co-firing in power plants, and small-scale combustion facilities, the resulting char can be used as high-quality fuel. It is also recognized that the char can be used as an adsorbent for the in situ remediation of soil and for the purification of water (Basu, 2013). Studies were performed to test various physical and chemical properties of the torrefied MSW feedstock and showed that the chars are more adsorbing when produced at 300°C which could have various agricultural purposes (Poudel et al., 2015; Yuan et al., 2015). The char product seemed to retain certain organic pollutants like dioxins and dioxin-like compounds, heavy metals with high boiling points (e.g., Pb and Zn) whereas low boiling points metals like Hg entered into the gas phase (Yu et al., 2016).

7.5 BIOLOGICAL TREATMENTS

7.5.1 Anaerobic Digestion

Anaerobic digestion, a series of natural bio-processes involves degradation of the organic fractions of waste by microbial consortium in the absence of oxygen. The major end products of the process are the renewable fuel (biogas) and digestate (Abudi et al., 2016). It majorly takes place in four different stages namely hydrolysis, acidogenesis, acetogenesis, and methanogenesis. Since the microorganisms participating may be specific for each step, the process depends on different environmental conditions such as temperature, organic loading rate, moisture content, pH, retention time and C/N ratio (Fernández-Rodríguez et al., 2013; Kang and Yuan, 2017). The treatment of anaerobic digestion is depicted in Figure 7.3.

i. **Temperature:** The operating temperature of the process is the most important factor to be considered since it is very much required for the survival and for the growth of microorganisms. Two types of microorganisms are present namely thermophilic (50–60°C) and mesophilic (30–40°C).Studies of anaerobic digestion at 35c and 50c showed that thermophillic conditions are more favoured.
ii. **Retention time:** The time required for the decomposition of organic fraction, called the retention time is determined by the chemical oxygen demand (COD) or biochemical oxygen demand (BOD) measures of the influents and effluents. More degradation is seen when retention time is longer.
iii. **Moisture content:** The moisture content required, one other important factor essential to control the turgidity of cell, to transport nutrients,enzymes, intermediates, products, and microorganisms, to help hydrolysis of complex organic matters; and to modify the shape and structure of enzymes and other macromolecules. For more degradation high moisture content is usually required.

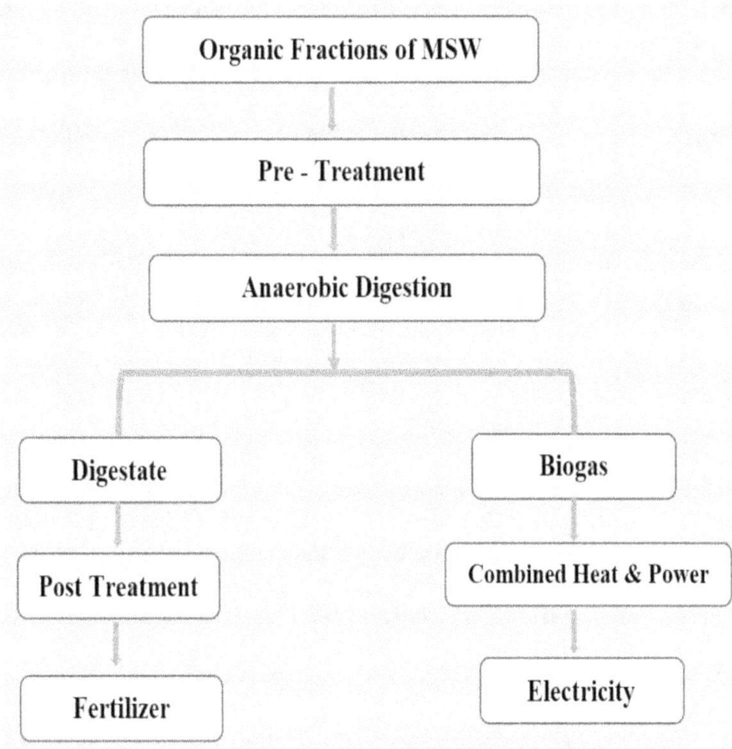

FIGURE 7.3 Process involved in anaerobic digestion of municipal solid waste.

iv. **pH:** pH is one another important factor since methanogens and methane production is sensitive to acidic conditions that are brought up in various stages resulting in the production of organic acids. pH lower than 5 can be lethal for methanogenic bacteria and can also cause digester failure.

v. **Organic loading rate:** The production rate of biogas is very much dependent on the organic loading rate. In case of higher volatile solids, more number of bacteria are required in the process.

vi. **Carbon and nitrogen content:** It is generally accepted that carbon serves as a source of energy and that nitrogen is necessary to promote microbial development. On the other hand, if nitrogen levels become the limiting factor, microbial populations will remain tiny and take longer to digest the available carbon; excess nitrogen might impede digestion.

The biogas comprises of CO_2 (30–50%), CH_4 (50–70%) and some impurities like H_2S, NH_3, halides, siloxane, and water vapours (Bhakov et al., 2014). Biogas thus produced can be used directly in combined heat and power (CHP) units, for cooking, or it can be refined for use in transportation (Karthikeyan et al., 2017). The complexity of organic waste may serve one major hinderance to the overall methanogenesis process (Kumar et al., 2018). It can be overcome by various pre-treatment processes such as (Mechanical, thermal, biological, chemical treatments, mainly for sorting, separation, sterilization and size reduction of the substrates. The pretreatment of MSW can be either optimized or avoided if waste is segregated properly at source. Anaerobic reaction takes place in three different stages and they are as follows:

Stage 1 – Hydrolysis

With the help of extracellular enzymes like hydrolases, the process of hydrolysis includes turning insoluble complex organic matter into soluble simpler molecules. Esterases, glycosidases (carbohydrates), and peptidases are a few examples. These enzymes are produced by hydrolytic bacteria such as *Clostridium*, *Peptococcus*, *Bacterioides*, *Vibrio*, *Bacillus*, *Staphylococcus*, and *Micrococcus* because of the sluggish depolymerization.

Stage 2 – Acidogenesis

In acidogenesis, the products of hydrolysis are then converted into volatile fatty acids, alcohols, carbon dioxide and lactate by fermentative bacteria. Different bacteria follow different fermentation pathways and some of them are *Saccharomyces*, *Butyribacterium*, *Clostridium*, *Lactobacillus* and *Streptococcus*. In the next stage the volatile fatty acids and alcohols are converted into acetate by acetogenic bacteria such as *Syntrophobacter wolinii* and *Smithella propionica.*

Stage 3 – Methanogenesis

The final stage, methanogenesis, involves the formation of methane gas from acetate and molecular hydrogen. Different species of Methanogens known as *Methanosaeta concilii* and *Methanothrix soehngenii* help in converting acetate to CH_4 and CO_2. *Methanobacterium bryantii*, *Methanobrevibacter arboriphilus*, and *Methanobacterium thermoautotrophicum* produce methane from H_2 and CO_2 and *Methanobacterium formicicum*, *Methanococcus voltae* and *Methanobrevibacter smithii* species aid in conversion of formate, H_2, and CO_2 to methane (Table 7.5).

The generated biogas can be utilized to feed natural gas systems, which is a better option, or as a source of heat for electricity generation. Biomethane can also be converted into biofuel for vehicles, which can be more profitable (Zou et al., 2016). The removal of CO_2, water, and H_2S can be accomplished by the use of numerous post-treatment techniques. Inhibiting the process during the methanogenic phase can also yield biohydrogen, which is regarded as cleaner biofuel when compared to

TABLE 7.5
Biogas Production from Different Biodegradable Components of MSW

Components	Biogas (Nm_3 /TS)	CH_4 (%)	CO_2 (%)
Carbohydrates	790–800	50	50
Raw Protein	700	70–71	29–30
Lignin	NIL	NIL	NIL
Raw Fat	1200–1250	67–68	32–33

natural gas and may be used in a blend to produce heat, energy, as well as for vehicle fuel and transportation. The constant utilization of chemical fertilizers can result in problems of soil degradation and environmental pollution which can be replaced by the digestate that is obtained from anaerobic digestion which rich in nutrients that can improve soil structures (Luana et al., 2019). Anaerobic digestion is a fairly complex process that can experience process instability. Because of this, it is important to continuously track the trend and development of the process stability (Figure 7.4). By adjusting various process parameters, it is possible to maximize the competence of the biogas production. When co-digestion is used in place of a single substrate (or mono-digestion), it can have various benefits, including improved nutritional balance (such as a higher C/N ratio), increased buffering capabilities, fewer inhibitory effects (such as the buildup of ammonia and VFA), and better process stability (Sarker et al., 2019; Wu et al., 2019).

7.5.2 Aerobic Composting

Composting, also known as aerobic degradation, is a biological solution for transforming MSW's organic material by using aerobic microbes that can function in the presence of oxygen. Along with CO2, H2O, nitrates, and sulphates, the aerobic process produces a stable solid material that is widely used as fertilizer for crops. The waste is broken down and digested by aerobic bacteria, which flourish in situations with plenty of oxygen. By adjusting the oxygen levels in the system, these processes can be sped up or controlled (Joshi and Ahmed, 2016). Composting in windrows, in vessels, and in aerated static piles are three common techniques used to accelerate the aerobic breakdown of organic matter. The aeration procedure enhances the porosity and oxygen availability required to quicken the rate of breakdown in each procedure.

i. **Windrow composting:** Long rows or piles of organic material are used in windrow composting and are turned automatically or manually at regular intervals to aerate them. The windrow's size, geometry, and rotating frequency regulate the heat production and oxygen passage through the core. To maintain the compost pile at the increased temperatures which is required for the bacteria to flourish and grow, it is crucial to take advantage of the exothermic nature of aerobic decomposition process.
ii. **Aerated static pile composting:** It does not turn the feedstock material like windrow composting does; instead, it arranges it in heaps rather than rows. When composting in a static pile, bulking agents like huge woodchips or other bulk materials are added to the pile for greater ventilation and to keep the decomposition process in the aerobic zone. Additionally, air blowers can be incorporated into the system for improved compost pile aeration.
iii. **In-vessel composting:** The best environmental control is achieved with in-vessel composting, which is done in a sealed drum or silo where variables like moisture, temperature, and aeration may be more precisely managed (Misra et al., 2018).

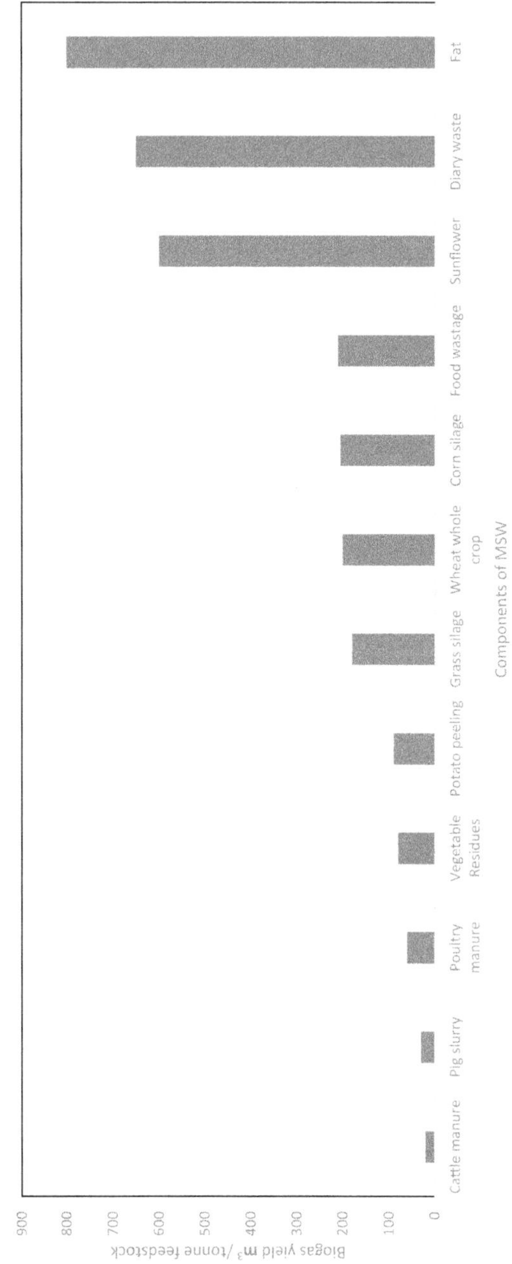

FIGURE 7.4 Biodegradable components and biogas yield of MSW.

7.5.3 Microbial Degradation of Plastics

Development of synthetic microbial consortium by manipulating the genetic structure of microorganism can be used to enhance their ability to degrade plastic wastes. The steps involved in the microbial degradation of plastics are:

- Bio-deterioration – a surface degradation process that alters mechanical, physical, and chemical properties of plastics. For example, the biofilm formed for *Rhodococcus ruber* C208 on polyethylene has shown higher levels of activity wherein it adhered to the polyethylene without any supply of external additional carbon even after being incubated for 60 days.
- Bio-fragmentation breaks down the polymeric plastics and converts it into respective oligomers, dimers or monomers by free radicals or ectoenzymes such as oxygenases released by the microorganisms; assimilation and mineralization are also some other options.

Though these studies are in their infancy, degradation of the non-biodegradable plastics using microbes seem to be achievable in the near future with advancements in science.

7.6 FUTURE SCOPE

Among all treatments that can be employed to make the "Waste to energy" discipline a more sustainable one, the biological treatments are seen to be more fitting and appropriate. The lower environmental impacts with less greenhouse gas emissions, its efficiency in energy recovery in an effort to lessen dependency on fossil fuels, its capability to treat more broader range of waste types, its more prominent waste reduction, lesser energy requirement, and relative cost effectiveness make the biological treatments more promising solutions to manage MSW in the long run. With regard to its future perspectives, increased application of biological treatments and its integration with other treatments can develop a more comprehensive system of MSW management, thereby maximizing the resource recovery and waste reduction. Advancements in biotechnology can be exploited in making the process faster and efficient in producing large amounts of energy in order to alternate the use of non-renewable energy. In addition optimization and improvement in the process methodology can serve to yield various by-products such as bio plastics and bio fertilizers. These treatments can be employed more in the developing countries where the major portion of MSW accounts are biodegradable. Thus, in comparison to the thermal treatments to manage MSW, biological treatments would serve a promising and sustainable solution to lead into a zero harmed environment in future.

7.7 CONCLUSION

The increase in the generation of MSW and energy demands are both the results of the growing population adapted to an urbanized modern world. Exploiting the potential of the MSW to recover energy is one of the most efficient methods to manage wastes and to meet the required energy needs. There are several strategies available for the treatments of these wastes and among them the biological treatment of MSW in the absence of oxygen has attracted more attention due to its tremendous potential to produce energy with minimal to no harm to the environment. Thermal treatments are effective in reducing the volume of wastes but are not completely reliable due to the harmful emissions released that can gradually destroy the existing green environment within no period of time. Thus, biological treatments would contribute the best to treat MSW and play a key element toward more sustainable management of municipal waste. One of the significant supporting factors for anaerobic digestion is that the major fraction of MSW is organic and is more suited for microbial digestion and later driven to energy plants. Anaerobically managed MSW serve enormous advantages over the conventional thermal treatments in reducing the emissions of greenhouse gases, producing soil amendments, and generating renewable sources of energy with continuous production of energy assured. Hence if the minimal flaws encountered during the process are overcome, the biological anaerobic digestion would become a promising option toward efficient management of MSW.

REFERENCES

Abudi, Z.N., Hu, Z., Sun, N., Xiao, B., Rajaa, N., Liu, C., Guo, D., 2016. Batch anaerobic co-digestion of OFMSW (organic fraction of municipal solid waste), TWAS (thickened waste activated sludge) and RS (rice straw): influence of TWAS and RS pretreatment and mixing ratio. *Energy*, 107, 131–140. https://doi.org/10.1016/j.energy.2016.03.141

Alam, O. and Qiao, X., 2020. An in-depth review on municipal solid waste management, treatment and disposal in Bangladesh. *Sustain. Cities Soc,* 52, 101775.

Ali, A., Bux Mahar, R., Ali Soomro, R. and Tufail Hussain Sherazi, S., 2017. Fe_3O_4 nanoparticles facilitated anaerobic digestion of organic fraction of municipal solid waste for enhancement of methane production. *Energ Source, Part A: Recovery, Utili, Environ Effects*, 39:16, 1815–1822. doi: 10.1080/15567036.2017.1384866

Ashraf, M.S., Ghouleh, Z. and Shao, Y., 2019. Production of eco-cement exclusively from municipal solid waste incineration residues. *Resources, Conservation and Recycling,* 149, 332–342.

Baran, B., Salih Mamis, M. and Baykant Alagoz, B., 2016. Utilization of energy from waste potential in Turkey as distributed secondary renewable energy source, *Renewable Energy*, c90, 493–500, ISSN 0960-1481. https://doi.org/10.1016/j.renene.2015.12.070

Basu, P., 2013. *Biomass gasification, pyrolysis and torrefaction – Practical design and theory* (2nd ed.). Elsevier Inc.

Bhakov, Z.K., Korazbekova, K.U. and Lakhanova, K.M., 2014. The kinetics of biogas production from codigestion of cattle manure. *Pak. J. Biol. Sci.,* 17, 1023–1029. doi: 10.3923/pjbs.2014.1023.1029

Bilgic, E., Yaman, S., Haykiri-Acma, H. and Kucukbayrak, S., 2016. Is torrefaction of polysaccharides-rich biomass equivalent to carbonization of lignin-rich biomass? *Bioresour Technol,* 200, 201–207

Cardoso Grangeiro, L., Gabriella Coêlho de Almeida, S., Sampaio de Mello, B., Tadeu Fuess, L., Sarti, A. and Dussán, K.J., 2019. Chapter 7–New trends in biogas production and utilization. In M. Rai and A.P. Ingle (Eds.), *Sustainable Bioenergy*. Elsevier, pp. 199–223, ISBN 9780128176542.

Chen, D., Yin, L., Wang, H. and He, P., 2014. Pyrolysis technologies for municipal solid waste: A review. *Waste Manag*, 34, 2466–2486. http://dx.doi.org/10.1016/j.wasman.2014.08.004

Chen, D., Zhang, Y., Xu, Y., Nie, Q., Yang, Z., Sheng, W. and Qian, G., 2022. Municipal solid waste incineration residues recycled for typical construction materials – a review. *RSC Advances*, 12:10, 6279–6291.

Chen, D.M.C., Bodirsky, B.L., Krueger, T., Mishra, A. and Popp, A., 2020. The world's growing municipal solid waste: trends and impacts. *Environ Res Lett*, 15:7, 074021.

Czajczyńska, D., Krzyżyńska, R., Jouhara, H. and Spencer, N., 2017. Use of pyrolytic gas from waste tire as a fuel: A review. *Energy*, 134, 1121–1131.

Diaz, L.F., Savage, G.M., Eggerth, L.L. and Golueke, C.G., 2020. *Composting and recycling: municipal solid waste*. CRC Press.

Fernández-Rodríguez, J., Pérez, M. and Romero, L.I., 2013. Comparison of mesophilic and thermophilic dry anaerobic digestion of OFMSW: Kinetic analysis. *Chem Eng J.* Oct 31;232, 59–64.

Gao, Q., Cieplik, M.K., Budarin, V.L., Gronnow, M. and Jansson, S., 2016. Mechanistic evaluation of polychlorinated dibenzo-p-dioxin, dibenzofuran and naphthalene isomer fingerprints in microwave pyrolysis of biomass, *Chemosphere*, 150, 168–175

Hameed, Z., Aslam, M., Khan, Z., Maqsood, K., Atabani, A.E., Ghauri, M., Khurram, M.S., Rehan, M. and Nizami, A.-S., 2021. Gasification of municipal solid waste blends with biomass for energy production and resources recovery: Current status, hybrid technologies and innovative prospects. *Renew. Sust. Energ. Rev, Elsevier*, 136:C, 110375.

Hoang, Q.N., Vanierschot, M., Blondeau, J., Croymans, T., Pittoors, R. and Van, J.C., 2021. Review of numerical studies on thermal treatment of municipal solid waste in packed bed combustion. *Fuel Communications*, 7, 100013,ISSN 2666-0520, https://doi.org/10.1016/j.jfueco.2021.100013

Jha, K.K. and Kannan, T., 2021. Recycling of plastic waste into fuel by pyrolysis-a review. *Mater. Today Proc.*, 37, 3718–3720.

Joseph, A.M., Snellings, R., Van den Heede, P., Matthys, S. and De Belie, N., 2018. The use of municipal solid waste incineration ash in various building materials: a Belgian point of view. *Mater*, 11:1, 141.

Joshi, R. and Ahmed, S., 2016. Status and challenges of municipal solid waste management in India: A review. *Cogent Environ Sci*, 2:1, 1139434.

Kaminsky, W. and Zorriqueta, I.-J., 2007. Catalytical and thermal pyrolysis of polyolefins. *J. Anal. Appl. Pyrolysis*, 79, 368–374.

Kang, A. J. and Yuan, Q., 2017. Enhanced anaerobic digestion of organic waste. In F. Mihai (Ed.), *Solid Waste Management in Rural Areas* (pp. 123–142). IntechOpen. 10.5772/intechopen.70148

Karthikeyan, O.P., Trably, E., Mehariya, S., Bernet, N., Wong, J.W.C. and Carrere, H., 2017. Pretreatment of food waste for methane and hydrogen recovery: a review. *Bioresour. Technol.* 249, 1025–1039. doi: 10.1016/j.biortech.2017.09.105

Khan, A.H., López-Maldonado, E.A., Alam, S.S., Khan, N.A., López, J.R.L., Herrera, P.F.M., Abutaleb, A., Ahmed, S. and Singh, L., 2022. Municipal solid waste generation and the current state of waste-to-energy potential: State of art review. *Energy Convers. Manag,* 267, 115905.

Kumar, S., Paritosh, K., Pareek, N., Chawade, A. and Vivekanand, V., 2018. De-construction of major Indian cereal crop residues through chemical pretreatment for improved biogas production: an overview. *Renew. Sustain. Energy Rev.,* 90, 160–170. doi: 10.1016/j.rser.2018.03.049

Lu, J.-S., Chang, Y., Poon, C.-S. and Lee D-J., 2020. Slow pyrolysis of municipal solid waste (MSW): a review. *Bioresour Technol*, 312, 123615.

Luo, X., Wu, T., Shi, K., Song, M. and Rao, Y., 2018. Biomass gasification: an overview of technological barriers and socio-environmental impact. In Yongseung Yun (Ed.), *Gasification for lowgrade feedstock* (pp. 3–17). InTech. https://doi.org/10.5772/intechopen.74191.

Malav, L.C., Yadav, K.K., Gupta, N., Kumar, S., Sharma, G.K., Krishnan, S., Rezania, S., Kamyab, H., Pham, Q.B., Yadav, S. and Bhattacharyya, S., 2020. A review on municipal solid waste as a renewable source for waste-to-energy project in India: Current practices, challenges, and future opportunities. *J. Cleaner Prod,* 277, 123227.

Miller, F.C., 2020. Composting of municipal solid waste and its components. In *Microbiology of solid waste*. CRC Press, pp. 115–154.

Ministry of Science & Technology, 2020. "Sustainable Processing of Municipal Solid Waste: 'Waste to Wealth' ", Posted On: 23 Oct 2020 6:43 PM by PIB Delhi, https://pib.gov.in/PressReleasePage.aspx?PRID=1667099

Miskolczi, N., Ates, F. and Borsodi, N., 2013. Comparison of real waste (MSW and MPW) pyrolysis in batch reactor over different catalysts. Part II: contaminants, char and pyrolysis oil properties. *Bioresour Technol*, 144, 370–379.

Misra, G.P., Kaushal, P., Bhaskarwar, A.K. and Grover, P.D., 2018. Requirement of preprocessing in a waste to energy (wte) plant based on indian municipal solid waste (MSW). *J Solid Waste Technol Manage*, 44: 2, 130–141.

Nanda, S. and Berruti, F., 2021a. A technical review of bioenergy and resource recovery from municipal solid waste. *J Hazard Mater,* 403, 123970.

Nanda, S. and Berruti, F., 2021b. Thermochemical conversion of plastic waste to fuels: a review. *Environ Chem Lett*, 19, 123–148. https://doi.org/10.1007/s10311-020-01094-7

Nidoni, P.G., 2017. Incineration process for solid waste management and effective utilization of byproducts. *Int Res J Eng Technol*, 4: 12, 378–382.

Nordin, A., Pommer, L., Nordwaeger, M. and Olofsson, I., 2013. Biomass conversion through torrefaction. In Erik Dahlquist (Ed.), *Technologies for converting biomass to useful energy*, CRC Press, pp. 217–244

Panigrahi, S. and Dubey, B.K., 2019. A critical review on operating parameters and strategies to improve the biogas yield from anaerobic digestion of organic fraction of municipal solid waste. *Renew Energy*, 143, 779–797, ISSN 0960-1481, https://doi.org/10.1016/j.renene.2019.05.040

Pham, T.P.T., Kaushik, R., Parshetti, G.K., Mahmood, R. and Balasubramanian, R., 2015. Food waste-to-energy conversion technologies: current status and future directions. *Waste Manag*, 38, 399–408. https://doi.org/10.1016/J. WASMAN.2014.12.004.

Poudel, J., Ohm, T.I. and Oh, S.C., 2015. A study on torrefaction of food waste, *Fuel*, 140, 275–281

Prajapati, P., Varjani, S., Singhania, R.R., Patel, A.K., Awasthi, M.K., Sindhu, R., Zhang, Z., Binod, P., Awasthi, S.K. and Chaturvedi, P., 2021. Critical review on technological advancements for effective waste management of municipal solid waste – Updates and way forward. *Environ Technol Innov,* 23, 101749.

Roy, H., Alam, S.R., Bin-Masud, R., Prantika, T.R., Pervez, M.N., Islam, M.S. and Naddeo, V., 2022. A review on characteristics, techniques, and waste-to-energy aspects of municipal solid waste management: Bangladesh perspective. *Sustainability*, 14:16, 10265.

Saidi, M., Gohari, M.H. and Ramezani, A.T., 2020. Hydrogen production from waste gasifcation followed by membrane fltration: a review. *Environ Chem Lett*, 18, 1529–1556. https://doi.org/10.1007/s1031 1-020-01030-9

Sarker, S., Lamb, J.J., Hjelme, D.R. and Lien, K.M., 2019. A review of the role of critical parameters in the design and operation of biogas production plants. *Appl. Sci.*, 9, 1915.

Sesotyo, P.A., Nur, M. and E Suseno, J., 2019. Plasma Gasification With Municipal Solid Waste As A Method Of Energy Self Sustained For Better Urban Built Environment: Modeling and Simulation. *IOP Conference Series: Earth and Environmental Science,* 396.

Shafiq, H., Azam, S.U. and Hussain, A., 2021. Steam gasification of municipal solid waste for hydrogen production using Aspen Plus® simulation. *Discov Chem Eng*, 1, 4. https://doi.org/10.1007/s43938-021-00004-9

Shah, A.V., Srivastava, V.K., Mohanty, S.S. and Varjani, S., 2021. Municipal solid waste as a sustainable resource for energy production: State-of-the-art review. *J Environ Chem Eng*, 9: 4, 105717.

Shah, G.M., Tufail, N., Bakhat, H.F., Ahmad, I., Shahid, M., Hammad, H.M., Nasim, W., Waqar, A., Rizwan, M. and Dong, R., 2019. Composting of municipal solid waste by ifferent methods improved the growth of vegetables and reduced the health risks of cadmium and lead. *Env Sci Pollut Res,* 26, 5463–5474. https://doi.org/10.1007/s11356-018-04068-z

Sharma, B.K., Moser, B.R., Vermillion, K.E., Doll, K.M. and Rajagopalan, N., 2014. Production, characterization and fuel properties of alternative diesel fuel from pyrolysis of waste plastic grocery bags. *Fuel Process. Technol.*, 122, 79–90.

Sharma, K.D. and Jain, S., 2020. Municipal solid waste generation, composition, and management: the global scenario. *Soc. Responsib. J*, 16: 6, 917–948.

Sipra, A.T., Gao, N. and Sarwar H., 2018. Municipal solid waste (MSW) pyrolysis for biofuel production: a review of effects of MSW components and catalysts. *Fuel Process Technol*, 175, 131–147.

Sliusar, N., Filkin, T., Huber-Humer, M. and Ritzkowski, M., 2022. Drone technology in municipal solid waste management and landfilling: A comprehensive review. *Waste Manag,* 139, 1–16.

Sudibyo, H., Majid, A.I., Pradana, Y.S., et al., 2017. Technological evaluation of municipal solid waste management system in Indonesia. *Energy Procedia,* 105, 263–269. https://doi.org/10.1016/j.egypro.2017.03.312

Taşkın, A. and Demir, N., 2020. Life cycle environmental and energy impact assessment of sustainable urban municipal solid waste collection and transportation strategies. *Sustain Cities Soc,* 61, 102339.

The World Bank, 2020. World bank country and leading groups. https://datahelpdesk.worldbank.org/knowledgebase/articles/906519-world-bank-country-and-lending-groups. Accessed 15 Apr 2020.

Vyas, S., Prajapati, P., Shah, A.V. and Varjani, S., 2022. Municipal solid waste management: Dynamics, risk assessment, ecological influence, advancements, constraints and perspectives. *Sci Total Environ*, 814, 152802.

Wang, Y., Shi, Y., Zhou, J., Zhao, J., Maraseni, T. and Qian, G., 2021. Implementation effect of municipal solid waste mandatory sorting policy in Shanghai. *J Environ Manag,* 298, 113512.

Waste Atlas (2018) What a waste: an updated look into the future of solid waste management. www.worldbank.org/en/news/ immersive-story/2018/09/20/what-a-waste-an-updated-look-intothe-future-of-solid-waste-management. Accessed 15 Apr 2020.

Werle, S., 2016. Wykorzystanie skoncentrowanego promieniowania słonecznego w procesie pirolizy biomasy. *Proc. ECOpole*, 10, 333–340.

Wu, D., Li, L., Zhao, X. and Peng, Y.Y., 2019. Anaerobic digestion: A review on process monitoring. *Renew. Sustain. Energy Rev.,* 103, 1–12.

Yadav, V. and Karmakar, S., 2020. Sustainable collection and transportation of municipal solid waste in urban centers. *Sustain Cities Soc,* 53, 101937.

Yin, K., Chan, W.P., Dou, X., Ahamed, A., Lisak, G. and Chang, V.W.C., 2020. Human exposure and risk assessment of recycling incineration bottom ash for land reclamation: a showcase coupling studies of leachability, transport modeling and bioaccumulation. *J Hazard Mater,* 385, 121600.

Yin, Y., Liu, Y.-J., Meng, S.-J., Kiran, E. U. and Liu, Y., 2016. Enzymatic pretreatment of activated sludge, food waste and their mixture for enhanced bioenergy recovery and waste volume reduction via anaerobic digestion. *Appl. Energy*, 179, 1131–1137. doi: 10.1016/j.apenergy.2016.07.083

Yu, J., Sun, L., Wang, B., Qiao, Y., Xiang, J., Hu, S. et al, 2016. Study on the behavior of heavy metals during thermal treatment of municipal solid waste (MSW) components. *Environ Sci Pollut Res*, 23, 253–265.

Yuan, G., Chen, D., Yin, L., Wang, Z., Zhao, L. and Wang, J.Y., 2014. High efficiency chlorine removal from polyvinyl chloride (PVC) pyrolysis with a gas–liquid fluidized bed reactor. *Waste Manag*, 34, 1045–1050.

Yuan, H., Wang, Y., Kobayashi, N., Zhao, D . and Xing, S., 2015. Study of fuel properties of torrefied municipal solid waste. *Energy Fuel,* 29, 4976–4980.

Zeng, K., Minh, D.P., Gauthier, D., Weiss-Hortala, E., Nzihou, A. and Flamant, G., 2015. The effect of temperature and heating rate on char properties obtained from solar pyrolysis of beech wood. *Bioresour. Technol.*, 182, 114–119.

Zhang, J., Kan, X., Shen, Y., Loh, K.-C., Wang, C.-H., Dai, Y. and Wah Tong, Y., 2018. A hybrid biological and thermal waste-to-energy system with heat energy recovery and utilization for solid organic waste treatment. *Energy*, 152, 214–222, ISSN 0360-5442, https://doi.org/10.1016/j.energy.2018.03.143

Zhu, Y., Zhang, Y., Luo, D., Chong, Z., Li, E. and Kong, X., 2021. A review of municipal solid waste in China: characteristics, compositions, influential factors and treatment technologies. *Environ. Dev. Sustain*, 23: 5, 6603–6622.

Zhu, Y., Zhao, Y., Zhao, C. and Gupta, R., 2020. Physicochemical characterization and heavy metals leaching potential of municipal solid waste incinerated bottom ash (MSWI-BA) when utilized in road construction. *Environ. Sci. Pollut. Res*, 27: 12, 14184–14197.

Zou, L., Ma, C., Liu, J., Li, M., Ye, M. and Qian, G., 2016. Pretreatment of food waste with high voltage pulse discharge towards methane production enhancement. *Bioresour. Technol.* 222, 82–88. doi: 10.1016/j.biortech.2016.09.104

8 Advancements in the Recovery and Refinement of Landfill Gas from Sanitary Landfills

Deepshikha Datta, Esha Mandal, Soheli Biswas, and Bimal Das

8.1 INTRODUCTION

The utilization of engineered landfills has become crucial in addressing the energy crisis and population growth. Various considerations have been adapted to influence the efficiency, recovery, and purification of landfill gas [1]. However, incorporating landfill gas as a renewable energy resource is still at par, when compared with other sources of energy as the advancements are a bit slow, keeping in mind the hazardous consequences of the landfills [2]. Landfill activity has been assessed using various numerical models, comprehensive spatial modelling, and geometry of designs, and also has been complemented with several biochemical processes, and other renewable energy sources to increase its efficacy as an energy source [3]. Yet another complication concerning the subject is the mitigation of unwanted harmful gases and unorganized waste, which pose a threat to humankind and the environment. Often the harmful gaseous emission from the waste decomposing in the landfill causes deadly diseases to mankind. The harmful wastes emitting radioactive radiation cause several mutations and also are responsible for life-threatening consequences [4]. Landfill gas (LFG) is a natural by-product of the decomposition of organic waste in landfills. LFG is composed primarily of methane and carbon dioxide, both of which are potent greenhouse gases that contribute to climate change [5]. The emission of LFGs can have a range of negative consequences. For example, when LFG is released into the atmosphere without being properly managed, it can contribute to air pollution and climate change. Methane, which makes up a significant portion of LFG, is about 28 times more effective at trapping heat in the atmosphere than carbon dioxide over a 100-year time horizon, making it a particularly potent greenhouse gas [6]. In addition, LFG can pose health and safety risks if it accumulates in enclosed spaces or is not properly ventilated. Methane

DOI: 10.1201/9781003364467-8

is also flammable and explosive in certain concentrations, which can create risks for workers and nearby communities [7]. Overall, the proper management of LFG emissions is important to mitigate their negative impacts on the environment and public health. This can involve techniques such as capturing LFG and using it for energy production or treating LFG to remove harmful contaminants before releasing it into the atmosphere. Also, the final product is recovered successfully using specific methodologies. Recent advancements have been proposed to contain them. Techniques like horizontal gas collection, hybrid landfill-gas power generation, and pH swing methods are some of the few which have shown sustained optimization in landfill gas being an efficient energy source [8].

Landfills are government-maintained lands or areas where wastes from different sectors are collected and deposited safely to reduce their exposure to the environment [9]. The landfills help in the reduction of the hazardous emission of methane and converting it to useful landfill gas, thus creating alternative fuel, generating electricity, preparing methanol, etc. Methane is the primary component of landfill gas, typically making up between 40–60% of the total gas composition. The methane in landfill gas is produced as a result of specific types of microorganisms known as methanogens, which convert organic matter into methane and carbon dioxide through a series of biochemical reactions. When organic waste is buried in a landfill, it is broken down by bacteria in a process known as anaerobic decomposition, which occurs in the absence of oxygen. During anaerobic decomposition, microorganisms break down organic materials such as food scraps, yard waste, and paper products, producing a mixture of gases including carbon dioxide, nitrogen, oxygen, and methane. This gas mixture is commonly referred to as landfill gas [10]. Once produced, landfill gas can be collected and processed for use as a renewable energy source. The methane in the gas can be combusted to generate electricity, heat buildings, or even be used as fuel for vehicles. By capturing and utilizing landfill gas, we can reduce greenhouse gas emissions and produce clean energy [11].

Landfill gases have shown an exclusive impact in modern times for every developed and developing country. The countries producing more hazardous wastes can now use the modern technologies being developed in the field of landfill gas optimization and its recovery [12]. Figure 8.1 describes a general landfill structure by showing different layers and divisions of functioning units such as liner, leachate collector, methane recovery system, etc. in it.

In the landfill system, the liners are generally responsible for leak prevention and ensuring proper enclosure of the waste materials. Leachate shown in Figure 8.1 is the liquid percolated from the solid waste and it needs to be removed [13]. The leachate collection and treatment plant work for the same by clearing up the leachate from the landfill and treating it outside. After the reactions and decompositions in the landfill, the explosive methane gas is produced and is recovered from the landfill by a methane recovery system [14]. As you have learned earlier, if the waste is exposed to the environment, it results in several harmful consequences and hence clay caps with different depths for different types of landfills are being used, thus preventing the direct exposure of the waste to the atmosphere. The types of waste differ with respect to time, type of landfills, climatic conditions, soil quality, drainage systems, etc., and

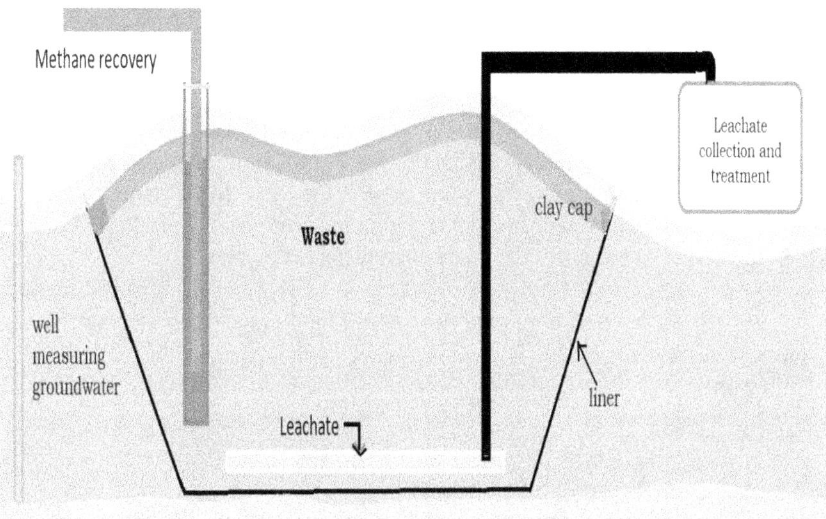

FIGURE 8.1 Basic components of landfills.

TABLE 8.1
Percentage of Deposition and Composting of the Wastes with Time

	Amount in percentage for different year (%)									
Materials	**1960**	**1970**	**1980**	**1990**	**2000**	**2005**	**2010**	**2015**	**2017**	**2018**
Paper and paperboard	17	15	25	28	43	50	63	67	66	68
Glass	2	1	5	20	23	21	27	28	25	25
Plastics	-	-	<1	2	6	6	8	9	9	9
Yard trimmings	-	-	-	12	52	62	58	61	69	63
Lead-acid batteries	-	76	70	97	93	96	99	99	99	99

varied a lot more in the past 50–60 years [15]. The statistics below show such a composition of waste being composted. Table 8.1 shows the percentage deposition of different types of waste in landfill and it is being observed that paper and paperboard has a maximum contribution.

This chapter investigates the types of landfills, affecting factors, and the current adapted techniques used in context to landfill gas-affiliated energy production and also aims to review the current advancements in the field of recovery and refinement of landfill gas. It also elaborates on the utilization of the landfill gas with consideration to the safety and economic prospect.

8.2 TYPES OF LANDFILLS

8.2.1 Municipal Solid Waste Landfills

This is the major type of landfill being practiced all over the globe. The residential wastes that are discarded end up being in municipal solid waste landfill which tends to have the best quality of safety, especially for the location of the landfill and the operating system [16]. As we come to know about modern MSW landfills, we find that it is restricted to some major layers and components as shown in Figure 8.2.

MSW landfills can be subdivided into three different parts, viz.,

- The upper layer consisting of topsoil, sand, and clay. Thus, sealing the trash after fulfilment of the MSW landfills prevents the exposure of the trash to the atmosphere.

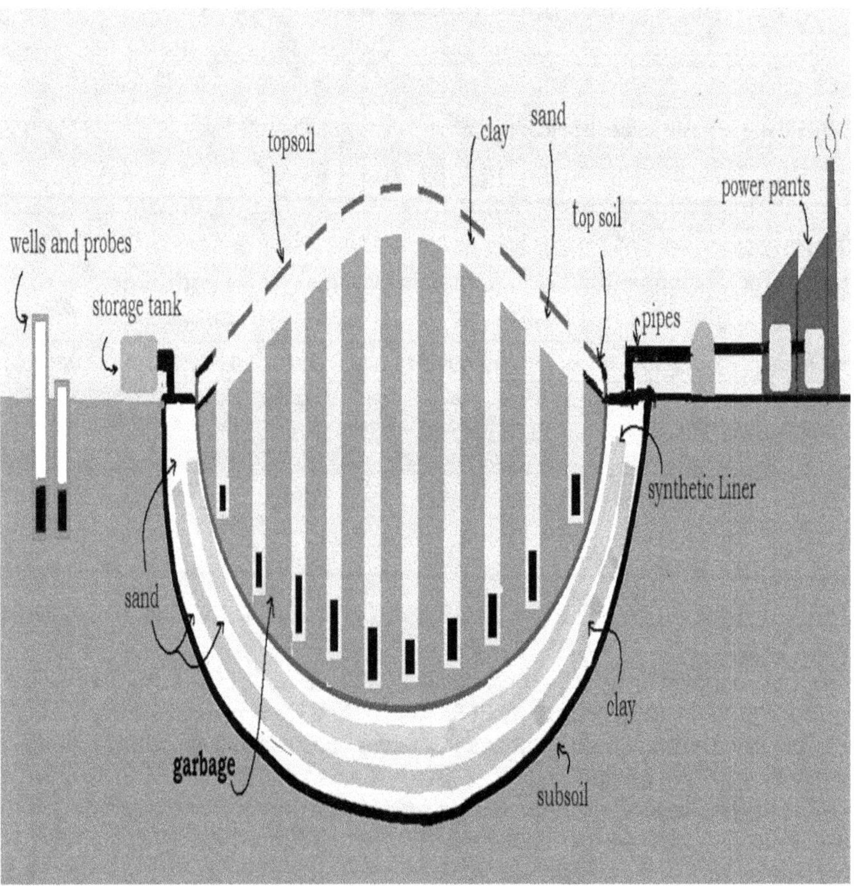

FIGURE 8.2 Municipal solid waste landfill system.

- The middle layer comprises the main component, i.e., the garbage, that is decomposed and methane is produced in due steps.
- The bottom layer incorporates synthetic liner, sand, and clay because clay and synthetic plastic lining are embraced to check the leaks in the system [17].

The leachate is pumped out for safe disposal and stored in the storage tank. Leakages tend to be more when the leachate is pumped. Therefore, structures like wells and probes are implemented in landfill [18]. The wells and probes as shown in Figure 8.2. are responsible for the detection of the leachate and the leakage of methane outside the landfills. Similarly, as and when the explosive methane gas is produced, it is collected by the pipes and transported to the power plants for the generation of electricity.

8.2.2 Industrial Waste Landfills

Industrial wastes are disposed of in this type of landfill. Most often, we find that construction debris is also disposed of here and are commonly termed as C&D (construction and demolition) landfills [19]. The renewable or reusable substances are retrieved and used up for further industrial or construction purposes. Wastes disposed of here are asphalt, concrete, metals, etc. [20]. The flow chart below shows the proceedings of the LFGs produced from the wastes. The gas is first collected and then primarily treated to remove the moisture content in it. The primary treatment is followed by secondary treatment, thus producing impurity-free LFG which is then processed and transferred to the gas engines and electricity-generating power plants. The harmful industrial wastes are hence converted into useful commodities or usable forms of energy [21]. Figure 8.3 shows the flow of landfill gas from industrial waste landfill.

8.2.3 Hazardous Waste Landfills

The most maintained and monitored landfills are specially designed to dispose of hazardous wastes. These wastes in the landfills are inspected multiple times to reduce the chances of hazards being exposed to the environment. The constructional details of hazardous waste landfills are described in Figure 8.4.

The wastes disposed of here are potential threats to the environment and public health. These landfills are filled more by chemical factories, waste treatment plants, certain agricultural industries and everyday used products such as cleansers, cosmetics, batteries, electronics, pharmaceuticals, etc. Dichlorodiphenyltrichloroethane (DDT) is reported as one of the most harmful wastes being disposed of in landfills, coming from agricultural industries. The landfill is such that it comprises different layers on the top as well as on the bottom. These types of landfills consist of more dangerous waste like certain worst chemicals that react with our hormones causing cancer, human and animal DNA spoilage, reproductive problems, certain birth defects, etc., and need to be protected all over to maintain their safety and regulation norms. Landfill products and waste have the potential to contain a wide range of chemicals, including some

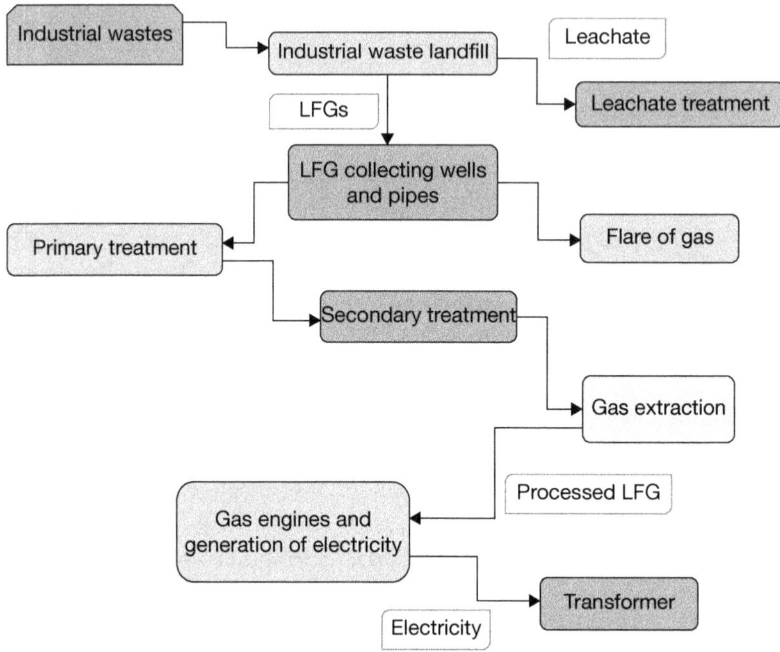

FIGURE 8.3 Flow of Land fill gas (LFG) from industrial waste landfill.

that have been linked to cancer. Hormones, such as estrogen and testosterone, are known to be present in some types of landfill waste, particularly medical and pharmaceutical waste [22]. While the presence of hormones in landfill waste may pose a potential health risk, the extent of that risk is not yet fully understood. Research has suggested that exposure to estrogen and other hormones may be linked to an increased risk of certain cancers, particularly breast cancer. In addition to hormones, landfill waste may also contain other chemicals that are known to be carcinogenic, such as benzene, vinyl chloride, and polychlorinated biphenyls (PCBs). These chemicals can leach into groundwater and soil, potentially contaminating nearby communities and posing a risk to human health [23].

Overall, it is important to properly manage and dispose of waste to reduce the potential health risks associated with exposure to harmful chemicals. This includes proper handling of medical and pharmaceutical waste to minimize the release of hormones and other hazardous substances into the environment. The HDPE cover and primary and secondary liners covering the landfill from different sides are responsible for checking the leakage in the landfill. The landfills are surrounded by wide-ranging clay caps. Thus providing a safe environment by preventing direct exposure of harmful waste to the environment. Leachate detection and removal are the most essential processes because it also contains perilous chemicals and other components, and thus cannot be left to overflow. The all-over system reduces the high risk of exposure to hazards and also produces essential explosive gases for the generation of electricity, fuel, etc. [23][24].

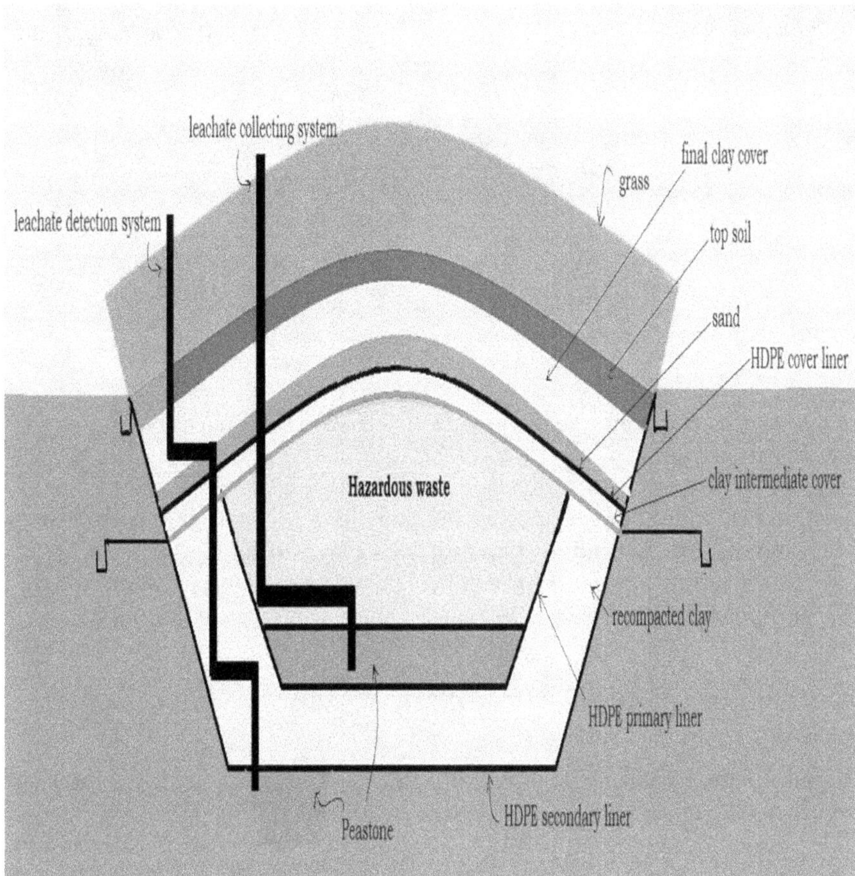

FIGURE 8.4 Hazardous waste landfill system.

8.2.4 Green Waste Landfills

These landfills are often unofficial. The organic substances that decompose naturally are decomposed in green waste landfills [25]. This in turn saves space in the MSW landfills often producing manure and other naturally decomposed reusable products. Figure 8.5 depicts the scheme of waste-to-product formation in green waste landfill. The wastes that are disposed of in this type of landfill are mulch, branches, biodegradable food material, etc. [26].

8.3 SANITARY LANDFILL

As a primary stream, Sanitary landfills are facilities for the disposal of mixed and unaltered or untreated municipal solid waste (MSW). Using engineering skills, confining sanitary wastes in the smallest possible area by depositing them on land and securing the lowest exposure to fresh air is termed landfills [27]. The sanitary landfills

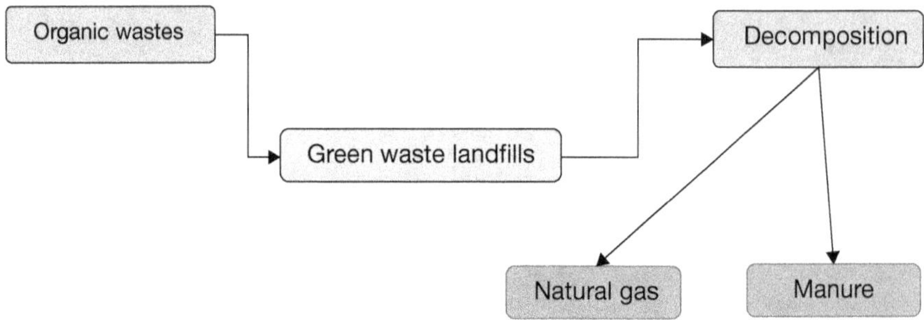

FIGURE 8.5 Waste to product formation in green waste landfill.

are those where liners are used to separate the trash. The different types of sanitary landfills are:

- Area landfill: in this type of landfill, a bulldozer is used to place solid waste substrate over the land and protect it from a layer of earth.
- Trench landfills: unlike area landfill, in this a layer of earth is dug out and waste is deposited in the trench [28].

8.3.1 Criteria or Property of Sanitary Landfills

Site Preparations

The following steps should be taken for the detailed design of the proposed landfill site and development of the treatment process [29][30]:

- Topographic survey of the site
- Detail leveling of the site
- Contour plan
- Cutting off 150 mm topsoil of the whole area and saving it for earth cover over the solid waste.
- Preparing the site for proper drainage of rainwater by cutting off the high-level ground and giving it a downslope of 1:400. Soil obtained from this cutting is used in providing the embankment of fencing of the area
- Construction of a road with greater width, to facilitate the trucks approaching the developmental site
- Providing infrastructure with all accessories at the site to facilitate the workers.

Parameters Adopted for Design

The following parameters are adequately considered in the design of the sanitary landfill site [30][31]:

- All the calculations are based on the population of the city
- It is considered that 30% of waste generated is landfilled, 60% is composted, and the remaining 10% is recycled.

- Waste will be compacted in four layers, one below the ground level and three above ground level. The depth of each layer will be 2 m.
- Excavated soil will be used as cover material.
- Cover for intermediate layers will be 0.15 m while the final layer will be covered by 0.6 m (2 ft) of cover material.
- Excavation for one year will be made in advance.
- A bond will be provided in each layer after one year.
- Waste will be compacted to a density of 1000–1300 kg/m3.
- Embankments will be provided at the periphery high enough to enclose the site with the fencing mounted on top.

8.3.2 Different Sanitary Landfills

Dry-type and wet-type landfills are two of the main types of landfills used for waste disposal.

Dry-type landfills are also known as "monofils" and are designed to accept only specific types of waste, such as construction and demolition debris, industrial waste, or certain types of hazardous waste. They typically have a liner system to prevent leachate from escaping and contaminating groundwater. Dry-type landfills can be located in areas with low rainfall and are typically designed with a lower leachate volume and lower methane production compared to wet-type landfill [32].

Wet-type landfills, also known as "bioreactor landfills," are designed to promote the rapid decomposition of waste through the addition of moisture and nutrients to support the growth of microorganisms. This type of landfill is intended for municipal solid waste, which typically contains high levels of organic materials. Wet-type landfills typically produce more leachate and methane compared to dry-type landfills, and require a more robust liner system to prevent contamination of groundwater and surrounding areas [29].

There are also other types of landfills, such as sanitary landfills, which are similar to wet-type landfills but with less emphasis on waste decomposition, and secure landfills, which are designed to contain hazardous waste and prevent it from contaminating the environment. It is important to note that the selection of a specific type of landfill depends on several factors, such as the type and volume of waste, location, environmental factors, and regulatory requirements. The goal is to design and operate a landfill that minimizes the impact on the environment and protects public health [33]. Besides this the other types of sanitary landfills with its specifications are explained in Table 8.2.

8.3.3 Sanitary Landfills Advantages and Drawbacks

Advantages

- **Excellent Energy Source:** Sanitary landfills act as excellent energy sources because they generate carbon dioxide and methane when the waste starts decomposing [36].

TABLE 8.2
Different Sanitary Landfills

1. Modern/ currently developed Droughty landfills	(a) Included processes–making straight line at decomplex base layer and decomplex upper covering process (b) Provide modern enact operator (i.e., highly engineered system), for the capture, collection, treatment of leachate and landfill gas (c) The buried waste is highly aggregated (d) Replenishment of leachate is not carried out
2. Modern/ currently developed Wet-type landfills	(a) Here also included processes–decomplex base layer and decomplex upper covering process (b) Leached out constituent and gases are arrested, expelled, accumulated by the modern enact operators (c) The excavated wastes are highly aggregated (d) Water-flow to stabilize them after the end of landfilling, which can be concluded by injecting air (which pre-controlled) (e) Landfills not covered with decomplex layer, fail to generate highly contaminated liquid or leachate and placed into moisture areas are also accounted in it
3. Upon the ground semi aerobic dumpsites	(a) These are stacked upon the land, by taking small space. (b) Base-layer has no linearity (c) It contains freely aggregated excavated wastes (d) Minimum sanitation is given at the disposal site; here contaminated soils are mostly used (e) Some soils are given as ultimate coverage
4. Billowy vulnerable land raises	(a) Upon-land wastage stacks are planned to provide ventilation within the landfill (b) Earthen barriers and a cheap clean water removal technique are given under the landfill (c) By a tractor, excavated wastes are aggregated incoherently (d) A technique to recirculate which has many branches is installed for the treatment of leachate and other wastewater from the facility, which includes body flushing component of the landfill, which is activated immediately after the closure of the landfill

Source: [29][30][31][34][35].

- **Eco-friendly:** Due to the efforts of environmentalists and conservationists who pushed for tight landfill legislation, regulations, and standards, sanitary landfills are eco-friendly [36].
- **Cleanliness and Waste Management:** When a community lacks a sanitary landfill or residents are aware that there is no effective waste management system in place, they will just dump waste in vacant lots or open dumps [36][37].
- **Good Storage Facility:** They also serve as a storage facility for more hazardous goods that must be kept away from the general population [36][37].

- **Low-Cost Option:** The waste in the sanitary landfill will just have to travel a short distance to the dump, lowering transportation expenditures [37].
- **Pollution Reduction:** This will also help to reduce pollution caused by garbage transportation [36].

Drawbacks

- **Demands Continuous Maintenance:** A landfill takes a lot of time and effort to construct and maintain. Furthermore, it only protects public health when properly administered and continuously maintained [38].
- Failure to take preventive steps could result in difficulties with the lining systems. As a result, the threat of groundwater contamination is constantly present [36][38].
- **Consumption of Huge Land:** The fundamental fault in the sanitary landfill concept is that it consumes a lot of land and resources to keep the garbage contained, as well as being potentially harmful to the environment [37][39].
- **Leachate:** Rainwater soaks into the landfill, resulting in foul-smelling liquid waste (leachate) that can transfer poisonous substance from the waste into the groundwater. This is why it's critical to properly line the landfill and avoid placing it near a river, stream, or lake [36][37].

8.4 LANDFILL GASES

These gases are the side products of the degradation of material (especially organic materials) in landfilled wastes that are produced naturally. LFG is made up of roughly 50% methane (CH_4) (natural gas is the main component), 50% (CO_2) and non-methane organic compounds are present in smaller quantities. CH_4 is a more powerful greenhouse gas than CO_2 and is effective in enclosing heat within the environment [40].

8.4.1 Landfill Gas Generation and Characteristics

Landfill gas generation can be produced by these processes: bacterial degradation, vaporization, and chemical reactions. The phases of landfill gas generation with its impact in various phases is shown in Figures 8.6 and 8.7 respectively. The typical landfill gas composition is shown in Table 8.3.

- **Bacterial degradation:** by bacterial degradation or decomposition, many landfill gas formations happen. In that case, bacteria present in garbage or waste break down those organic wastes naturally and soil covers the landfill [40]. There are many wastes like household garbage, vegetable waste, many industrial wastes, and wooden waste or products made up of paper. As mentioned above, four phases of decomposing happens, and the composition of the gases also changes [41][42].
- **Vaporization**: process where compounds alter their liquid-state or solid-state into vapor-state to generate landfill gases. This is vaporization, also known as volatilization. Vaporization of particular chemicals dumped in the can create non-methane organic landfill gases [41].

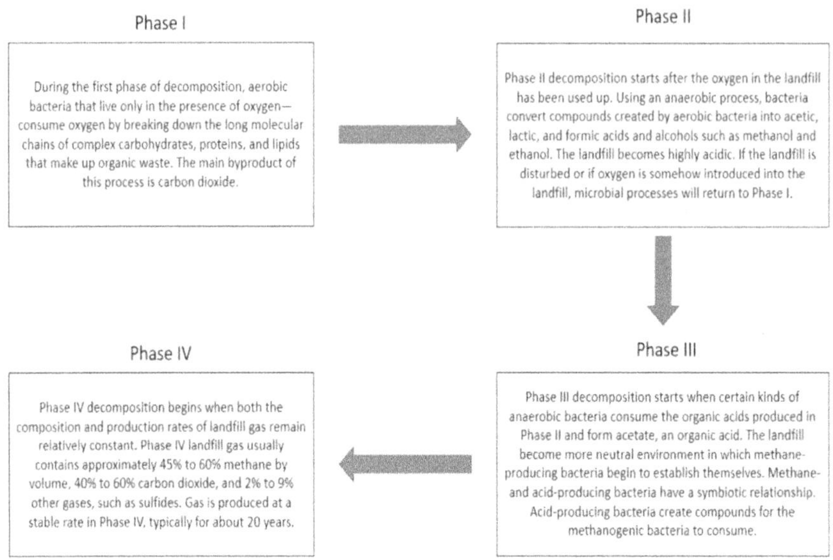

FIGURE 8.6 Phases of landfill gas generation.

Source: [40].

- **Chemical reactions:** with particular reactions by the chemicals present in that garbage, landfill gases like non-methane organic can be made, e.g. if chlorine reacts with ammonia in the waste pile, it produces a dangerous gas such as nitrogen trichloride that causes injuries to health [42].

8.4.2 Landfill Gas Composition and Quality

Landfill gas can be made up of a mixture of various gases. Its natural compositions are 50–70% CH_4 and 35–50% CO_2. Landfill gas also includes N, O, NH_3, H, CO sulfides, and NMOCs e.g., C_2HCL_3, C_6H_6, and H_2C=CH-Cl in a smaller amount. Table 8.1 lists "typical" landfill gases, their percentage by volume, and their characteristics [43].

8.5 METHANOGENIC DECOMPOSITION

A landfill, deposited with organic waste and provided with anaerobic conditions, is more prone to methanogenic decomposition. The different microorganisms involved in decomposition of organic waste in landfill condition is shown in Table 8.4. The methane-producing bacteria, so-called methanogens, are responsible for the decomposition of organic waste thus producing methane gas, majorly contributing to landfill gases. The methanogens utilize CO_2, acetate, and other organic acids for their

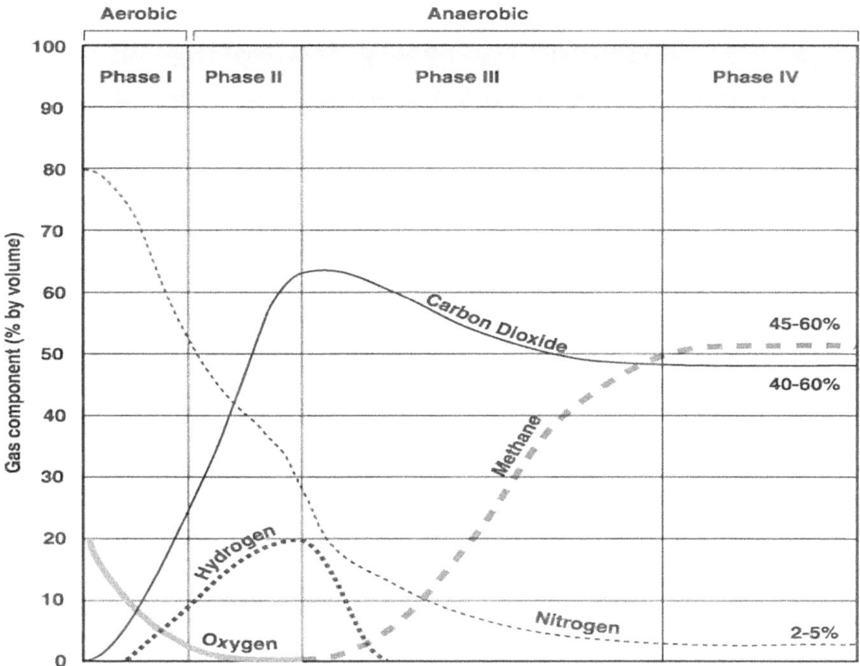

FIGURE 8.7 Gases impact in every phase.

Source: [42].

functioning in landfills. The LFGs produced by this method are provided with a high amount of methane and are highly efficient [44].

Methane (CH_4) is the primary gas produced during landfill gas production, making up around 50% of the total volume of landfill gas [45]. It could be observed that production of methane takes place by anaerobic digestion in Stage 4 decomposition by methanogenesis as shown in Figure 8.8. While methane is not toxic, it is flammable and can pose a significant safety risk if it accumulates in an enclosed space. Another gas produced during landfill gas production is carbon dioxide (CO_2), which can also be harmful in high concentrations. Additionally, trace amounts of other gases such as hydrogen sulfide (H_2S), ammonia (NH_3), and volatile organic compounds (VOCs) may be present in landfill gas and can pose health risks if inhaled in high concentrations [46].

However, the composition of landfill gas can vary depending on the type of waste in the landfill, the age of the landfill, and other factors. As a result, it is important to monitor and control the emissions from landfills to minimize the release of potentially dangerous gases into the atmosphere. This is typically done through the installation of gas collection and control systems that capture and treat the landfill gas before it is released into the environment [47][48].

TABLE 8.3
Typical Landfill Gases Composition and Their Characteristics

Sl. No.	Constituents	Volume percent	Features
1.	Methane	45–60	Colourless Odourless Burns in air with pale flame Also called marsh gas
2.	Carbon dioxide	40–60	Colourless Odourless Slightly acidic Non–flammable gas
3.	Nitrogen	2–5.1	Odourless Tasteless Colourless Liquid N_2 is non-toxic
4.	Hydrogen	0–0.23	Odourless Colourless
5.	Carbon monoxide	0–0.021	Odourless and flammable Colorless Highly poisonous
6.	Ammonia	0.1–1.1	Colourless Having pungent smell Highly soluble
7.	Oxygen	0.1–1.1	Odourless Colourless Tasteless
8.	NMOCs	0.01–0.6	May be naturally occurring or synthesized Commonly includes ethyl benzene, trichloroethylene, 1,2–cis dichloroethylene, carbonyl sulphide, tetrachloroethylene, methyl ethyl ketone, etc.

Source: [41][42][43].

8.6 LANDFILL GAS RECOVERY PROCESSES

8.6.1 Passive Collection

Landfill gas collection systems are separated into passive and active systems. A passive system collects gas simply by inserting vents into the landfill and allowing LFG to travel ambiently up through the vents [49]. The vents may be vertical or horizontal within the waste and may extend lengths up to 75% of the depth of the landfill waste. For a passive system to be effective, the landfill must have adequate internal pressure to push air upwards [50]. Generally, a passive system releases the collected gas into the atmosphere; thus, passive systems are not frequently used in large landfills since many LFG components are greenhouse gases regulated by the Clean Air Act or other

TABLE 8.4
Different Microorganisms Involved in Decomposition of Organic Waste in Landfill Condition

No. of phase	Bacteria involved	Substrate	Condition of landfill	Product
1	Aerobic bacteria	Carbohydrates, proteins, lipids	In presence of abundant oxygen	Monomers or simple products of the substrates
2	Anaerobic bacteria	Monomers or simple products of the carbohydrates, proteins, lipids	In presence insufficient oxygen. Highly acidic after reaction	Acetic acid, formic acid, lactic acid, methanol and ethanol. N_2 and P conc. Increases in the landfill, enhancing diverse microbial growth
3	Anaerobic bacteria Initial Methanogenic bacteria	Acids and alcohols from phase 2. CO_2 and acetate	Absence of oxygen. More neutral pH after reaction	Organic acids like acetate. Methane is also produced
4	Generally anaerobic bacteria Stable methanogenic bacteria	All the products from previous phases	Mostly in absence of oxygen	Landfill gases are produced. Decomposition continues

legislation. Passive systems are often practical for lower-volume landfills or landfills that have been closed for a significant period and are no longer producing large volumes of gas [51].

8.6.2 Active Collection

Active systems of landfill gas collection generate a vacuum inside extraction wells to extract landfill gas from inside the system [49]. The vacuum is created using blowers that are sized and distributed proportionally to the amount of gas that is being collected through the overall system [51]. The extraction wells all connect to a larger header pipe, which may be located just below the surface in a sand trench. Extraction wells may extend vertically or horizontally within the landfill. Calculations must be completed to ensure that the distribution of wells adequately extracts gas from all areas of the landfill [52].

8.7 LANDFILL GAS REFINING AND TREATMENT

The gases propagated within a landfill can be accumulated and used in numerous ways. The landfill gas can be utilized directly within the site with boiling or other

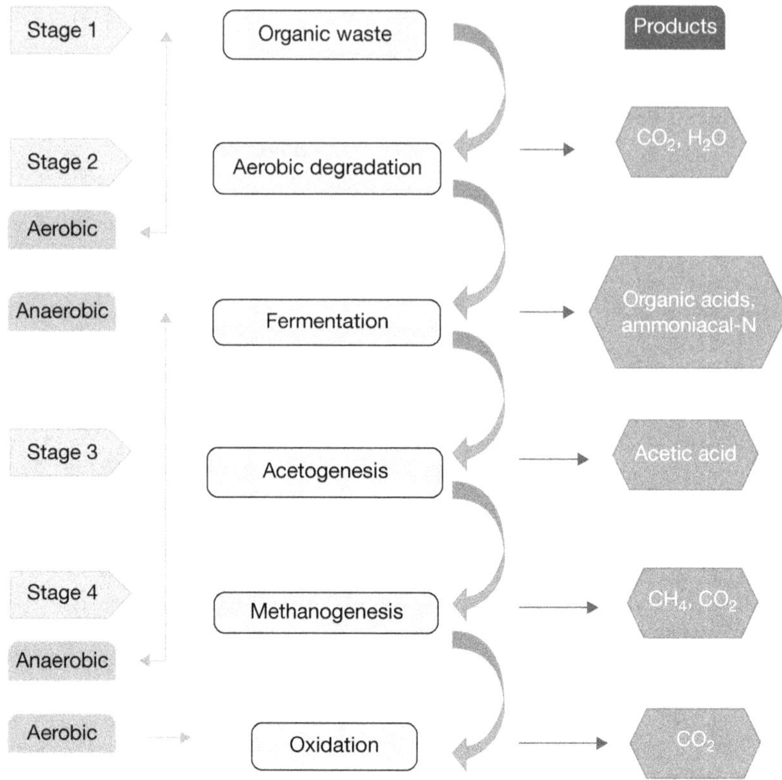

FIGURE 8.8 Stages of gas production by microbial actions in landfills.

burning operators, which provides temperature. Electric current is produced within it using turbines. Then these gases are transmitted through the pipeline pathways by eliminating several contagious materials. Landfill gas can also be used to evaporate leachate, another by-product of the landfill process.

8.7.1 Primary Treatment

Electricity generation is the most widely used landfill gas-to-energy (LFGE) technology in the United States [53]. Microturbines, reciprocating engines, internal combustion turbines, steam, and combined cycle power plants are successfully used to convert landfill gases into electricity. Medium BTU is used to express the amount of heat energy that can be derived from a given quantity of the fuel. One cubic foot of natural gas contains about 1,000 BTUs of energy. This information is important for determining the efficiency of heating systems and appliances that use these fuels, as well as for comparing the cost and performance of different types of fuels, gas sales are the second most widely used LFGE technology in the USA. Medium BTU gas projects typically provide limited treatment before transporting the GHG (Green House Gas) through a dedicated pipeline to the end user. In medium BTU electric and gas projects, primary treatment steps are applied when necessary to condition the GHG before use. The primary treatment steps, including dewatering, filtration, hydrogen sulfide removal, and VOC (volatile organic compound) removal, are discussed below [54].

- **Dehydration**

Typically, LFG is saturated with water vapor within the relatively warm landfill. When extracted and subjected to relatively cool temperatures outside of the landfill, some of the water vapor in the LFG will condense and form liquid water or condensate. Condensate must be managed, or prevented from forming, throughout the entire LFGE process. Of most importance, liquid water cannot be allowed at the burner tip or an engine's fuel valve [54].

The simplest and most cost-effective method to prevent condensate from forming, after pressurization or compression of LFG, is temperature control. Condensation downstream of the cooler can be avoided by setting the temperature of the LFG exiting the cooler above its dew point temperature (so that the temperature of the LFG in the downstream piping does not cool below its dew point), or by installing an LFG re-heater or heat tracing on the downstream piping (to prevent the LFG temperature from falling below the cooler's LFG outlet temperature [55].

A typical dehydration system for LFGE projects removes most of the moisture in the LFG before delivery to the pipeline or onsite use. Typical dehydration systems lower the dew point of the LFG to 40°F and then re-heat the LFG to at least 20°F above the dew point. Typical equipment includes a reheat gas-to-gas heat exchanger, a chilled glycol-to-LFG heat exchanger, and a condensate knock-out [56].

- **Filtration**

Condensate knockouts are typically located before the inlet of the compression equipment. Besides managing condensate, most knockouts will contain a stainless-steel demister. The demister element typically is designed to remove 10 microns and above particulates, in addition to water droplets. For engine projects, a coalescing filter is also used to provide particulate removal down to 0.3 microns [54][56].

- **Removal of Hydrogen Sulfide**

There are numerous hydrogen sulfide (H2S) removal systems that are commercially available. Typically, the goal for the removal of H2S is to reduce corrosion of power generation equipment, to pre-treat LFG for high-BTU gas processing, and/or to reduce sulfur oxides (SOx) emissions, associated with LFG combustion. H2S treatment systems must be evaluated on a life-cycle basis to ascertain whether a high-capital cost or low-capital cost system is the best fit for an individual landfill [56].

H2S removal technologies are typically grouped per the following process categories:

- Physical Adsorption
- Solid Chemical Scavenging

- Liquid Chemical Scavenging
- Solvent Absorption
- Liquid Redox
- Biological Processes

- **VOC Removal**

If landfill gas is the source of VOC contamination, it is possible to halt the source rather than use the pump-and-treat remediation method [54][56].

8.7.2 Advanced Treatment

Advanced treatment is required to produce high-BTU gas for injection into natural gas pipelines or the production of alternative fuels. Advanced treatment steps provide additional LFG processing and may employ multiple clean-up processes. The type of advanced treatment depends on the constituents that need to be removed for potential end users [56].

8.8 DIFFERENT WASTE AND ITS EFFECT ON THE ENVIRONMENT

8.8.1 Different Wastes Sent for Landfilling

8.8.1.1 Solid Waste

Solid waste is the main contributor to the creation of landfills worldwide. Homes, restaurants, schools, offices, and public places produce many used materials, rubbish, and garbage. The waste products include paper, wood, plastic, and obsolete electronic products, which are dumped in landfills [57].

Because most of the waste products are not biodegradable, they lay in landfills for years. The result is worse in poorly managed landfills, as the waste damages the land and the environment.

- **Domestic wastes:** household waste, food waste, garbage (kitchen waste), rubbish (paper, polythene, plastics, glass pieces), old toys, old clothes, old mattresses, etc.
- **Community wastes:** educational institutions waste, government offices, markets, public cleansing, bulky wastes, hospitals, construction works, etc. [57].

8.8.1.2 Agricultural Waste

Waste materials from crops, animal manure, and farm waste are also collected and dumped in landfills. Most of these agricultural remains release highly toxic substances as they decompose and contaminate water resources and land.

Agricultural wastes: Farm waste, livestock yards, crop residues, bagasse from sugarcane, outdated pesticides and fertilizers, manure, weedicides, fungicides, slaughter waste, plastics and containers, organics, etc. [58].

8.8.1.3 Manufacturing, Industry, and Construction Waste

Construction activities, industrial processes, and power plants produce lots of by-products and solid residues. The main waste products are generated by power plants, oil refineries, pharmaceuticals, agricultural products, and construction works. Unfortunately, this solid waste eventually finds its way to different landfills from time to time [58].

- **Commercial wastes:** bulky wastes from shops, offices, hotels, non-government markets, stores, tires, electronics, plastic bags, bottles, buckets, packaging materials, paper fibers, thermocol, discarded electric waste, etc.
- **Industrial wastes:** paper and pulp wastes, oil refineries, tanneries, distilleries, thermal power plants, chemical industries, metal smelters, coal, ash, acids, chemicals, textiles, plastics, nuclear wastes, unused metal sheets, metal scraps, rubber, leather, toxic effluents, fibers (residues), heavy metals, solvents, resins, sludge, abrasive, etc.
- **Construction wastes**: Demolition, excavation, renovation works, road works, site clearance, wood, glass, metal, plastic, concrete, etc.

Some major examples are:

Mining wastes: Waste rock, tailings, mine water, chemicals, and others, etc.

Radioactive wastes: Nuclear explosions, nuclear testing, use of radioactive substances in medical and scientific research, products poisoned by radioactive materials that are used in diagnosis or other radio therapeutic items, etc.

Municipal wastes: Household discharge, street sweeping, sewage treatment plant waste, waste from schools and other institutions, public toilets, etc.

Biomedical wastes: Hospitals, clinics, laboratories, etc.

Healthcare wastes: Infectious waste, pathological waste, sharps waste, chemical waste, pharmaceutical waste, cytotoxic waste, general waste, etc. (Syringes, needles, disposal scalpels and blades, expired, unused, and contaminated drugs and vaccines, swabs, bandages, gloves, disposable medical devices, urine bags, sanitary napkins, napkins, diapers, mammalian body parts or tissues, other liquefied things or fluid, contaminated or infected animal corpses, disinfectants, sterilant, heavy metals, broken thermometers, batteries, chemicals, etc.)

Electronic wastes (E-wastes): Discarded electronic devices like computers, TV, music systems, transistors, tape recorders, mobile phones, computer cabinets, motherboards, CDs, cassettes, moue, wires, cords, switches, chargers, batteries, circuits, etc. [57][58].

8.8.2 Effect of Different Types of Waste on the Environment

8.8.2.1 Air Pollution a Major Concern

Landfills emit over ten toxic gases, with methane being the most fatal. Methane is naturally produced in activities involving organic matter decay from different sources.

According to EPA (Environmental Protection Agency), MSW (Municipal solid waste) landfills accounted for over 15% of methane emissions in the USA in 2018. Compared to greenhouse gas emissions during the same year, these methane amounts are equal to emissions released by over 20 million passenger vehicles driven over the year [59].

The methane released from unmanaged decomposing organic matter in landfills can also trap solar radiation 20 times better than carbon dioxide. That results in increased global temperatures, especially in urban areas where most landfills are located. Besides methane, other agricultural and household chemicals like bleach find their way into landfills and generate toxic gases that highly affect the neighborhood's air quality.

8.8.2.2 Groundwater Contamination

The contamination of groundwater by leachates is a significant concern arising from the increase in landfills globally. That's because landfills are packed with hazardous waste that inevitably continues to deteriorate groundwater, especially when it's located near a lake, river, or ocean. These toxic materials range from household cleaners to industrial solvents and electronic waste that contain mercury, lead, and cadmium. Most of these landfill toxins penetrate the soil and reach fresh waterways, which sadly end up in the foods and water we consume. This pollution affects animal and plant life, as research shows that 82% of all landfills have leaks [60].

8.8.2.3 Soil and Land Pollution

Landfills render the land and soil where it's set unusable. It also affects the adjacent land because toxic chemicals can spread through the soil surrounding the area over time. These toxins destroy the soil's upper layer, distort soil fertility, and affect the plants' lives. That upsets the land's ecosystem and can result in health complications if the soil is used for agricultural purposes [60].

8.8.2.4 Health Concerns

Unmanaged landfills pose the risk of severe health issues with low birth rates and weights, congenital disabilities, and increased cancer cases, especially for those residing near these plants. For example, TCE (Trichloroethylene) is a common carcinogenic substance that originates from landfill leachate and is associated with several adverse health effects. That includes immunotoxicity, neurotoxicity, several forms of cancer, developmental toxicity, and liver issues [60].

People living near landfills have also reported other symptoms and discomfort issues like fatigue, headaches, and sleepiness. These effects result from the toxic chemicals emitted by landfill waste. That includes everything from harmful gases to water contaminants that result in adverse health effects. Over 50% of people living close to landfills suffer from lung and heart diseases at one point or the other [61].

8.8.2.5 Landfill Fires

Landfill gases and wastes can easily start fires, especially during summer. Once the fires ignite, stopping them can be challenging even with proper equipment, and it further contributes to air pollution. These fires can also destroy neighboring human and animal habitats if not put out immediately.

Methane gas is highly combustible, and it can wreak havoc when in abundant supply. In landfills, the combustion of this gas also worsens as it burns together with other chemicals that increase toxins in the area [61].

8.8.2.6 Economic Costs

The social and economic cost of managing a landfill can be very high. From managing groundwater contamination to minimizing the number of harmful gasses emitted, there's a lot that needs lots of finance to avoid future complications. Additionally, a municipality must incur additional costs in complying with environmental regulatory policies [61].

Since most material packed in landfills doesn't easily decompose, it can be challenging to effectively design integrated waste management facilities and strategies with limited resources. These high capital investments drain the municipality funds and might encourage some communities to leave landfills unmanaged.

8.9 ADVANCED LANDFILLING TECHNIQUE

Landfill techniques may be considered under seven headings as shown below. The different types of landfill techniques are shown in Figure 8.9.

(a) Location and engineering
(b) Phasing and cellular infilling
(c) Waste emplacement methods
(d) Waste pre-treatment
(e) Environmental monitoring
(f) Gas Control
(g) Leachate management

8.9.1 Location and Engineering

Site-specific factors determine the acceptability of a particular waste disposal strategy at any location. Theoretically, a designed landfill for any waste can be located anywhere, with a sufficiently high level of technology [66]. In practice, the perceived risk of non-containment is such that many countries limit landfills for hazardous waste and possibly MSW to fewer sensitive locations, such as non-aquifers, and may also stipulate a minimum unsaturated depth below the landfill. In other cases, acceptability depends on the results of a risk assessment, which examines the impact on groundwater quality of the worst possible leakage rates [67].

The main components of landfill engineering are usually a containment liner, a protective liner layer, a leachate drainage layer, and a top cover [66]. The most common containment methods are mineral liners (such as clay), polymeric flexible membrane liners (FML) such as high-density polyethylene (HDPE), or composite liners consisting of a mineral liner and FML in close contact. Other materials are also used, such as bentonite-reinforced soil (BES) and asphalt concrete [68].

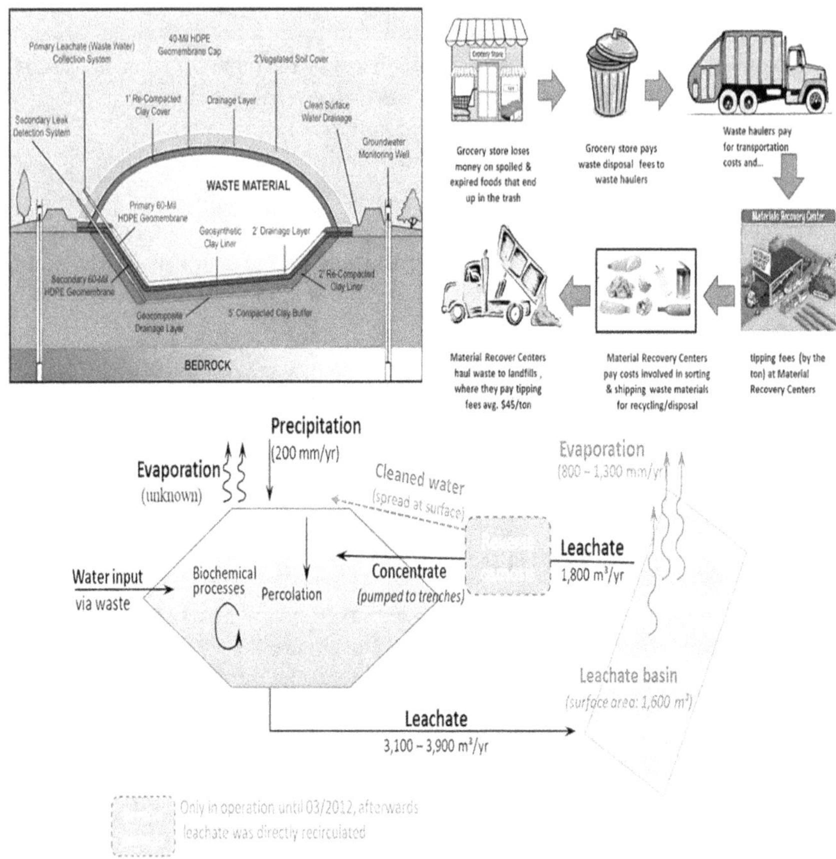

FIGURE 8.9 Landfill techniques (a) Location and engineering, and Phasing and cellular infilling [62]; (b) Waste emplacement methods, and Waste pre-treatment [63]; (c) Environmental monitoring, and Gas control [64]; (d) Leachate management [65].

8.9.2 Phasing and Cellular Infilling

Landfills are often filled in stages. This is usually done for purely logistical reasons. Because of the size of some landfills, it is cost effective to prepare and backfill parts of the site in sequence [67]. In addition, the active phases are sometimes further subdivided into smaller cells, the area of which can usually vary from 0.5 to 5 ha. Often these cells can be designed to be hydraulically isolated from each other [69].

There are two main reasons for cellular in filling:

1. To allow the segregation of different waste types within a single landfill.
2. To minimize the active area and thus minimize leachate formation, by allowing clean rainwater to be discharged from unfilled areas while individual cells are filled. This can increase the physical complexity of leachate removal

arrangements and if the cells receive different waste types, each cell may produce leachate with different characteristics. This may in turn influence the design of leachate treatment and disposal facilities [67].

8.9.3 Waste Emplacement Methods and Pre-Treatment

Wastes are usually compacted at the time of deposit. This is done to gain maximum economic benefit from the void space and to minimize later problems caused by excessive settlement. The degree of compaction achieved depends on the equipment used, the nature of the wastes, and the placement techniques [66].

Equipment may vary from small, tracked bulldozers, to specialized steel-wheeled compactors. The latter is claimed to be able to achieve in situ waste densities of more than 1 ton/m^3 with MSW. Experience suggests that to achieve this, it is necessary to place wastes in thin layers, not more than 1m thick, and to make many passes with the compactor. At many landfills, waste is placed in much thicker lifts of 2.5m or more and receives relatively few passes by the compactor. Densities of ~0.7–0.8t/m3 are more typical in such situations [67].

Some operators of MSW landfills add moisture, or wet organic wastes such as sewage sludge, at the time of waste emplacement, to encourage rapid degradation, and in particular to encourage the early establishment of methanogenesis. There is ample experimental and field evidence to show that this can be effective [63].

8.9.4 Environment Monitoring

Monitoring is an essential part of landfill management and has two important functions: [66][67]

- It is necessary to confirm the degradation and stabilization of the wastes within the landfill.
- It is necessary to detect any unacceptable impact of the landfill on the external environment so that action can be taken. Monitoring can be divided into several distinct aspects, as follows:
- Gas–Landfill gas quality within the site; soil gas quality outside the site; air quality in and around the site;
- Leachate–Leachate level within the site; leachate flow rate leaving the site; leachate quality within the site;
- Leachate quality leaving the site;
- Water–Groundwater quality outside the site; surface water quality outside the site;
- Settlement–Settlement of wastes after infilling.

8.9.5 Gas Control

At most landfills receiving degradable wastes such as MSW and many non-hazardous industrial wastes, it is necessary to extract landfill gas to prevent it from migrating away from the landfill. Landfill gas (LFG), a mixture of methane and carbon dioxide, has the potential to cause harm to human health, via explosion or asphyxiation, and to cause environmental damage such as crop failure [67]. Examples of all three have occurred both within and outside landfills. The techniques for extracting and controlling LFG are now reasonably well established and in common use. Vertical gas extraction wells are usually installed after infilling has ceased in a particular area. Gas is extracted, usually under applied suction, and routed either to a flare or to a gas utilization scheme. It is now quite common to generate electrical power from LFG and to recover heat. In some cases, LFG has been used directly as a fuel source in brick kilns, cement manufacture, and for heating greenhouses [70].

In conjunction with extraction wells, it is often necessary to install passive control systems, in the form of barriers and venting trenches around the perimeter of landfills.

8.9.6 Leachate Management

Leachate management refers to the process of controlling, treating, and disposing of leachate, which is a liquid that results from the percolation of water through waste materials. Leachate typically contains high concentrations of pollutants, including organic matter, heavy metals, and other harmful substances, and therefore requires proper management to prevent environmental contamination and health risks. Leachates may broadly be divided into a few types as shown in Table 8.5 [67][70].

Treatment techniques depend on the nature of the leachate and the discharge criteria. The common methods of managing leachate are as follows [65].

1. Collection and treatment: Leachate is collected from the bottom of a landfill or waste storage facility and is treated to remove harmful substances before being discharged or reused.
2. Evaporation and condensation: This method involves using natural evaporation or mechanical systems to remove water from the leachate, leaving behind concentrated pollutants that can be treated and disposed of.
3. Bioreactor landfills: In this method, the landfill is designed to encourage the growth of microorganisms that break down organic matter in the leachate and reduce its concentration of pollutants.
4. Land application: Leachate can be applied to land for irrigation or fertilization purposes, but only after being treated to meet local regulations and ensure environmental safety.

It is technically feasible to treat the leachate to almost any required discharge quality. Aerobic biological treatment is the basis of the vast majority of sewage treatment plants, but many other methods are used to remove components that are not adequately removed by biological methods. The degree of treatment and the most appropriate methods depending on the location. The time required for active leachate management depends on the rate at which pollutants are flushed from the landfill. With conventional, low-permeability upper caps and containment strategies, the time is likely to be several centuries for wastes with high pollution potential, such as MSW [66].

TABLE 8.5
Types of Leachates

Types of Leachates	Concentration of components	Elements
1) Hazardous waste leachate	Highly variable concentrations of a wide range of components.	Extremely high concentration of substances such as salts, halogenated organics, and trace elements can occur.
2) Municipal solid waste leachate	High initial concentrations of organic matter (COD >20,000 mg/l and a BOD/COD ratio >0.5) falling to low concentrations (COD in the range of 2,000 mg/l and a BOD/COD ratio <0.25) within a period of 2-10 years.	High concentrations of nitrogen (>1000 mg/l) of which more than 90% is Ammonia-N. This type of leachate is relatively consistent for landfills receiving MSW, mixed non-hazardous industrial and commercial waste and for many uncontrolled dumps.
3) Non-hazardous, low-organic waste leachate	Relatively low content of organic matter (COD does not exceed 4,000 mg/l and it has a typical BOD/COD ratio of <0.2) and a low content of nitrogen (typically total N is in the range of 200 mgN/l, but can be as high as 500 mgN/l).	Relatively low trace element concentrations are observed. This type of leachate comes from landfills receiving only non-hazardous waste exclusive of MSW.
4) Inorganic waste leachate	Relatively high initial concentrations of salts (chlorides plus sulphates in the range of 15,000 mg/l) and a low content of organic matter (typically COD <1,000 mg/l) and low content of nitrogen (total-N <100 mg/l).	Trace element concentrations are often negligible. This type of leachate is typical of landfills for MSW incineration ash.

8.10 LANDFILL GAS UTILIZATION

Landfills occupy the third-highest position in the methane-producing sector after fossil fuel and agriculture. Landfill gas utilization involves the collection, recovery processing, and treatment of methane and other gases produced in the landfill from the decomposition of the wastes in it and thus supplying the treated gas in different sectors of society and different forms [71]. Figure 8.10 shows the collection of landfill gas employing the collecting well in the first phase, processing the LFG first with primary treatment for the removal of moisture followed by secondary treatment to remove the hydrogen sulfide and other impurities in the second phase and treatment

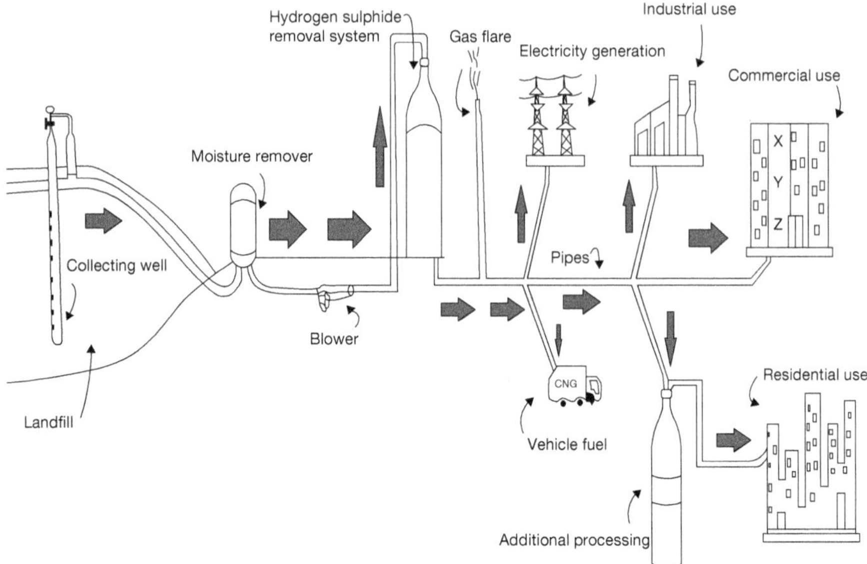

FIGURE 8.10 Utilization of landfill gases.

in the gas flare in the third phase and at last the processed gas is supplied to electricity generating plants, industries, commercial buildings, residential complexes, CNG drove vehicles, etc. [72].

8.10.1 Electricity Generation

LFGs with a marvelous impact on modern society acquire almost half of the industry of the generation of electricity [71]. There are multiple technologies for generating electricity, as listed below:

- Microturbines: the compressor present in the system draws the air into a recuperator, creating an air-to-air heat exchanger, recovering the heat from the collected exhaust gas.
- Internal reciprocation of combustion engines: this type of engine employs gaseous expansions and the resultant pressure increase in the core of the confined cylinder to move pistons of different sizes and thus rotate the shaft, producing mechanical energy which is then converted to electrical energy and supplied.
- Fuel cells: these act similarly to batteries and are electrochemical cells. The LFGs are produced to act as the fuel here and allow the fuel cell to produce electricity employing the electrodes and electrolytes.
- Turbines: being a mechanical rotary device, it generally converts mechanical energy into electrical energy.

In the above-mentioned technologies, the reciprocation of combustion engines affords very high efficiency at a low cost.

8.10.2 Direct Usage of Medium Btu-gas

As you know, several natural gases need supportive agents for their combustion. Often, the LFGs act as the supportive agents for the usage of several other fuels like fuel oil or natural gas, and taking into consideration the direct use of landfill gas, they can be directly treated in the boiler, kiln, and other thermal applications [73].

8.10.3 Steam Generation

Using the updated technologies, the LFGs are now used in combined heat and power production, as such, using a cogeneration process to generate both electricity as well as thermal energy, generally in the form of steam. The steam is then used in several heat and mass transfer operations [74].

8.10.4 Alternative Fuel

In the generation of the uprising cost of every fuel system, LFGs are slowly capturing the domestic as well as the commercial market. The landfill gases are being used as pipeline fuels and with more purification may be used as a vehicle fuel with high efficiency and eco-friendly [75].

8.10.5 Methanol Synthesis

The wastes in the landfills are decomposed anaerobically by the methanogenic bacteria and thus producing methane and subsequently methanol, which is then purified and filtered for further use [76].

8.11 LANDFILL GAS COST AND ECONOMY

As and when the landfill gases are recovered, the cost is also determined based on the efficiency, density, and efficacy of the landfill gas and the process using which the gas is being produced [77]. The landfill gas recovery contributes to the global economy in several ways [78] by

- Generating revenue by selling the gas that has been recovered.
- Creating jobs for several functioning in the recovery plants, as such, involving positions for:
  - Design: as studied earlier in this chapter, the waste deposited in the landfills is hazardous and hence are needed to be safely covered with the highest efficiency for supplies and collecting systems to work in it. Hence, designers are employed to implement the proper design with the highest efficiency and safety.
  - Construction: after the design is accomplished and approved, constructors with proper skills are hired to construct the design made by the designer in the real field. The liners, drainage, and collecting systems are the most essential systems to be constructed in a landfill.

- Operation: not only constructing the landfill but also operating it properly gives the desired result, as such, waste deposition, collecting the leachate, maintaining the level of waste, and treatment of waste and leachate must be guided by a skilled operator in a landfill.

Thus, creating employment for: [79][80]

(a) Engineers: Generally the operators, and designers involved in landfill operation and designing are well-skilled engineers.
(b) Impedimenta vendors: Commodities required for landfill construction are vented from specific sectors. Thus, creating and running several business opportunities.
(c) Labour: In underdeveloped and developing countries people with less education and more skills and capacity to work are hired as laborers in the landfills, generally working under the guidance of the operator in case of operation.
(d) Construction firms: For constructing a properly equipped landfill system, high-profile construction firms are involved. Thus, increasing business in the construction field thus employing thousands of workers, and adding up to the economy of the country.
   - Having a recovery plant in the locality reduces the cost price of the locality thus making it available to every class of the population.
   - This also increases local employment and sales.

8.12 FACTORS THAT AFFECT LANDFILL GAS GENERATION

Due to the large variability in biodegradation settings, waste composition, and site characteristics, the rate of degradation and volume of landfill gas produced per unit of waste can vary greatly, and accurately predicting how these parameters affect total gas production can be difficult.

8.12.1 Waste Composition

When organic waste is present in a landfill, bacteria produced additional landfill gases (examples- CO_2, CH_4, N, H_2S) in the time of putrefaction/degradation. Increment in the transmission of chemical compounds provokes the production of other nonmethanogenic compounds [81][82].

8.12.2 Age of Refuse

In general, newly excavated or combusted wastages secrete more of these gases by microbial degradation, vaporization, and chemical formulation than older wastages (buried more than 10 years). Top gas usually appears as an outcome 5 to 7 years after landfilling [81].

8.12.3 Moisture Content

Wastewater moisture is one of the most important components for successful organic decomposition and gasification. A typical landfill moisture content is 20–30%, although there is some variation depending on site conditions. When the humidity increases to 60–80%, there is an increase in the biodegradation of the waste along with the maximum formation of CH_4 [83]. This increase in gas production is mainly due to the improved distribution of nutrients and microorganisms as a result of the leachate flowing through a wider part of the waste matrix. Increased moisture content can provide positive results, but it should be noted that uncontrolled addition of water, such as excessive rainfall infiltration into the landfill, can have negative effects such as interruption of CH_4 production and increased leachate volume. Filtrate generated at landfills must be collected and processed following regulations [84].

8.12.4 Nutrient Content

Adequate availability of nutrients such as carbon, nitrogen, oxygen, hydrogen, and phosphorus are crucial in ensuring proper digestion of organic. The ratio of 30 biodegradable carbon to nitrogen ratio was found to be ideal for maximizing CH_4 production. On the other hand, a surplus of sulphate may prevent gas generation [82][83].

8.12.5 Temperature

Temperature can have a great impact on landfill gas production, especially for shallower landfills that are vulnerable to seasonal conditions such as heavy rain or snowfall [81]. Methanogens can carry out normal functions in the range of 59°F to 113°F, and the optimal temperature for CH_4 production was found to be between 90°F to 95°F. Below 50°F, a drastic reduction in degradation rate was observed [82]. While the higher temperature does improve volatilization and stimulate chemical reactions, injecting the heat into a landfill might be challenging and costly, and thus the focus should be to seal the heat generated through reactions in the landfill by carefully designing thermal insulating soil covers [83].

8.12.6 pH Level

For maximum CH_4 production, the landfill waste should maintain a neutral pH level of around 7, and methanogenic activity could be severely inhibited if the pH drops below 6 or rises above 8. Due to the acids produced during the anaerobic reaction, the pH may decrease to an undesirable level for gas production [81]. Such a phenomenon can be exacerbated with the addition of digested sludge or old refuse, which accelerates and stimulates more microbial activity but also produces more acids that will inhibit methanogenesis. To mitigate this, a layer of dry buffer material can be placed along with landfill waste, or the waste can be pre-treated before burial to control the acid release in the landfill [84].

8.12.7 Waste Characteristics

Depending on the population and mean standard of living near each landfill, landfill waste composition can differ significantly, meaning the fraction of organic waste, which dominates landfill gas production, will also fluctuate greatly. Wastes like food, paper, and garden cuttings are easily biodegradable, thus contributing mostly to earlier gas production, whereas materials like plastic, wood, and rubber are harder to biodegrade and will biodegrade much more slowly [82]. For accelerated gas production and more rapid stabilization of the landfill, waste can be shredded before placement such that the surface area to mass ratio is increased, resulting in a greater chance of interaction between microbes and waste. However, with the reduction in waste particle size, the surplus of acid release is of serious concern, which could slow down methanogenic activities [81][83].

8.12.8 Waste Compaction

The level of waste compaction can also influence landfill gas production. With higher compaction, wastes are physically closer to nutrients and microorganisms, promoting more degradation and increased CH_4 production [81]. Yet high compaction may not be ideal for some landfills, such as a landfill with high moisture content or a bioreactor landfill with leachate circulation, as the compacted waste could be difficult for leachate to flow through and therefore result in a build-up of pore pressure [82]. Thus, a looser compaction level might be more suitable for waste with high moisture content to ensure an easier circulation of leachate [83]. Furthermore, the timing of compaction is also important since compaction at the beginning of methane production with only a small population of microorganisms could inhibit methane production, but compaction during the active methane production phase could enhance the decomposition rate [84].

8.13 LANDFILL GAS SAFETY AND HEALTH

Landfill odors are a frequent source of complaints from community members. People may also be concerned about the health effects associated with these odors and other emissions from the landfill [85]. People in communities near landfills are often concerned about odors emanating from landfills. They say these odors are the origin of unwanted health and hygiene effects or show indications of a health problem such as headaches and vomiting. At low concentrations, typically associated with landfill gas, it is unclear whether the component itself or its odors are causing the reaction. Generally, these effects disappear when the smell can no longer be detected [86].

Landfill gas odors are created by microbial or chemically formulated techniques. These odors can migrate into the surrounding community [87]. Potential sources of landfill odor include sulfides, ammonia, and some NMOCs when present in high enough concentrations. Landfill odors can also occur when certain types of waste are disposed of, such as manure and fermented grains [88].

- **Sulfides:** Hydrogen sulfide, dimethyl sulfide, and mercaptans are the three general sulfides accountable for landfill odors. These gases can produce a very strong rotten smell even though they are present in very low amounts. Of these three sulfides, hydrogen sulfide is diffused from landfills at the highest levels and concentrations [87].
- **Ammonia:** This is odoriferous gas, produces during landfilling. This is an important compound because it maintains everyone like plants and animal's life cycle. Though ammonia is usually used as a sanitation liquid in household affairs, it is well known for its fragrance [88].
- **NMOCs:** Some Non-Methane Organic Compounds, like $H_2C{=}CHCl$, C_6H_6, etc., other compounds have a chance to produce smells. But in the environment, these are secreted at a very minimum level and don't spread a severe/dangerous odor issue [86].

Many people may find the odors emanating from a landfill offensive or unpleasant. In response to the smell, some people may experience nausea or headaches [86]. Although such responses are undesirable, medical attention is usually not required. Often, symptoms such as headache and nausea go away when the smell goes away [87]. However, the consequences for everyday life can be more long-lasting. Families living near the Connecticut landfill described the frequent odors as extremely disruptive. One family reported waking up before dawn to a nauseous draft that persisted for 2 or more hours. Loss of sleep and frustration due to frequent unpleasant odors have greatly increased the level of stress in family life. Although landfill odors may not be associated with long-term negative health effects or illness for most people, the additional disruption and stress associated with daily activities can significantly affect the quality of life [88].

8.14 CONCLUSION

Landfills are a necessary part of modern waste management, but they come with significant environmental and health risks. This chapter highlights the various types of landfills, with their different properties and purposes. Sanitary landfills are designed to reduce the risks associated with waste disposal, but they still generate significant amounts of landfill gas, which can have serious consequences if not managed properly. Landfill gas recovery and utilization technologies have developed significantly in recent years, providing opportunities for economic and environmental benefits. However, the success of these technologies depends on factors such as waste composition, age, and moisture content, which affect landfill gas generation. The recovery and refinement of landfill gas from sanitary landfills have come a long way since their inception. With the advancement in technology, it has become possible to not only reduce the harmful emissions from landfill gas but also generate electricity and heat from it. The utilization of landfill gas as a renewable energy source has become increasingly popular due to its potential to reduce greenhouse gas emissions and provide a reliable source of energy. However, there is still room for improvement in terms of efficiency, cost effectiveness, and safety. It is important to monitor the environment around landfills, including groundwater and soil, to ensure that they do not become contaminated. Finally, landfill gas safety and health concerns should be addressed to protect workers and nearby residents from exposure to hazardous gases. Overall, while there are challenges associated with landfilling, continued research and development can help mitigate the environmental and health risks associated with waste disposal. The advancements in the recovery and refinement of landfill gas hold promising potential for addressing the energy needs of our society while reducing our impact on the environment.

REFERENCES

1. Popov, V. "A new landfill system for cheaper landfill gas purification." *Renewable Energy* 30, no. 7 (2005): 1021–1029.
2. Slack, R. J., Gronow, J. R. and Voulvoulis, N. "Household hazardous waste in municipal landfills: contaminants in leachate." *Science of the Total Environment*, 337, no. 1–3 (2005): 119–137.
3. Gielen, D., Boshell, F., Saygin, D., Bazilian, M.D., Wagner, N. and Gorini, R. "The role of renewable energy in the global energy transformation." *Energy Strategy Reviews,* 24 (2019): 38–50.
4. Zuberi, M.J.S. and Ali, S.F. "Greenhouse effect reduction by recovering energy from waste landfills in Pakistan." *Renewable and Sustainable Energy Reviews,* 44 (2015): 117–131.
5. Howarth, R.W., Santoro, R. and Ingraffea, A. "Methane and the greenhouse-gas footprint of natural gas from shale formations: A letter." *Climatic Change,* 106 (2011): 679–690.
6. Mohajan, H.K. "Dangerous effects of methane gas in atmosphere." *International Journal of Economic and Political Integration*, 2, no. 1 (2012): 3–10.
7. Wang, D., Qian, X., Ji, T., Jing, Q., Zhang, Q. and Yuan, M. "Flammability limit and explosion energy of methane in enclosed pipeline under multi-phase conditions." *Energy*, 217 (2021): 119355.
8. Rajaram, V., Zia Siddiqui, F. and Khan, Mohd E. *From landfill gas to energy: Technologies and challenges*. CRC press, 2011.
9. Gee, T. "Status and trends of waste management in Singapore." *Waste Management in the Coastal Areas of the ASEAN Region: Roles of Governments, Banking Institutions, Donor Agencies, Private Sector and Communities,* 10 (1992): 57.
10. Bijmans, M.F.M., Buisman, C.J.N., Meulepas, R.J.W. and Lens, P.N.L. "Sulfate reduction for inorganic waste and process water treatment." In M. Moo-Young (Ed.), *Comprehensive biotechnology* (pp. 384–395). Elsevier, 2019.
11. Yip, C. H. and Chua, K.H. "An overview on the feasibility of harvesting landfill gas from MSW to recover energy." *ICCBT*, 28 (2008): 303.
12. Rettenberger, G. "Utilization of landfill gas and safety measures." In Raffaello Cossu and Rainer Stegmann (Eds.), *Solid Waste Landfilling* (pp. 463–476). Elsevier, 2018. https://doi.org/10.1016/B978-0-12-407721-8.00023-1
13. Chavan, D., Arya, S. and Kumar, S. "Open dumping of organic waste: Associated fire, environmental pollution and health hazards." In Chaudhery Hussain and Subrata Hait (Eds.), *Advanced organic waste management* (pp. 15–31). Elsevier, 2022, https://doi.org/10.1016/B978-0-323-85792-5.00014-9.
14. Leonardo, P., Ciarapica, F.E. and Bevilacqua, M. "Environmental assessment of a landfill leachate treatment plant: Impacts and research for more sustainable chemical alternatives." *Journal of Cleaner Production*, 183 (2018): 1021–1033.

15. Ishchenko, V. and Vasylkivskyi, I. "Environmental pollution with heavy metals: case study of the Household Waste." *Studies in Systems, Decision and Control,* 198 (2020): 161–175.
16. Rim-Rukeh, A. "An assessment of the contribution of municipal solid waste dump sites fire to atmospheric pollution." *Open Journal of Air Pollution*, 3, no. 3 (2014): 53.
17. Mudau, A. "A laboratory investigation of the effects of water content and waste composition on leachate and gas generation from simulated MSW." PhD diss., University of the Witwatersrand, Faculty of Engineering and the Built Environment, School of Civil and Environmental Engineering, 2012.
18. Morris, J.W.F. and Barlaz, M.A. "A performance-based system for the long-term management of municipal waste landfills." *Waste Management*, 31, no. 4 (2011): 649–662.
19. Lee, G.F. and Jones-Lee, A. *Impact of municipal and industrial non-hazardous waste landfills on public health and the Environment: An overview*. Vol. 530. G. Fred Lee & Associates, 1994.
20. Benfenati, E., Barcelò, D., Johnson, I., Galassi, S. and Levsen, K. "Emerging organic contaminants in leachates from industrial waste landfills and industrial effluent." *TrAC Trends in Analytical Chemistry*, 22, no. 10 (2003): 757–765.
21. Tang, Y.Y., Tang, K.H.D., Maharjan, A.K., Aziz, A.A. and Bunrith, S. "Malaysia moving towards a sustainability municipal waste management." *Industrial and Domestic Waste Management*, 1, no. 1 (2021): 26–40.
22. Gautam, P., Kumar, S. and Lokhandwala, S. "Advanced oxidation processes for treatment of leachate from hazardous waste landfill: A critical review." *Journal of Cleaner Production*, 237 (2019): 117639.
23. Hilger, H.A., Wollum, A.G. and Barlaz, M.A. *Landfill methane oxidation response to vegetation, fertilization, and liming*. Vol. 29, no. 1. American Society of Agronomy, Crop Science Society of America, and Soil Science Society of America, 2000.
24. Misra, V. and Pandey, S.D. "Hazardous waste, impact on health and environment for development of better waste management strategies in future in India." *Environment International*, 31, no. 3 (2005): 417–431.
25. Carden Jr, J.L. "Hazardous waste management." In *Introduction to environmental health* (pp. 179–214). Springer-Verlag, 1985.
26. Ali, G., Nitivattananon, V., Abbas, S. and Sabir, M. "Green waste to biogas: Renewable energy possibilities for Thailand's green markets." *Renewable and Sustainable Energy Reviews*, 16, no. 7 (2012): 5423–5429.
27. Siddiqua, A., Hahladakis, J.N. and Ahmed KA Al-Attiya, W. "An overview of the environmental pollution and health effects associated with waste landfilling and open dumping." *Environmental Science and Pollution Research,* 27(2022): 58514–58536. https://doi.org/10.1007/s11356-022-21578-z
28. Chian, E.S.K. and DeWalle, F.B. "Sanitary landfill leachates and their treatment." *Journal of the Environmental Engineering Division*, 102, no. 2 (1976): 411–431.
29. El-Fadel, M. and Khoury, R. "Modeling settlement in MSW landfills: A critical review." *Critical reviews in environmental science and technology*, 30, no. 3 (2000): 327–361.
30. Madon, I., Drev, D. and Likar, J. "Long-term groundwater protection efficiency of different types of sanitary landfills: Model description." *MethodsX*, 7 (2020): 100810.
31. Mohareb, A.K., Warith, M.A. and Diaz, R. "Modelling greenhouse gas emissions for municipal solid waste management strategies in Ottawa, Ontario, Canada." *Resources, Conservation and Recycling*, 52, no. 11 (2008): 1241–1251.
32. Cossu, R. and Stegmann, R. *Solid waste landfilling: Concepts, processes, technology*. Elsevier, 2018.

33. Madon, I., Drev, D. and Likar, J. "Long-term risk assessments comparing environmental performance of different types of sanitary landfills." *Waste Management*, 96 (2019): 96–107.
34. Goss, G., Alessi, D., Allen, D., Gehman, J., Brisbois, J., Kletke, S., Zolfaghari, A. et al. "Unconventional wastewater management: a comparative review and analysis of hydraulic fracturing wastewater management practices across four North American basins." *Canadian Water Network* (2015). http://cwn-rce.ca/project/unconventional-wastewater-management-a-comparative-review-and-analysis-of-hydraulic-fracturing-wastewater-management-practices-across-four-north-american-basins/
35. Reinhart, D.R. "Full-scale experiences with leachate recirculating landfills: case studies." *Waste Management & Research*, 14, no. 4 (1996): 347–365.
36. Madon, I., Drev, D. and Likar, J. "Long-term risk assessments comparing environmental performance of different types of sanitary landfills." *Waste Management,* 96 (2019): 96–107.
37. Dos Santos, I.F.S., Gonçalves, A.T.T., Borges, P.B., Barros, R.M. and da Silva Lima, R. "Combined use of biogas from sanitary landfill and wastewater treatment plants for distributed energy generation in Brazil." *Resources, Conservation and Recycling,* 136 (2018): 376–388.
38. Reinhart, D.R., McCreanor, P.T. and Townsend, T. "The bioreactor landfill: Its status and future." *Waste Management & Research*, 20, no. 2 (2002): 172–186.
39. Hird, M.J. "Waste, landfills, and an environmental ethic of vulnerability." *Ethics & the Environment*, 18, no. 1 (2013): 105–124.
40. Yaashikaa, P.R., Kumar, P.S., Nhung, T.C., Hemavathy, R.V., Jawahar, M.J., Neshaanthini, J.P. and Rangasamy, G. "A review on landfill system for municipal solid wastes: Insight into leachate, gas emissions, environmental and economic analysis." *Chemosphere*, 309 (2022): 136627.
41. Huang, Q., Yufei, Y.A.N.G., Xiangrui, P.A.N.G. and Qi, W.A.N.G. "Evolution on qualities of leachate and landfill gas in the semi-aerobic landfill." *Journal of Environmental Sciences*, 20, no. 4 (2008): 499–504.
42. Beede, D.N. and Bloom, D.E. "The economics of municipal solid waste." *The World Bank Research Observer*, 10, no. 2 (1995): 113–150.
43. Misra, V., and Pandey, S.D. "Hazardous waste, impact on health and environment for development of better waste management strategies in future in India." *Environment International*, 31, no. 3 (2005): 417–431.
44. Venkataramani, E.S., Ahlert, R.C., Corbo, P. and Irvine, R.L. "Biological treatment of landfill leachates." *Critical Reviews in Environmental Control*, 14, no. 4 (1984): 333–376.
45. Purmessur, B. and Surroop, D. "Power generation using landfill gas generated from new cell at the existing landfill site." *Journal of Environmental Chemical Engineering*, 7, no. 3 (2019): 103060.
46. Kurniawan, T.A., Lo, W., Chan, G. and Sillanpää, M.E.T. "Biological processes for treatment of landfill leachate." *Journal of Environmental Monitoring*, 12, no. 11 (2010): 2032–2047.
47. Sanphoti, N., Towprayoon, S., Chaiprasert, P. and Nopharatana, A. "Enhancing waste decomposition and methane production in simulated landfill using combined anaerobic reactors." *Water Science and Technology*, 53, no. 8 (2006): 243–251.
48. Kattel, E., Kivi, A., Klein, K., Tenno, T., Dulova, N. and Trapido, M. "Hazardous waste landfill leachate treatment by combined chemical and biological techniques." *Desalination and Water Treatment*, 57, no. 28 (2016): 13236–13245.

49. Christensen, T.H., Cossu, R. and Stegmann, R. *Landfilling of waste: Biogas*. CRC Press, 2020.
50. Surroop, D. and Mohee, R. "Power generation from landfill gas." In *2nd International Conference on Environmental Engineering and Applications IPCBEE*, vol. 17, pp. 237–241. 2011.
51. Hird, Myra J. "Waste, landfills, and an environmental ethic of vulnerability." *Ethics & the Environment*, 18, no. 1 (2013): 105–124.
52. Yaashikaa, P.R., Kumar, P.S., Nhung, T.C., Hemavathy, R.V., Jawahar, M.J., Neshaanthini, J.P. and Rangasamy, G. "A review on landfill system for municipal solid wastes: Insight into leachate, gas emissions, environmental and economic analysis." *Chemosphere* (2022): 136627.
53. Huang, Q., Yufei, Y.A.N.G., Xiangrui, P.A.N.G. and Qi, W.A.N.G. "Evolution on qualities of leachate and landfill gas in the semi-aerobic landfill." *Journal of Environmental Sciences*, 20, no. 4 (2008): 499–504.
54. Li, S., Yoo, H.K., Macauley, M., Palmer, K. and Shih, J.-S. "Assessing the role of renewable energy policies in landfill gas to energy projects." *Energy economics*, 49 (2015): 687–697.
55. Beede, D.N. and Bloom, D.E. "The economics of municipal solid waste." *The World Bank Research Observer*, 10, no. 2 (1995): 113–150.
56. Blanchet, M. J. *Treatment and Utilization of Landfill Gas: Mountain View Project Feasibility Study*. Environmental Protection Agency, Office of Solid Waste, 1977.
57. Lisk, D.J. "Environmental effects of landfills." *Science of the Total Environment*, 100 (1991): 415–468.
58. Jaramillo, P. and Matthews, H.S. "Landfill-gas-to-energy projects: Analysis of net private and social benefits." *Environmental Science & Technology*, 39, no. 19 (2005): 7365–7373. https://doi.org/10.1021/es050633j
59. Ezekwe, C.I., Agbakoba, A. and Igbagara, P.W. "Source Gas Emission And Ambient Air Quality Around The Eneka Co-Disposal Landfill In Port Harcourt, Nigeria." *Gas*, 2, No. 1 (2016): 11–23.
60. Joshi, B.R. and Hubbard, R.M. "Forensics analysis of solid state drive (SSD)." In *2016 Universal Technology Management Conference (UTMC)* (Vol. 2016, pp. 1–12), 2016. researchgate. net.
61. Shin, H.C., Park, J.W., Kim, H.S. and Shin, E.S. "Environmental and economic assessment of landfill gas electricity generation in Korea using LEAP model." *Energy Policy*, 33, no. 10 (2005): 1261–1270.
62. Eklund, B., Anderson, E.P., Walker, B.L. and Burrows, D.B. "Characterization of landfill gas composition at the fresh kills municipal solid-waste landfill." *Environmental Science & Technology*, 32, no. 15 (1998): 2233–2237.
63. Meegoda, J.N., Hettiarachchi, H. and Hettiaratchi, P. "Landfill design and operation." In *Sustainable Solid Waste Management* (pp. 577–604). American Society of Civil Engineers (ASCE), 2016. https://doi.org/10.1061/9780784414101.ch18
64. Tränkler, J., Visvanathan, C. and Kuruparan, P. "Mechanical biological waste treatment in South-East Asia—process features and treated waste performance." In *Proceedings APLAS, 3rd Asia Pacific Landfill Symposium*, pp. 191–198. 2004.
65. Raghab, S.M., Abd El Meguid, A.M. and Hegazi, H.A. "Treatment of leachate from municipal solid waste landfill." *HBRC Journal*, 9, no. 2 (2013): 187–192.
66. Shin, K.W., Kim, S-H., Kim, J.-H., Hwang, S.D. and Kang, J.-C. "Toxic effects of ammonia exposure on growth performance, hematological parameters, and plasma components in rockfish, Sebastes schlegelii, during thermal stress." *Fisheries and Aquatic Sciences*, 19 (2016): 1–8.

67. Welander, U. *Characterization and treatment of municipal landfill leachates.* [Doctoral Thesis (compilation), Biotechnology]. Department of Biotechnology, Lund University, 1998.
68. Chilton, J. ed. *Groundwater in the urban environment* (Vol. 2). AA Balkema, 1999.
69. Allen, A. "Containment landfills: the myth of sustainability." *Engineering Geology*, 60, no. 1–4 (2001): 3–19.
70. Li, Y., Park, S.Y. and Zhu, J. "Solid-state anaerobic digestion for methane production from organic waste." *Renewable and Sustainable Energy Reviews*, 15, no. 1 (2011): 821–826.
71. Home, Landfill Gas Primer. "Landfill gas basics." In *Landfill Gas Primer – An Overview for Environmental Health Professionals*. Available online at www.atsdr.cdc.gov/HAC/landfill/PDFs/Landfill_2001_ch2mod.pdf [Accessed on 13 February 2023]
72. Panwar, N.L., Kaushik, S.C. and Kothari, S. "Role of renewable energy sources in environmental protection: A review." *Renewable and sustainable energy reviews*, 15, no. 3 (2011): 1513–1524.
73. Panwar, N.L., Kaushik, S.C. and Kothari, S. "Role of renewable energy sources in environmental protection: A review." *Renewable and sustainable energy reviews* 15, no. 3 (2011): 1513–1524.
74. Morgan, S.M. and Yang, Q. "Use of landfill gas for electricity generation." *Practice Periodical of Hazardous, Toxic, and Radioactive Waste Management,* 5, no. 1 (2001): 14–24.
75. Meier, J.G., Hung, W.S.Y. and Sood, V.M. "Development and application of industrial gas turbines for medium-Btu gaseous fuels." Journal of Engineering for Gas Turbines and Power 108, no. 1 (1986): 182–190. https://doi.org/10.1115/1.3239869
76. Ahmed, S.I., Johari, A., Hashim, H., Mat, R., Lim, J.S., Ngadi, N. and Ali, A. "Optimal landfill gas utilization for renewable energy production." *Environmental Progress & Sustainable Energy*, 34, no. 1 (2015): 289–296.
77. Tomić, T. and Schneider, D.R. "Circular economy in waste management–Socio-economic effect of changes in waste management system structure." *Journal of Environmental Management*, 267 (2020): 110564.
78. Johari, A., Ahmed, S.I., Hashim, H., Alkali, H. and Ramli, M. "Economic and environmental benefits of landfill gas from municipal solid waste in Malaysia." *Renewable and Sustainable Energy Reviews*, 16, no. 5 (2012): 2907–2912.
79. Shuput, T.A. "Landfill gas recovery: Should your community consider it." *Energy Technol.(Wash., DC);(United States)* 12, no. CONF-850301- (1985).
80. Panwar, N.L., Kaushik, S.C. and Kothari, S. "Role of renewable energy sources in environmental protection: A review." *Renewable and Sustainable Energy Reviews*, 15, no. 3 (2011): 1513–1524.
81. Eklund, B., Anderson, E.P., Walker, B.L. and Burrows, D.B. "Characterization of landfill gas composition at the fresh kills municipal solid-waste landfill." *Environmental Science & Technology*, 32, no. 15 (1998): 2233–2237.
82. Barlaz, M.A., Chanton, J.P. and Green, R.B. "Controls on landfill gas collection efficiency: instantaneous and lifetime performance." *Journal of the Air & Waste Management Association*, 59, no. 12 (2009): 1399–1404.
83. Khalil, M.J., Gupta, R. and Sharma, K. "Microbiological degradation of municipal solid waste in landfills for LFG generation." *International Journal of Engineering and Technical Research,* 2321 (2014): 10–14.
84. Rettenberger, G. "Collection and disposal of landfill gas." In Raffaello Cossu and Rainer Stegmann (Eds.), *Solid Waste Landfilling* (pp.449–462). Elsevier, 2018.

85. Yuan, L. and Abichou, T. “Methane emission estimation and control through the life cycle of MSW landfills.” In *GeoFlorida 2010: Advances in Analysis, Modeling & Design* (pp. 2888–2895), 2010. https://doi.org/10.1061/41095(365)294
86. Kamarrudin, N., Zulkafli, N.H., Sikirman, A., Mahayuddin, N.M., Sigau, B.A., Hamid, K.H.K. and Akhbar, S. “Concentration and toxicological study on sanitary landfill gases at drilling point closed cell.” In *2013 IEEE Business Engineering and Industrial Applications Colloquium (BEIAC)* (pp. 333–338). IEEE, 2013, April.
87. Ko, J.H., Xu, Q. and Jang, Y.C. “Emissions and control of hydrogen sulfide at landfills: A review.” *Critical Reviews in Environmental Science and Technology*, 45, no. 19 (2015): 2043–2083.
88. Abraham, S., Joslyn, S. and Suffet, I.H. “Treatment of odor by a seashell biofilter at a wastewater treatment plant.” *Journal of the Air & Waste Management Association*, 65, no. 10 (2015): 1217–1228.

9 Bioprocessing of Organic Municipal Solid Waste for Biomethane and Biohydrogen Production

Musa Manga, Swaib Semiyaga, Sarah Lebu, Anne Nakagiri, Charles B. Niwagaba, Aaron Salzberg, and Chimdi C. Muoghalu

9.1 INTRODUCTION

Globally, about 2.01 billion tons of municipal solid waste (MSW) are generated annually and this is expected to increase extensively to about 3.4 billion tons by 2050 (Watts et al., 2018). Urbanization and industrialization are partly responsible for increasing quantities of solid waste whose management and disposal is a major challenge. The organic portion of MSW contributes to global warming, public health risks, and ecosystem imbalance (Lim et al., 2016; Manga, 2017). With increasing population growth, the aforementioned environmental problems will become a major crisis in the coming years, making it imperative that actions are taken toward a more sustainable society. Waste-to-energy technologies like gasification, pyrolysis, anaerobic digestion, and incineration which combine waste treatment with energy production can be used for effective management of MSW as well as decrease interdependence on fossil fuels (Panigrahi and Dubey, 2019; Semiyaga et al., 2022; Muoghalu et al., 2023). Among the waste-to-energy technologies, the bio-based process–anaerobic digestion (AD) has the lowest negative environmental impact (Zainal et al., 2018). Energy products obtained from the AD process are biomethane and biohydrogen. Apart from being a waste-to-energy process, AD is an important pillar of the circular economy as it mitigates greenhouse gas emissions, recycles nutrients from biomass to make fertilizers, and curtails environmental contamination and spread of harmful diseases from landfilling (Wu et al., 2015). In addition, with increased emphasis on the need to achieve sustainability goals by shifting from fossil fuel to green energy that are carbon neutral, the AD process which produces bioenergy plays an important in this paradigm shift (Nguyen et al., 2019).

Biomethane and biohydrogen are environmentally friendly forms of energy and can be used in the generation of electricity. Biomethane is generated in anaerobic digestion when all the stages of the process – hydrolysis, acidogenesis, acetogenesis,

DOI: 10.1201/9781003364467-9

and methanogenesis – are allowed to occur (Manga et al., 2023). Biohydrogen on the other hand is an intermediate product of the AD process and is produced when the methanogenic stage of AD is inhibited (Stamatelatou et al., 2014). The production of either biomethane or biohydrogen from OFMSW has been investigated by several studies (Panigrahi and Dubey, 2019; Show et al., 2011). However, research on simultaneous generation of both biomethane and biohydrogen from OFMSW is limited. The generation of biomethane and biohydrogen from OFMSW can be carried out with two-phase AD process has more advantages than single-stage AD process as it brings about increased yield of both biomethane and biohydrogen, has shorter HRT, less erratic operation from improved buffering capacity and better resistance to organic shock loads (Ariunbaatar et al., 2015). Also, the biomethane and biohydrogen produced can be combined together to form biohythane which is a pure fuel that can be used in the automobile industry. Biohythane can be a better alternative to compressed natural gas (CNG) by adding hydrogen which enhances the flammability range of methane, reducing the burning time of engines without the emission of nitrous oxide into the environment (Hans and Kumar, 2019). For commercial production and utilization of biomethane and biohydrogen, proper evaluation of the process parameters as well as problems associated with two-stage AD process is required. To this end, this chapter presents an extensive review of the AD process, process parameters, pretreatment techniques, bioreactor design as well as constraints and future prospects in biomethane and biohydrogen generation from the AD process.

9.2 CHARACTERISTICS OF OFMSW

9.2.1 General Description for OFMSW

The feedstock used in anaerobic digestion influences the yield, quality and production rate of bioenergy, and process cost (Ghimire et al., 2015). Thus, it is pertinent that the composition of the substrate (in this case OFMSW) to be used in the process is well understood. OFMSW refers to a heterogeneous mixture of organic wastes from residential, business, institutional, and partly industrialized areas. OFMSW comprises food waste from residential and commercial kitchens, yard waste (i.e., flowers and grass) and paper waste with food waste being the major fraction. OFMSW can be obtained as source-separated organic waste or as mixed waste (Panigrahi and Dubey, 2019; Manga et al., 2021; Manga et al, 2019). The quantity and composition of OFMSW varies with climatic condition, geographic location, characteristics of a population, season, collection method and frequency, and legislation (Panigrahi and Dubey, 2019).

9.2.2 Physical Characteristics of OFMSW

Organic fraction of municipal solid waste is made up of a wide range of particle sizes. Particle size influences the biodegradation rate during the AD process (Panigrahi and Dubey, 2019). Larger particle sizes have smaller available surface areas, and this slows the biodegradability of OFMSW, which in turn prolongs the hydrolysis stage of AD (Ye and Berson, 2014). Hence, prior to the AD process, it is pertinent that the

OFMSW is macerated and pulverized to particle sizes between 1–5cm (Vögeli et al., 2014). Very small particle sizes (i.e., less than 1 cm) can lead to acid accumulation in the digester (Izumi et al., 2010). Common techniques that are used in particle size reduction for OFMSW are shredding, milling, grinding, and maceration.

Another major physical parameter which impacts the AD process is feedstock density. For OFMSW, the density is a function of the technology employed in separating OFMSW from MSW (Xu et al., 2016). Forster-Carneiro et al. (2008) posited that high-density OFMSW is more biodegradable than that with low density.

9.2.3 Chemical and Compositional Characteristics of OFMSW

The bioenergy potential of OFMSW is largely influenced by its chemical characteristics particularly the pH, total solids (TS), and volatile solids (VS). In addition, elemental constituents like carbon (C), hydrogen (H), nitrogen (N), and sulfur (S) influence the yield of bioenergy from OFMSW. When OFMSW is used for the AD process, the biohydrogen and biomethane yield from the process is a function of micro molecules (like carbohydrate, protein, and lipids) and macro molecules (like cellulose, hemicellulose, and lignin) that are present in OFMSW (Panigrahi and Dubey, 2019). Most macro and micro molecules of OFMSW are suitable for biomethane production but biohydrogen is produced in large quantities from the carbohydrate (sugar) content of OFMSW (Stamatelatou et al., 2014). Soluble and readily accessible sugars (like glucose, hexose, and sucrose) are the major fraction of OFMSW which can be converted into biohydrogen (Guo et al., 2014). OFMSW contains materials like paper, wood, food scraps, and yard waste, which are rich in carbon and thus suitable substrate for biohydrogen generation.

9.3 ANAEROBIC DIGESTION FOR BIOMETHANE AND BIOHYDROGEN PRODUCTION

Anaerobic digestion (AD) is a complex process involving biochemical reactions which lead to organic matter decomposition by a consortium of microorganisms in an oxygen-limited environment to generate biogas (Mathew et al., 2015). Generally, AD occurs in a sequence of four stages: hydrolysis, acidogenesis, acetogenesis, and methanogenesis (Mathew et al., 2015). In hydrolysis, complex organic compounds – that is carbohydrates (polysaccharides), proteins, and lipids – are decomposed into simple sugars, amino acids, and fatty acids, respectively through enzymatic action (Adekunle and Okolie, 2015). This is followed by acidogenesis where the products of the hydrolysis stage are degraded by acidogenic bacteria to form volatile fatty acids (VFAs), hydrogen, carbon-dioxide, and alcohols. In the third stage of acetogenesis, acetogenic bacteria process VFAs into acetate, more hydrogen gas (H_2), and carbon dioxide (CO_2). The last stage, methanogenesis, is carried out by acetotrophic and hydrogenotrophic methanogens. Acetotrophic methanogens convert acetate to methane and CO_2 while the hydrogenotrophic methanogens convert H_2 and CO_2 to methane (André et al., 2016).

Anaerobic digestion is primarily used in the production of biomethane, but it can be optimized to generate both biomethane and biohydrogen (Khan et al., 2016). If the production process is targeted toward the generation of biomethane, then all the stages of anaerobic digestion will occur in one bioreactor/digester (Hans and Kumar, 2019). This one-stage process usually leads to the generation of 60–65% biomethane and about 30–40% of CO_2 with biohydrogen and other elements in trace amounts (Jafari and Zilouei, 2016). When generating both biohydrogen and biomethane, a two-stage AD is usually applied. In two-stage anaerobic digestion, a bioreactor is used for hydrolysis, acidogenesis and acetogenesis which forms the first stage. The products of the first bioreactor are then placed in the second bioreactor used for methanogenesis which forms the second stage (Franke-Whittle et al., 2014). Inhibitory metabolites and VFAs generated from the first bioreactor are provided for methanogens to enhance biomethane generation in the second stage (Kvesitadze et al., 2012). Not only does this technique enhance the generation of biohydrogen in the first stage, it also increases the biomethane yield and energy density in the second stage (Kvesitadze et al., 2012; Malave' et al., 2015; Massanet-Nicolau et al., 2013). For instance, Kvesitadze et al. (2012) carried out anaerobic digestion of solid waste with both one-and two-stage processes. It was noted that the two-stage AD process had higher H_2 and CH_4 contents – 23 and 26% increase respectively – than the one-stage process.

Besides increased yield of biohydrogen and biomethane, the two-stage AD process takes a shorter period of time (18–20 days) compared to one stage (28–30 days) (Hans and Kumar, 2019). This is due to the fact that hydraulic retention time (HRT) (usually 1–3 days) is adequate during acidogenesis to generate H_2 and organic acids which can be quickly extracted and the effluents can be fed to the methanogenic bioreactor which usually produces CH_4 within 10–15 days (Kongjan et al., 2011). Thus, using the two-stage AD process has more advantages than using one-stage AD as it has shorter HRT, higher reduction of COD, reduced environmental concerns, higher energy efficiency and recovery (Hans and Kumar, 2019). The biomethane and biohydrogen can be simultaneously used for energy. This makes the system have less erratic operation from improved buffering capacity, increased tolerance to organic loading shocks, and failure from VFA accumulation (Ariunbaatar et al., 2015).

Table 9.1 is a summary of the properties of biohydrogen and biomethane produced from the AD process while Figure 9.1 shows the typical pathway for bioenergy production from OFMSW.

9.4 FACTORS WHICH INFLUENCE THE AD TO PRODUCE BIOMETHANE AND BIOHYDROGEN FROM OFMSW

The production of biomethane and biohydrogen from AD process is usually affected by different environmental, operational, and chemical conditions. These factors include system pH, temperature, HRT, organic loading rate, nutrients, and hydrogen partial pressure.

9.4.1 System pH

The system pH is a key factor as it directly influences the digestion process of OFMSW and its products (Panigrahi and Dubey, 2019). The metabolic pathway,

TABLE 9.1
Properties of Hydrogen and Methane from the AD Process

Parameter	Hydrogen (H_2)	Methane (CH_4)
Molecular weight	2.00	16.00
Flammability limit in air (% vol.)	4–75	5–15
Minimum ignition energy (MJ)	0.02	0.20
Ignition temperature (K)	858.00	918.00
Flame temperature in air (K)	2318.00	2148.00
Burning speed (cm/s)	270.00	37.00
Mass lower heating value (MJ/kg)	120.00	50.00
Volumetric lower heating value (MJ/m^3)	10.81	35.37
Octane number	>130	107.50
Laminar burning velocity in NTP air (cm/s)	265–325	37–45
Stoichiometric air-to-fuel ratio	34.20	17.19
Adiabatic exponent	1.14	1.32
Quenching distance (cm)	0.06	0.20

Source: Liu et al. (2013).

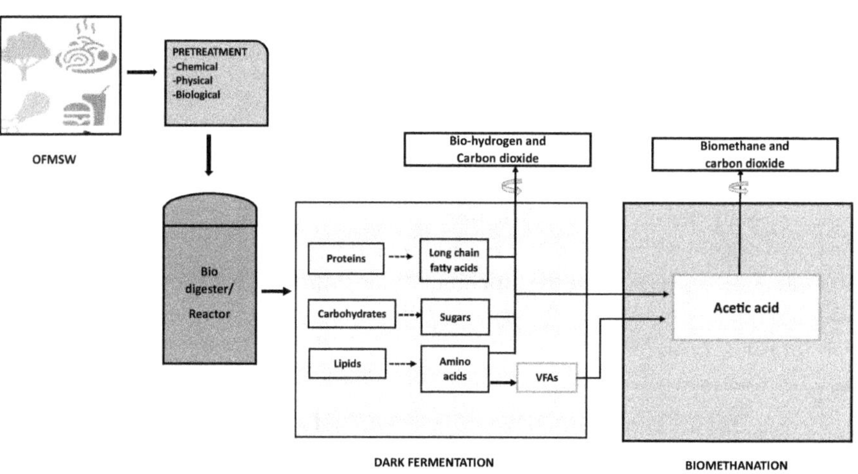

FIGURE 9.1 Pathway for Production of Biomethane and Biohydrogen from OFMSW.

microbial growth, and energy yields from the AD of OFMSW are highly influenced by pH (Chozhavendhan et al., 2020). The different species of bacteria used in the AD of OFMSW have pH ranges at which they operate best (Appels et al., 2008). Quite a number of studies have reported that the optimum pH needed to achieve maximum biohydrogen yield ranges between 5.2 and 6.0 (Chang and Lin, 2004; Ren et al., 2009; Zhang et al., 2008). Methanogens which are used during the methanogenesis stage usually require neutral pH range of 6.5–7.2 (Appels et al., 2008). Acidic pH range negatively impacts the hydrogenase enzyme activity and inhibits the production of biohydrogen while very high pH (> 9) conditions eliminate methanogens (Ngo et al., 2012). Thus, in order to ensure optimal biohydrogen and biomethane production during AD of OFMSW, the pH needs to be continuously monitored and controlled. OFMSW is acidic (pH: 4.5–5.9) and pH control for it is usually carried out by decreasing the organic loading rate or using bases like sodium, potassium and calcium hydroxide, sodium and calcium carbonate at the startup of anaerobic digestion (Chozhavendhan et al., 2020; Panigrahi and Dubey, 2019).

9.4.2 Temperature

Temperature affects substrate degradation, microbial growth, and metabolic activity during AD of OFMSW (Chozhavendhan et al., 2020). Typical temperature range for anaerobic digesters is 20 to 65°C. The temperature used in anaerobic digesters can be in the mesophilic (20°C–40 °C) or thermophilic (40°C–65°C) ranges. The results of several studies show that higher production of bioenergy (biomethane and biohydrogen) occurs when thermophilic operation is used (Cavinato et al., 2013; Intanoo et al., 2016). Increased energy yield from thermophilic digestion can be attributed to increased system entropy, faster acclimatization rate for thermophilic inoculum, enhanced disintegration of complex biopolymers such as lignocelluloses present in the feedstock, and increased enzyme activity (Zhong et al., 2015). Also, high temperature can improve the production of biohydrogen through inhibition of methanogenesis by hindering growth of hydrogen-consuming methanogens (Khan et al., 2016). Although thermophilic AD leads to higher energy yield, one major disadvantage with it is the need for high investment and maintenance costs (Beevi et al., 2015).

9.4.3 Hydraulic Retention Time

The hydraulic retention time (HRT) is the average period of time organic material stays in the bioreactor, that is the period of time it is in contact with microorganisms (Jain et al., 2015). HRT greatly affects microbial metabolism and bioenergy generation from AD process (Rawoof et al., 2021). HRT is usually affected by factors such as feedstock type, inoculum activity, pH, and organic loading rate. Commonly used HRT for biohydrogen production from OFMSW usually ranges between 1–3 days (Abe et al., 2019) while it takes longer time (> 20 days) for biomethane production. HRT can be used as a tool to ensure continuous biohydrogen production in the first stage of the two-stage AD process. Since the specific growth rate for methane-producing bacteria which are hydrogen consumers ($0.0167–0.02h^{-1}$) is less than that of hydrogen-producing bacteria ($0.172h^{-1}$), using low HRT will limit the development of methane-generating bacteria involved in methane generation. However, the HRT should not be too low as this can cause untimely removal of biomass from the reactor, in turn reducing digestion stability (Panigrahi and Dubey, 2019).

9.4.4 Organic Loading Rate

Organic loading rate (OLR) is the amount of organic feedstock delivered per unit volume of the bioreactor for biodegradation within a specified time period (Chozhavendhan et al., 2020). OLR has notable impact on bioenergy generation production as OLR which exceeds the decomposition rate or hydrolytic capacity of the digester can lead to process instability and reduce bioenergy generation (Mao et al., 2015). Most AD plants for OFMSW apply OLR of 4.4–22 g VS/L/day (Panigrahi and Dubey, 2019). High OLR leads to lower bioenergy generation as a result of increased creation of inhibitory compounds like VFAs, nutrient deficiency, and shock on microorganisms (Mu et al., 2018). The optimal OLR (7.5 VS/L/day) used in the production of bioenergy is usually based on bioreactor type, temperature, pH, temperature, and concentration of the feedstock (Jiang et al., 2020). Apart from its effect on bioenergy yield, OLR has significant influence on the functioning of circulating and feeding pumps which easily wear out when loaded with high solids content (Panigrahi and Dubey, 2019).

9.4.5 Nutrients

Micro- and macro-nutrients are instrumental in the anaerobic digestion of OFMSW. Macro nutrients such as phosphorus (P), nitrogen (N), and sulfur (S) are usually applied as bulking agents (Panigrahi and Dubey, 2019). In addition to use as bulking agents, they usually supplement carbohydrate-based feedstock for optimal cell cultivation and biohydrogen production (Show et al., 2011). Micro-nutrients such as iron (Fe), magnesium (Mg), nickel (Ni), zinc (Zn), copper (Cu), cobalt (Co) and chromium (Cr) usually act as crucial cofactors in enzymatic reactions (Panigrahi and Dubey, 2019).

A major consideration with respect to nutrient is the carbon to nitrogen (C/N) ratio. The C/N ratio gives an indication of the nutrient level of the feedstock, ammonia nitrogen evolved and the amount of VFAs that accumulate in the bioreactor (Panigrahi and Dubey, 2019). The AD process is highly sensitive to C/N ratio, which is usually from 20:1 and 30:1 with 25:1 as best ratio for the development of bacteria (Khalid et al., 2011). OFMSW contains materials like paper, yard waste, wood and food scraps, thereby making its C/N ratio high. This high C/N ratio results in slower protein disintegration and low ammonia formation within the digester, thus co-digestion with different kinds of nitrogenous waste like manure can help improve the AD process (Velásquez Piñas et al., 2018).

9.4.6 Hydrogen Partial Pressure

Hydrogen partial pressure at liquid state influences biohydrogen generation from the AD process as hydrogen concentration in the reactor determines the amount of biohydrogen generated (Rawoof et al., 2021). Biohydrogen generation is less favorable with increase in hydrogen partial pressure, thus it is pertinent that surplus hydrogen is removed from the AD system to sustain production (Tiwari et al., 2006). The removal of excess hydrogen can be achieved through sparging of carbon dioxide, nitrogen, or creation of a vacuum (Mizuno et al., 2000). In sparging, gas is bubbled through the slurry/liquid in the digester to remove the excess hydrogen gas present in the digester. Higher biohydrogen yield is obtained with external gas sparging (involving use of CO2 and nitrogen) than internal biogas sparging (Show et al., 2011). It is also pertinent that the sparging gas does not contain carbon monoxide as carbon monoxide inhibits the production of hydrogenase (Lee et al., 2010).

9.5 TECHNIQUES TO IMPROVE THE YIELD OF BIOMETHANE AND BIOHYDROGEN FROM THE AD PROCESS

9.5.1 Pretreatment of Substrates

The pretreatment of substrates prior to use in the AD process helps in improving decomposition and enhances the bioenergy yield faster (Rawoof et al., 2021). Pretreatment is very important for hydrolysis, the slowest step in anaerobic digestion (Brodeur et al., 2011). The effect of pretreatment on feedstock digestion is based on type of pretreatment and feedstock type (Satari et al., 2019). Frequently applied pretreatment methods include physical, chemical, biological, and thermal treatment.

9.5.1.1 Physical Pretreatment

This involves the application of mechanical forces to treat the substrate rather than using enzymes and chemicals. Examples of physical pretreatment used in the anaerobic digestion process include mechanical and ultrasonic pretreatment.

(a) **Mechanical pretreatment:** Owing to the fact that OFMSW contains different particle sizes, mechanical pretreatment is needed prior to AD to homogenize the particle size. Mechanical pretreatment helps to improve feedstock decomposition by decreasing its particle size, crystallinity, and polymerization, hence, increasing the pore volume of the substrate as well as its surface area to volume ratio (Jain et al., 2015). Common techniques used for mechanically pretreating OFMSW include milling, grinding, crushing, and irradiation (Esposito et al., 2011). In the mechanical pretreatment of OFMSW, the range of particles for AD is 1–5 cm; with the optimum particle size for generation of bioenergy being 3.5 cm and it should be taken into consideration as extremely small particle sizes can lead to foaming during anaerobic digestion (Motte et al., 2015; Esposito et al., 2011).

(b) **Ultrasound pretreatment:** Ultrasound waves are cyclic acoustic pressure waves (compression and rarefaction) which are based on a frequency more than 20 kHz. Ultrasonic pretreatment breaks the structure of the substrate through powerful vibrations of the ultrasonic waves, and this enhances the AD process (Motte et al., 2015). Applying ultrasound to OFMSW which contains lignocellulosic feedstock breaks down the structure by disintegrating the hydrogen bond present (Kotay and Das, 2009). Ultrasonic pretreatment of organic materials usually involves two frequency ranges: low frequency (20–100 kHz) which results in cavitation and high frequency (>100 kHz) which induces chemical reactions (Mischopoulou et al., 2016). Most times, ultrasound pretreatment of solid waste is conducted at reduced frequency (20–40 kHz) (Cesaro et al., 2014).

9.5.1.2 Thermal Pretreatment

This technique enhances the efficiency of anaerobic digestion by disintegrating recalcitrant organics (Ariunbaatar et al., 2015) breaking up floccules of macromolecules, and increasing dewaterability (Wang et al., 2010). Key parameters which influence thermal pretreatment are temperature, time, with the mode of heat transfer with temperature being the most critical (Panigrahi and Dubey, 2019). The most widely used temperature for thermal treatment of OFMSW is 121°C for a duration range of 20 to 30 minutes (Chozhavendhan et al., 2020). Generally, there are two types of thermal treatment techniques: techniques where only temperature is regulated, i.e., hot air oven, microwave, hot water bath; and techniques where both temperature and pressure are regulated (hydrothermal pretreatment), i.e., autoclave treatment. Hydrothermal techniques like autoclaving do not use chemicals and thus retain the nutrients in OFMSW while eliminating pathogens in the feedstock (Chozhavendhan et al., 2020). Generally, the amount of soluble chemical oxygen demand (COD), VFA, and amount of biogas produced are used for measuring the effect of thermal pretreatment (Jin et al., 2016).

9.5.1.3 Chemical Pretreatment

Chemical pretreatments involves application of organic and inorganic solvents in the hydrolysis of lignocellulosic components of OFMSW (Chozhavendhan et al., 2020). This is one of the most preferred treatment due to its easy scale-up and low energy demand. The main types of chemical pretreatments are acid, alkaline, and oxidative methods.

Acid pretreatment: This method is suitable substrates rich in lignocellulose (Mancini et al., 2018). This is usually carried out with the aim of disintegrating lignin, cellulose, and hemicellulose, into monosaccharides. It can be achieved via breaking down the covalent hydrogen bonds and van der Waals forces binding the particles of the organic material (Sarto et al., 2019). Both dilute and concentrated acids can be employed in the acidification of the substrate, however dilute acid is more preferred based on the fact that it is less toxic, less corrosive, cheaper, and makes use of smaller quantities of neutralizing chemicals (Panigrahi and Dubey, 2019). Examples of dilute acids that are commonly used in acid pretreatment of lignocellulosic fractions of MSW are dilute nitric acid (HNO_3), hydrochloric acid (HCl), sulfuric acid (H_2SO_4), phosphoric acid (H_3PO_4), with H_2SO_4 being the most commonly used because of its cheapness and availability.

Alkali pretreatment: Alkaline pretreatment is for making cellulose and hemicellulose in OFMSW easily degradable by hydrolytic (Panigrahi and Dubey, 2019). Commonly used alkali for the treatment of OFMSW are sodium hydroxide (NaOH), sodium carbonate (Na_2CO_3), potassium hydroxide (KOH), magnesium hydroxide ($Mg(OH)_2$), calcium hydroxide ($Ca(OH)_2$), calcium carbonate ($CaCO_3$), and ammonium hydroxide (NH_4OH). One key benefit of using alkali pretreatment is that when maintained, it leads to preventing hydrogen sink phenomena during the acidogenesis stage (Mao et al., 2015). However, when compared to acid pretreatment, alkaline pretreatment needs higher quantities of chemical dosage as some chemicals are consumed by the biomass which makes it more expensive than treating with acid.

9.5.1.4 Biological Pretreatment

Biological pretreatment is usually applied for distorting covalent cross linkages and non-covalent forces holding lignin and hemicellulose together thereby improving the digestibility of complex waste. Though treating biologically is environmentally friendly as it involves low chemical usage, it is too slow and costly. This is because the biological method needs high retention time which makes it a time-consuming process (Chozhavendhan et al., 2020). Parameters which influence biological pretreatment are incubation temperature and period, moisture content, medium pH, and feedstock particle size. Commonly used biological pretreatment techniques for OFMSW are enzymatic, pre-composting, application of mature compost, aeration, and use of fungal species (Fdez.-Güelfo et al., 2011).

9.5.2 Co-Digestion

Co-digestion involves digesting multiple organic materials simultaneously (Rawoof et al., 2021). This method has more benefits than mono-digestion because it brings about a balance of the C/N ratio, curtails inhibition (from ammonia and VFA), and enhances energy production kinetics (Xie et al., 2016). It presents a viable method for dealing with nutritional deficiencies which reduce the efficiency of anaerobic digestion (Yang et al., 2015). Hence, prior to co-digestion of OFMSW, it is necessary that its composition is well known so as to know what materials it can be co-digested with. If its lignin content is very high, then it can be co-digested with nitrogen-rich substrates like manure to reduce inhibitory effects (Karthikeyan and Visvanathan, 2013).

9.5.3 Application of Additives

Chemical, biological, and metal additives can be applied to improve microbial metabolism and yield of biomethane and biohydrogen. Additives provide nutritional supplement for the growth of microorganisms and also expedite intercellular electron transport (Sun et al., 2019). Biological additives such as bacterial strains, fungi, and enzymes are used in bio-augmentation (Bouabidi et al., 2019). Adding chemicals like metal cations and micronutrients enriches microorganisms and in turn increase in bioenergy generation. The application of metals in small amounts can alter enzymatic activity during AD and enhance the efficiency of the process (Sun et al., 2019). Metal additives applied in anaerobic digestion as monomers (i.e., gold, silver, copper, iron, nickel, and lead), metal ions (i.e., Ca^{2+}, Mg^{2+}, Cu^{2+}), metal oxides or all of them combined together. One major advantage using additives has over other methods is that it is more cost effective (Yang et al., 2015).

9.6 DIGESTERS/BIOREACTORS USED IN BIOMETHANE AND BIOHYDROGEN PRODUCTION

9.6.1 Conventional Two-Phase Bioreactors

The commonly used bioreactors for the two-phase AD of OFMSW include: continuous stirred-tank reactors (CSTRs), sequencing batch reactors (SBRs), membrane bioreactors (MBRs), fixed bed bioreactors (FBR), fluidized bed bioreactors, and up flow anaerobic sludge blanket (UASB) reactors (Show et al., 2011). In CSTRs, constant stirring occurs, and this leads to complete mixing and suspension of microorganisms in the mixed liquor in the reactor. With this technique, good substrate to microorganism contact is established and mass transfer occurs easily. A fixed bed reactor makes use of microcarriers inside the reactor for biomass retention. In FBR, the mixing velocity is not as high as CSTRs, and this reduces the mass transfer rate and in turn reduces the rate of substrate conversion and bioenergy production (Lee et al., 2010. MBRs retain microorganisms in the reactor by combining the features of membrane filtration and that of activated sludge process. The main limitations of these reactors are membrane fouling and high operational costs (Buitrón et al., 2019).

9.6.2 Hybrid Bioreactors

The hybrid bioreactors operate on the same principle as two-stage bioreactors where degradation of the feedstock to H2 and organic acids occurs in the first step of the process. However, both biohydrogen and biomethane can be obtained in the second stage unlike the conventional two-stage bioreactors where only methane is obtained in the second stage (Koutrouli et al., 2009). The extraction of large quantities of hydrogen from the second stage is usually achieved through photo-fermentation (Kapdan and Kargi, 2006) or the use of fuel cell (García-Peña et al., 2009). Increased hydrogen yield is obtained by combining dark and photo-fermentation in these kinds of reactor (Nath and Das, 2006). Although research so far has shown the viability of hybrid systems, one major technical barrier in the use of photo-fermentation is low photosynthetic efficiencies and in turn low biohydrogen yields. An alternative method of increasing the yield of biohydrogen in the second stage is through the use of microbial electrolysis cells (MECs) where electricity is used in microbial fuel cell to supply energy required for changing organic acids to hydrogen (Nath and Das, 2006). Successful pilot-scale operation of hybrid systems for fermentation of kitchen waste has been documented with increased production rates of 5.4 L H_2/L. d and 6.1 L CH_4/L. d and improved COD reduction of 80% (Ueno et al., 2007).

9.6.3 Recirculated Two-Phase Anaerobic Digesters

Recirculated two-phase anaerobic digestion (R-TPAD) is a modification of the two-phase AD, which combines the functions of dark fermentation and methane generation (Liu et al., 2013). The R-TPAD system consists of two different-sized CSTR reactors; the smaller one is used for dark fermentation while the larger one is used for methane production. Impellers are used in mechanically mixing the contents at different levels. What makes R-TPAD different from conventional two-phase AD is that the alkalinity and microorganisms required for consistent bioenergy generation are recycled from the effluent of the second which leads to increased production of biohydrogen (Chu et al., 2008; Jiang et al., 2018). For example, Cavinato et al. (2012) investigated the pilot-scale digestion of OFMSW with R-TPAD system and recorded specific gas production rate of 0.m^3/kgTVS.

Table 9.2 is a summary of different bioreactors and the optimum operating conditions.

9.7 MICROBIAL COMMUNITIES INVOLVED IN BIOHYDROGEN AND BIOMETHANE GENERATION

Anaerobic digestion process is usually implemented by a complex consortium of microorganisms; hydrolysis, acidogenesis, and acetogenesis is usually implemented by bacteria while methanogenesis is usually conducted by archaea (Stams and Plugge, 2009). Table 9.3 shows the various classes of microorganisms in different phases of the AD process. These groups of organisms function in syntropy and their metabolic pathways are intra-dependent (Nguyen et al., 2019). Thus, any change in their metabolic pathways bring about considerable changes in the quantity and composition of microorganisms found in the anaerobic digester. This was noted by Si et al. (2016), who observed that change from only biomethane production to biohythane (biohydrogen and biomethane) production led to significantly modified microbial composition in the bioreactor. They noted that the amount of acidogenic bacteria (i.e., family Clostridiaceae) was lower in the biohythane system than in the biomethane system. However, the quantity of acetogens (Syntrophaceae, Syntrophomonadaceae and Desulfovibrionaceae) and acetate-oxidizing bacteria (Spirochaetes) increased. These results imply that biomethane reactors in biohythane system have better acetogenic and acetate-utilizing microorganisms.

The formation of the abovementioned group of microorganisms needed in the AD process takes a long period of time. Thus, to kickstart and speed up the AD process, microbial cultures which can be pure or mixed are usually applied as inoculum. The microbial culture is pure when it contains only a specific group of microorganisms for a specific purpose. For instance, if the aim of anaerobic digestion is just for biohydrogen production, then pure microbial strains like *Enterobacter aerogenes* and *Clostridium acetobutylicum* can be used (Hasibar et al., 2020). However, if the aim of anaerobic digestion is for methane production, then mixed cultures which contain the aforementioned strains together with methanotrophic bacteria such as Methylomicrobium *alcaliphilum*, *Methylosinus trichosporium*, and *Methylococcus capsulatus* can be used (Henard et al., 2018).

9.8 CONSTRAINTS AND AREAS OF FUTURE RESEARCH

Two major challenges in the production of biohydrogen are the comparatively small hydrogen yield and production rate. Hydrogen yield can be enhanced by applying fitting microbial strains, modifying the process, improving digester design and use of genetic and molecular engineering techniques to enhance the metabolism of microorganisms. In addition, molecular techniques can be used in generating microbial strains to disintegrate refractory organics.

Although the two-stage AD system brings about increased yield of both biomethane and biohydrogen, unlike the single-stage AD process where the system uses its self-buffering capacity to balance acidity from VFAs generation, the first stage (dark fermentation) of the two-stage AD is usually unstable due to continuous production of acidity (VFA). Extremely low pH (<5) can hinder biohydrogen generation. Application of additives like acids, bases, and buffers can help keep the pH in the required range. However, this is usually expensive, and it reduces the economics and the sustainability of the process. One technique that can be applied to regulate the

TABLE 9.2
Biohydrogen and Biomethane Yield from Different Types of Two-phase AD bioreactors

Reactor configuration	Waste type	VS/TS ratio (%)	Inoculum	Operating conditions for hydrogen reactor
CSTR reactors for both phases of the digestion process	Olive pulp	96.4	Olive pulp anaerobic sludge	Temperature: 35 °C HRT: 7.5-30h pH: 5 ± 0.1
Hybrid system consisting of dark fermentation reactor + MFC + MECs	Municipal biowaste (food waste + fruit–vegetable waste + dewatered sewage sludge (DSS		Anaerobic seed sludge	Temperature: 60 °C HRT: 32h DF reactor pH: 7.0
R-TPAD with two CSTRs	Cassava residue	91	Anaerobic mixed microflora from sewage sludge digester	Temperature: 55 °C HRT: 10 days pH: 5.3
2 stage anaerobic batch film sequencing reactors	Cassava wastewater		PBR anaerobic sludge	Temperature: 30°C HRT: 2-4 h pH: 6.0
2 stage DMBR	Food waste			Temperature: 55 °C HRT: 8h
Laboratory scale; batch operated serum bottles.	Microalgae		Anaerobic sludge	
Laboratory scale; batch operated serum bottles.	Cornstalk		Anaerobic sludge from UASB reactor for treating wastewater	Temperature: 55 °C PH: 5.0
UASB reactors for both stages of the digestion process	Wheat straw hydrolysate	75	Granular sludge and digested manure	Temperature: 70 °C HRT: 1 day pH: 5.5
R-TPAD with two CSTRs	Food waste	82.3	Anaerobic digester sludge	Temperature: 55 °C HRT: 3.3–6days pH: 5.5 OLR: 16-21 kgVS/ m^3 d

Operating conditions for methane reactor	COD removal efficiency	Biohydrogen yield	Biomethane yield	Reference
Temperature: 35 °C HRT: 5-20d OLR: 150–600 gTCOD/L/d pH: 7.51–7.62	4.4–10.2%	0.13–0.231 L/L/d	0.96–1.69 L/L/d	Koutrouli et al. (2009)
Temperature: 35 °C HRT: 40 days pH: 7.10		0.48 L/L /d (based on MEC volume)		Wang et al. (2010)
		0.31 L/L/day	0.81 L/L/day	Jiang et al. (2018)
Temperature: 30 °C HRT: 21-38 days		0.7 L/L/d	2.71 L/L/d	Mari et al. (2020)
Temperature: 35 °C HRT: 6 days		7.09 L/L/d	0.99 L/L/d	Jung et al. (2022)
Temperature: 37 °C PH: 7.0	81.8% COD removal	46 ± 2.4 mL/g-VS	393.6 ± 19.5 mL/g-VS	Yang et al. (2011)
Temperature: 37 °C PH: 7.0		64 mL/g VS	115 mL/g VS	Lu et al. (2009)
Temperature: 55 °C HRT: 5.2 days pH: 7.0		89 mL/g VS	307 mL/g VS	Kongjan et al. (2011)
Temperature: 55 °C HRT: 12.6 days pH: 5.5 OLR: 5–10 kgVS/m^3 d		overall SGP of 780 mL/gVS		Cavinato et al. (2012)

(*continued*)

TABLE 9.2 (Continued)
Biohydrogen and Biomethane Yield from Different Types of Two-phase AD bioreactors

Reactor configuration	Waste type	VS/TS ratio (%)	Inoculum	Operating conditions for hydrogen reactor
2 stage static batch reactors	Giant reed, Maize, Olive pomace, Wheat bran	76.2	Anaerobic digestion sludge from agricultural residue, poultry and cattle manure biogas plant	Temperature: 37 °C pH: No pH control
2 stage automatic methane potential test system (AMPTS)	Microalgae (*Chlorella* sp.)	90.0	Anaerobic sludge from sewage treatment plant	Temperature: 37°C pH: 6.0
2 stage CSTRs	Food waste	86.7	Anaerobic digester sludge treating maize straw	Temperature: 55 °C HRT: 5 days pH: 5.5
2 stage glass reactors	Food waste	96.7	Anaerobic digestion sludge	Temperature: 55 °C HRT:48h pH: 6.0
CSTR for hydrogen reactor and ABR for methane reactor	Food waste			Temperature: 55 °C HRT: 2.87 days pH: 5.4–5.6
CSTR for hydrogenogenic stage and PBR for methanogenic stage	Artificial garbage slurry containing milled paper	83	Sludge compost for biohydrogen reactor and anaerobic sludge for methane reactor	Temperature: 55 °C HRT: 0.5d pH: 6.0
CSTR for biohydrogen AHBC for biomethane	POME Palm oil mill effluent	78.69	Anaerobic seed sludge from reactor processing POME and SLS	Temperature: 55 °C HRT: 3days
CSTR for biohydrogen and UASB for methane	Wheat straw		Granular anaerobic sludge	Temperature: 70 °C pH: 6.9

Operating conditions for methane reactor	COD removal efficiency	Biohydrogen yield	Biomethane yield	Reference
Temperature: 37 °C pH: No pH control		<0.1 mL/gVS 13.8 mL/gVS <0.1 mL/gVS 18.9 mL/gVS	195.7 mL/gVS 216.7 mL/gVS 136.6 mL/gVS 243.5 mL/gVS	Corneli et al. (2016)
Temperature: 37°C pH: 7.0		8.29 ± 0.33 mL/gVS	434.38 ± 5.72 mL/gVS	Wu et al. (2020)
Temperature: 35 °C HRT: 9 days pH: 7.3–7.8		135mL/g VS	510 mL/g VS	Algapani et al. (2019)
Temperature: 35 °C HRT: 30 days pH: 8.0		174.6 mL/gVS	113.3–264.1 mL/gVS	Cheng et al. (2020)
Temperature: 55 °C HRT: 14.4 days pH: 7.0		1.9–2.6 mol/mol hexose	1.0e1.2 g-COD/g-COD_{Cr}	Kobayashi et al. (2012)
HRT: 22h–24h pH:7.5	82% COD removal and 96% VSS decomposition	282 mL/g COD	389 mL/g COD	Ueno et al. (2007)
Temperature: 35 °C HRT: 35days	66.27%	86 mL/g-COD	250.33 mL/g-COD	Sani et al. (2021)
Temperature: 37 °C pH: 7.52	93% reduction in COD	Biohythane gas with the composition of 46-57% H_2, 43-54% CH_4 and 0.4% CO_2		Willquist et al. (2012)

TABLE 9.3
Microorganisms Involved in Different Phases of the AD Process

Phase	Family	Phylum	Class	Metabolic functions	Reference
Hydrolysis	Clostridiaceae	Firmicutes	Clostridia	Degradation of cellulose and hemicellulose	Chirania et al. (2022)
	Bacteroidaceae	Bacteroidetes	Bacteroidia	Carbohydrate degradation	Chirania et al. (2022)
Acidogenesis	Erysipelotrichaceae	Firmicutes	Erysipelotrichia	Glucose fermentation with major metabolic end products as butyrate, lactate and formate	De Maesschalck et al. (2014)
	Clostridiaceae	Firmicutes	Clostridia	Glucose fermentation with end products as hydrogen, butyrate, acetate, and lactate	Lin et al. (2007)
	Bacteroidaceae	Bacteroidetes	Bacteroidia	Glucose fermentation with end products as acetate, CO_2, and hydrogen	Khan et al. (1980)
	Veillonellaceae	Firmicutes	Negativicutes	Glucose fermentation with end products as acetate, propionate, isobutyrate, butyrate and isovalerate	Marchandin and Jumas-Bilak (2014)
	Synergistaceae	Synergistetes	Synergistia	Fermentation of glucose and organic acids	Honda et al. (2013)
	Thermotogaceae	Thermotogae	Thermotogae	Fermentation of carbohydrates and peptides	Hao and Wang (2015)
	Ruminococcaceae	Firmicutes	Clostridia	Fermentation of glucose with end products as hydrogen and VFAs	Liu et al. (2015)
Acetogenesis	Syntrophaceae	Proteobacteria	Deltaproteobacteria	Propionate and butyrate-utilizing bacteria	Moertelmaier et al. (2014)
	Syntrophomonadaceae	Firmicutes	Clostridia	Bacteria involved in syntrophic association with hydrogenotrophic methanogens	Morotomi et al. (2011)
	Desulfovibrionaceae	Proteobacteria	Deltaproteobacteria	Utilization of lactate and pyruvate to form acetate, hydrogen and CO_2	Zhao et al. (2012)
Methanogenesis	Methanosaetaceae	Euryarchaeota	Methanomicrobia	Acetoclastic methanogens	Liu and Whitman, 2008)
	Methanobacteriaceae	Euryarchaeota	Methanobacteria	Hydrogenotrophic methanogens	Liu and Whitman (2008)
	Methanospirillaceae	Euryarchaeota	Methanomicrobia	Hydrogenotrophic methanogens	Liu and Whitman (2008)

pH of the system is the use of RTPAD. However, high ammonia concentration and methanogens present in the recirculated AD reject wastewater is usually a course of concern. Hence, there is need for more studies not just to assess the influence of recirculating the AD reject wastewater on biohydrogen generation but also to improve the quality of the wastewater that is being circulated.

In addition, the reuse of digestate from the two-stage AD process is another major challenge. For conventional single-phase AD process, the digestate obtained as the end product can be recycled back into the bioreactor for buffering. However, for two-stage anaerobic digestion, reusing the digestate in the first stage will have inimical effect on the microbial consortium in the bioreactor due to the presence of methane-producing microorganisms in the digestate. Hence, there is need for more research to obtain possible solutions for improving the quality of the digestate produced in the second stage of two-phase anaerobic digestion so that it can be recycled.

Further, OFMSW is heterogeneous in nature, has complex structure and high solids content, and this is a limitation in the production of bioenergy from its AD as there is a lack of effective digesters for handling its high solids content. Digesters which are commonly used for OFMSW are those that have been developed for the treatment of biosolids which have less solids content (4% to 6% total solids) than OFMSW. Hence, there is need for research on reactor configurations which incorporate new mixing and conveying mechanisms inside the digesters to avoid settlement of solids and improve contact between the substrate and microorganisms.

9.9 CONCLUSION

Rapid population growth, urbanization, and industrialization are partly responsible for large quantities of MSW and increased demand for energy. The sequential biohydrogen and biomethane generation from OFMSW using the two-stage AD process has the dual advantage of managing the waste while generating energy in a sustainable manner. However, despite the fact that the two-stage AD process brings about increased yield of biohydrogen and biogas at low HRT and hence smaller and inexpensive reactors, it still has operational challenges such as high pretreatment costs, inhibitory compounds, and low process stability and this has impaired the scaling up of this technology. Thus, it is pertinent that more research is carried out on the design of two-stage reactors for optimized biohydrogen and biomethane generation from OFMSW. Apart from optimizing the design of bioreactors, genetic and metabolic engineering techniques can be used to identify and develop microbial communities that can be used to improve the yield of bioenergy from AD of OFMSW.

REFERENCES

Abe, J.O. et al. (2019) 'Hydrogen energy, economy and storage: Review and recommendation', *International Journal of Hydrogen Energy*, 44(29), pp. 15072–15086. Available at: https://doi.org/10.1016/j.ijhydene.2019.04.068

Adekunle, K. F., and Okolie, J. A. (2015). A review of biochemical process of anaerobic digestion. *Advances in Bioscience and Biotechnology*, 6(03), 205–212. Available at: http://dx.doi.org/10.4236/abb.2015.63020

Algapani, D. E. et al. (2019). Bio-hydrogen and bio-methane production from food waste in a two-stage anaerobic digestion process with digestate recirculation. *Renewable Energy*, 130, 1108–1115. Available at: https://doi.org/10.1016/j.renene.2018.08.079

André, L. et al. (2016) 'Methane production improvement by modulation of solid phase immersion in dry batch anaerobic digestion process: Dynamic of methanogen populations', *Bioresource Technology*, 207, pp. 353–360. Available at: https://doi.org/10.1016/j.biortech.2016.02.033

Appels, L. et al. (2008) 'Principles and potential of the anaerobic digestion of waste-activated sludge', *Progress in Energy and Combustion Science*, 34(6), pp. 755–781. Available at: https://doi.org/10.1016/j.pecs.2008.06.002

Ariunbaatar, J., Scotto Di Perta, E., et al. (2015) 'Effect of ammoniacal nitrogen on one-stage and two-stage anaerobic digestion of food waste', *Waste Management,* 38(1), pp. 388–398. Available at: https://doi.org/10.1016/j.wasman.2014.12.001

Bouabidi, Z.B., El-Naas, M.H. and Zhang, Z. (2019) 'Immobilization of microbial cells for the biotreatment of wastewater: A review', *Environmental Chemistry Letters*, 17(1), pp. 241–257. Available at: https://doi.org/10.1007/s10311-018-0795-7

Brodeur, G. et al. (2011). Chemical and physicochemical pretreatment of lignocellulosic biomass: a review. *Enzyme Research*, 2011. 17, 1–17. Available at: https://doi:10.4061/2011/787532

Buitrón, G., Muñoz-Páez, K.M. and Hernández-Mendoza, C.E. (2019) 'Biohydrogen production using a granular sludge membrane bioreactor', *Fuel,* 241(July 2018), pp. 954–961. Available at: https://doi.org/10.1016/j.fuel.2018.12.104

Cavinato, C. et al. (2012) 'Bio-hythane production from food waste by dark fermentation coupled with anaerobic digestion process: A long-term pilot scale experience', *International Journal of Hydrogen Energy*, 37(15), pp. 11549–11555. Available at: https://doi.org/10.1016/j.ijhydene.2012.03.065

Cesaro, A. et al. (2014) 'Enhanced anaerobic digestion by ultrasonic pretreatment of organic residues for energy production', *Journal of Cleaner Production,* 74, pp. 119–124. Available at: https://doi.org/10.1016/j.jclepro.2014.03.030

Chang, F.-Y., and Lin, C.-Y. (2004). Biohydrogen production using an up-flow anaerobic sludge blanket reactor. *International Journal of Hydrogen Energy*, *29*(1), 33–39. Available at: https://doi.org/10.1016/S0360-3199(03)00082-X

Cheng, H. et al. (2020). Advanced methanogenic performance and fouling mechanism investigation of a high-solid anaerobic membrane bioreactor (AnMBR) for the co-digestion of food waste and sewage sludge. *Water Research*, 187, 116436. Available at: https://doi.org/10.1016/j.watres.2020.116436

Chirania, P. et al. (2022) 'Metaproteomics reveals enzymatic strategies deployed by anaerobic microbiomes to maintain lignocellulose deconstruction at high solids', *Nature Communications*, 13(1), pp. 1–13. Available at: https://doi.org/10.1038/s41467-022-31433-x

Chozhavendhan, S. et al. (2020) 'A review on feedstock, pretreatment methods, influencing factors, production and purification processes of bio-hydrogen production', *Case Studies in Chemical and Environmental Engineering*, 2(July), p. 100038. Available at: https://doi.org/10.1016/j.cscee.2020.100038

Chu, C.-F. et al. (2008). A pH- and temperature-phased two-stage process for hydrogen and methane production from food waste. *International Journal of Hydrogen Energy*, 33(18), 4739–4746. Available at: https://doi.org/10.1016/j.ijhydene.2008.06.060

Corneli, E. et al.. (2016). Energy conversion of biomass crops and agroindustrial residues by combined biohydrogen/biomethane system and anaerobic digestion. *Bioresource Technology*, *211*, 509–518. Available at: https://doi.org/10.1016/j.biortech.2016.03.134

De Maesschalck, C. et al. (2014) '*Faecalicoccus acidiformans* gen. nov., Sp. Nov., Isolated from the chicken caecum, And reclassification of *streptococcus pleomorphus* (Barnes et al. 1977), *Eubacterium biforme* (Eggerth 1935) and *eubacterium cylindroides* (Cato et al. 1974) as *faecalicoccus pleomorphus* comb. nov., *Holdemanella biformis* gen. nov., Comb. nov. and *faecalitalea cylindroides* gen. nov., Comb. nov., Respectively, Within the family *erysipelotrichaceae*', *International Journal of Systematic and Evolutionary Microbiology,* 64, pp. 3877–3884. Available at: https://doi.org/10.1099/ijs.0.064626-0

Esposito, G. et al. (2011) 'Modelling the effect of the OLR and OFMSW particle size on the performances of an anaerobic co-digestion reactor', *Process Biochemistry*, 46(2), pp. 557–565. Available at: https://doi.org/10.1016/j.procbio.2010.10.010

Fdez.-Güelfo, L.A. et al. (2011) 'Biological pretreatment applied to industrial organic fraction of municipal solid wastes (OFMSW): Effect on anaerobic digestion', *Chemical Engineering Journal*, 172(1), pp. 321–325. Available at: https://doi.org/10.1016/j.cej.2011.06.010

Forster-Carneiro, T., Pérez, M. and Romero, L.I. (2008) 'Anaerobic digestion of municipal solid wastes: Dry thermophilic performance', *Bioresource Technology*, 99(17), pp. 8180–8184. Available at: https://doi.org/10.1016/j.biortech.2008.03.021

Franke-Whittle, I.H. et al. (2014) 'Investigation into the effect of high concentrations of volatile fatty acids in anaerobic digestion on methanogenic communities', *Waste Management*, 34(11), pp. 2080–2089. Available at: https://doi.org/10.1016/j.wasman.2014.07.020

García-Peña, E.I. et al. (2009) 'Semi-continuous biohydrogen production as an approach to generate electricity', *Bioresource Technology*, 100(24), pp. 6369–6377. Available at: https://doi.org/10.1016/j.biortech.2009.07.033

Ghimire, A. et al. (2015) 'A review on dark fermentative biohydrogen production from organic biomass: Process parameters and use of by-products', *Applied Energy*, 144, pp. 73–95. Available at: https://doi.org/10.1016/j.apenergy.2015.01.045

Guo, X.M. et al. (2014) 'Predictive and explicative models of fermentative hydrogen production from solid organic waste: Role of butyrate and lactate pathways', *International Journal of Hydrogen Energy*, 39(14), pp. 7476–7485. Available at: https://doi.org/10.1016/j.ijhydene.2013.08.079

Hans, M. and Kumar, S. (2019) 'Biohythane production in two-stage anaerobic digestion system', *International Journal of Hydrogen Energy*, 44, pp. 17363–17380. Available at: https://doi.org/10.1016/j.ijhydene.2018.10.022

Hao, J. and Wang, H. (2015) 'Volatile fatty acids productions by mesophilic and thermophilic sludge fermentation: Biological responses to fermentation temperature', *Bioresource Technology*, 175, pp. 367–373. Available at: https://doi.org/10.1016/j.biortech.2014.10.106

Hasibar, B. et al. (2020) 'Increasing biohydrogen production with the use of a co-culture inside a microbial electrolysis cell', *Biochemical Engineering Journal*, 164(September), pp. 1–6. Available at: https://doi.org/10.1016/j.bej.2020.107802

Henard, C.A. et al. (2018) 'Biogas biocatalysis: Methanotrophic bacterial cultivation, metabolite profiling, and bioconversion to lactic acid', *Frontiers in Microbiology*, 9(October), pp. 1–8. Available at: https://doi.org/10.3389/fmicb.2018.02610

Honda, T., Fujita, T. and Tonouchi, A. (2013) 'Aminivibrio pyruvatiphilus gen. nov., sp. Nov., an anaerobic, amino-acid-degrading bacterium from soil of a Japanese rice field', *International Journal of Systematic and Evolutionary Microbiology*, 63(PART10), pp. 3679–3686. Available at: https://doi.org/10.1099/ijs.0.052225-0

Intanoo, P., Chaimongkol, P., and Chavadej, S. (2016). Hydrogen and methane production from cassava wastewater using two-stage upflow anaerobic sludge blanket reactors (UASB) with an emphasis on maximum hydrogen production. *International Journal of Hydrogen Energy*, 41(14), 6107–6114. Available at: https://doi.org/10.1016/j.ijhydene.2015.10.125

Izumi, K. et al. (2010) 'Effects of particle size on anaerobic digestion of food waste', *International Biodeterioration and Biodegradation*, 64(7), pp. 601–608. Available at: https://doi.org/10.1016/j.ibiod.2010.06.013

Jafari, O. and Zilouei, H. (2016). Enhanced biohydrogen and subsequent biomethane production from sugarcane bagasse using nano-titanium dioxide pretreatment. *Bioresource Technology*, 214, 670–678. Available at: https://doi.org/10.1016/j.biortech.2016.05.007

Jain, S. et al. (2015) 'A comprehensive review on operating parameters and different pretreatment methodologies for anaerobic digestion of municipal solid waste', *Renewable and Sustainable Energy Reviews,* 52, pp. 142–154. Available at: https://doi.org/10.1016/j.rser.2015.07.091

Jiang, H. et al. (2018). Bio-hythane production from cassava residue by two-stage fermentative process with recirculation. *Bioresource Technology*, 247, 769–775. Available at: https://doi.org/10.1016/j.biortech.2017.09.102

Jiang, J. et al. (2020) 'Effect of organic loading rate and temperature on the anaerobic digestion of municipal solid waste: Process performance and energy recovery', *Frontiers in Energy Research*, 8(May), pp. 1–10. Available at: https://doi.org/10.3389/fenrg.2020.00089

Jin, Y., Li, Y. and Li, J. (2016) 'Influence of thermal pretreatment on physical and chemical properties of kitchen waste and the efficiency of anaerobic digestion', *Journal of Environmental Management,* 180, pp. 291–300. Available at: https://doi.org/10.1016/j.jenvman.2016.05.047

Jung, H., Kim, D., Choi, H., and Lee, C. (2022). A review of technologies for in-situ sulfide control in anaerobic digestion. *Renewable and Sustainable Energy Reviews*, 157, 112068. Available at: https://doi.org/10.1016/j.rser.2021.112068

Kapdan, I.K. and Kargi, F. (2006) 'Bio-hydrogen production from waste materials', *Enzyme and Microbial Technology*, 38(5), pp. 569–582. Available at: https://doi.org/10.1016/j.enzmictec.2005.09.015

Karthikeyan, O.P. and Visvanathan, C. (2013) 'Bio-energy recovery from high-solid organic substrates by dry anaerobic bio-conversion processes: A review', *Reviews in Environmental Science and Biotechnology*, 12(3), pp. 257–284. Available at: https://doi.org/10.1007/s11157-012-9304-9

Khalid, A. et al. (2011) 'The anaerobic digestion of solid organic waste', *Waste Management*, 31(8), pp. 1737–1744. Available at: https://doi.org/10.1016/j.wasman.2011.03.021

Khan, A.W. et al. (1980) 'Degradation of cellulose by a newly isolated mesophilic anaerobe, bacteroidaceae family', *FEMS Microbiology Letters*, 7(1), pp. 47–50. Available at: https://doi.org/10.1111/j.1574-6941.1980.tb01574.x

Khan, M.A. et al. (2016) 'Optimization of process parameters for production of volatile fatty acid, biohydrogen and methane from anaerobic digestion', *Bioresource Technology*, 219, pp. 738–748. Available at: https://doi.org/10.1016/j.biortech.2016.08.073

Kobayashi, T. et al. (2012) 'Effect of sludge recirculation on characteristics of hydrogen production in a two-stage hydrogen-methane fermentation process treating food wastes', *International Journal of Hydrogen Energy*, 37(7), pp. 5602–5611. Available at: https://doi.org/10.1016/j.ijhydene.2011.12.123

Kongjan, P., O-Thong, S. and Angelidaki, I. (2011) 'Performance and microbial community analysis of two-stage process with extreme thermophilic hydrogen and thermophilic methane production from hydrolysate in UASB reactors', *Bioresource Technology*, 102(5), pp. 4028–4035. Available at: https://doi.org/10.1016/j.biortech.2010.12.009

Koutrouli, E.C. et al. (2009) 'Hydrogen and methane production through two-stage mesophilic anaerobic digestion of olive pulp', *Bioresource Technology*, 100(15), pp. 3718–3723. Available at: https://doi.org/10.1016/j.biortech.2009.01.037

Kvesitadze, G. et al. (2012) 'Two-stage anaerobic process for bio-hydrogen and bio-methane combined production from biodegradable solid wastes', *Energy*, 37(1), pp. 94–102. Available at: https://doi.org/10.1016/j.energy.2011.08.039

Lee, H.S., Vermaas, W.F.J. and Rittmann, B.E. (2010) 'Biological hydrogen production: Prospects and challenges', *Trends in Biotechnology*, 28(5), pp. 262–271. Available at: https://doi.org/10.1016/j.tibtech.2010.01.007

Lim, S.L., Lee, L.H. and Wu, T.Y. (2016) 'Sustainability of using composting and vermicomposting technologies for organic solid waste biotransformation: Recent overview, greenhouse gases emissions and economic analysis', *Journal of Cleaner Production,* 111, pp. 262–278. Available at: https://doi.org/10.1016/j.jclepro.2015.08.083

Lin, P.Y. et al. (2007) 'Biological hydrogen production of the genus Clostridium: Metabolic study and mathematical model simulation', *International Journal of Hydrogen Energy*, 32(12), pp. 1728–1735. Available at: https://doi.org/10.1016/j.ijhydene.2006.12.009

Liu, Y. and Whitman, W.B. (2008) 'Metabolic, phylogenetic, and ecological diversity of the methanogenic archaea', *Annals of the New York Academy of Sciences*, 1125, pp. 171–189. Available at: https://doi.org/10.1196/annals.1419.019

Liu, Z. et al. (2013) 'States and challenges for high-value biohythane production from waste biomass by dark fermentation technology', *Bioresource Technology*, 135, pp. 292–303. Available at: https://doi.org/10.1016/j.biortech.2012.10.027

Liu, Z. et al. (2015) 'Effects of furan derivatives on biohydrogen fermentation from wet steam-exploded cornstalk and its microbial community', *Bioresource Technology*, 175, pp. 152–159. Available at: https://doi.org/10.1016/j.biortech.2014.10.067

Malave', A.C.L. et al. (2015) 'Multistep anaerobic digestion (MAD) as a tool to increase energy production via H2 + CH4', *International Journal of Hydrogen Energy*, 40(15), pp. 5050–5061. Available at: https://doi.org/10.1016/j.ijhydene.2015.02.068

Mancini, G. et al. (2018) 'Increased biogas production from wheat straw by chemical pretreatments', *Renewable Energy*, 119, pp. 608–614. Available at: https://doi.org/10.1016/j.renene.2017.12.045

Manga, M. (2017). 'The Feasibility of Co-Composting as an Upscale Treatment Method for Faecal Sludge in Urban Africa', PhD thesis, School of Civil Engineering, University of Leeds: Leeds, UK. Available online: https://etheses.whiterose.ac.uk/16997 (accessed on 3 November 2022).

Manga, M., Aragón-Briceño, C., Boutikos, P., Semiyaga, S., Olabinjo, O., & Muoghalu, C. C. (2023). Biochar and Its Potential Application for the Improvement of the Anaerobic Digestion Process: A Critical Review. *Energies*, 16(10), 4051. Available at: https://doi.org/10.3390/en16104051

Manga, M.; Camargo-Valero, M.A.; Anthonj, C.; Evans, B.E. (2021) 'Fate of faecal pathogen indicators during faecal sludge composting with different bulking agents in tropical climate', *International Journal of Hygiene and Environmental Health*, 232, 113670, Available at: https://doi.org/10.1016/j.ijheh.2020.113670.

Manga, M., Camargo-Valero, M.A., and Evans, B.E. (2019) 'Inactivation of viable Ascaris eggs during faecal sludge co-composting with chicken feathers and market waste', *Desalination and Water Treatment-Science and Engineering* 2019, 163, 347–357, Available at: https://doi.org/10.5004/dwt.2019.24494

Mao, C. et al. (2015) 'Review on research achievements of biogas from anaerobic digestion', *Renewable and Sustainable Energy Reviews*, 45, pp. 540–555. Available at: https://doi.org/10.1016/j.rser.2015.02.032

Marchandin H., Jumas-Bilak E. (2014) 'The family Veillonellaceae'. In: Rosenberg E, DeLong EF, Lory S, Stackebrandt E, Thompson F (eds) *The prokaryotes: Firmicutes and Tenericutes*. Springer, Berlin Heidelberg, Berlin, Heidelberg, pp 433–453.

Mari, A. G. et al. (2020). Biohydrogen and biomethane production from cassava wastewater in a two-stage anaerobic sequencing batch biofilm reactor. *International Journal of Hydrogen Energy*, 45(8), 5165–5174. Available at: https://doi.org/10.1016/j.ijhydene.2019.07.054

Massanet-Nicolau, J., Dinsdale, R., Guwy, A., and Shipley, G. (2013). Use of real time gas production data for more accurate comparison of continuous single-stage and two-stage fermentation. *Bioresource Technology*, 129, 561–567. Available at: https://doi.org/10.1016/j.biortech.2012.11.102

Mathew, A.K. et al. (2015) 'Biogas production from locally available aquatic weeds of Santiniketan through anaerobic digestion', *Clean Technologies and Environmental Policy*, 17(6), pp. 1681–1688. Available at: https://doi.org/10.1007/s10098-014-0877-6

Mischopoulou, M. et al. (2016). Effect of ultrasonic and ozonation pretreatment on methane production potential of raw molasses wastewater. *Renewable Energy*, 96, 1078–1085. Available at: https://doi.org/10.1016/j.renene.2015.11.060

Mizuno, O. et al. (2000) 'Characteristics of hydrogen production from bean curd manufacturing waste by anaerobic microflora', *Water Science and Technology*, 42(3–4), pp. 345–350. Available at: https://doi.org/10.2166/wst.2000.0401

Moertelmaier, C. et al. (2014) 'Fatty acid metabolism and population dynamics in a wet biowaste digester during re-start after revision', *Bioresource Technology*, 166, pp. 479–484. Available at: https://doi.org/10.1016/j.biortech.2014.05.085

Morotomi, M., Nagai, F. and Watanabe, Y. (2011) 'Description of Christensenella minuta gen. nov., sp. Nov., isolated from human faeces, which forms a distinct branch in the order Clostridiales, and proposal of Christensenellaceae fam. Nov', *International Journal of Systematic and Evolutionary Microbiology*, 62(1), pp. 144–149. Available at: https://doi.org/10.1099/ijs.0.026989-0

Motte, J. C. et al. (2015). Dynamic observation of the biodegradation of lignocellulosic tissue under solid-state anaerobic conditions. *Bioresource Technology*, 191, 322–326. Available at: https://doi.org/10.1016/j.biortech.2015.04.130

Mu, L. et al. (2018) 'Semi-continuous anaerobic digestion of extruded OFMSW: Process performance and energetics evaluation', *Bioresource Technology*, 247(September 2017), pp. 103–115. Available at: https://doi.org/10.1016/j.biortech.2017.09.085

Muoghalu, C., Owusu, A., Nakagiri, A., Semiyaga, S., Labu, S., Iorhemen, O.T., Manga, M. (2023) 'Biochar as a novel technology for treatment of onsite domestic wastewater: A critical review', *Frontiers in Environmental Science 11*, 202, Available at: https://doi.org/10.3389/fenvs.2023.1095920

Nath, K. and Das, D. (2006) 'Amelioration of biohydrogen production by a two-stage fermentation process', *Industrial Biotechnology*, 2(1), pp. 44–47. Available at: https://doi.org/10.1089/ind.2006.2.44

Ngo, T.A., Nguyen, T.H. and Bui, H.T.V. (2012) 'Thermophilic fermentative hydrogen production from xylose by Thermotoga neapolitana DSM 4359', *Renewable Energy*, 37(1), pp. 174–179. Available at: https://doi.org/10.1016/j.renene.2011.06.015

Nguyen, L.N., Nguyen, A.Q. and Nghiem, L.D. (2019) 'Microbial Community in Anaerobic Digestion System: Progression in Microbial Ecology', in X.-T. Bui et al. (eds) *Water and wastewater treatment technologies*. Singapore: Springer Nature, pp. 331–355.

Panigrahi, S. and Dubey, B.K. (2019) 'A critical review on operating parameters and strategies to improve the biogas yield from anaerobic digestion of organic fraction of municipal solid waste', *Renewable Energy,* 143, pp. 779–797. Available at: https://doi.org/10.1016/j.renene.2019.05.040

Rawoof, S.A.A. et al. (2021) 'Sequential production of hydrogen and methane by anaerobic digestion of organic wastes: a review', *Environmental Chemistry Letters*, 19(2), pp. 1043–1063. Available at: https://doi.org/10.1007/s10311-020-01122-6

Sajeena Beevi, B., Madhu, G. and Sahoo, D.K. (2015) 'Performance and kinetic study of semi-dry thermophilic anaerobic digestion of organic fraction of municipal solid waste', *Waste Management,* 36, pp. 93–97. Available at: https://doi.org/10.1016/j.wasman.2014.09.024

Sani, K. et al. (2021). Effectiveness of using two-stage anaerobic digestion to recover bioenergy from high strength palm oil mill effluents with simultaneous treatment. *Journal of Water Process Engineering*, 39, 101661. Available at: https://doi.org/10.1016/j.jwpe.2020.101661

Sarto, S., Hildayati, R. and Syaichurrozi, I. (2019) 'Effect of chemical pretreatment using sulfuric acid on biogas production from water hyacinth and kinetics', *Renewable Energy,* 132, pp. 335–350. Available at: https://doi.org/10.1016/j.renene.2018.07.121

Satari, B., Karimi, K. and Kumar, R. (2019) *Cellulose solvent-based pretreatment for enhanced second-generation biofuel production: A review, Sustainable Energy and Fuels*. Royal Society of Chemistry. Available at: https://doi.org/10.1039/c8se00287h

Semiyaga, S., Nakagiri, A., Niwagaba, C.B. and Manga, M. (2022) 'Application of Anaerobic Digestion in Decentralized Faecal Sludge Treatment Plants'. In *Anaerobic Biodigesters for Human Waste Treatment* (pp. 263–281). Springer, Singapore. Available at: https://doi.org/10.1007/978-981-19-4921-0_14

Show, K.Y., Lee, D.J. and Chang, J.S. (2011) 'Bioreactor and process design for biohydrogen production', *Bioresource Technology*, 102(18), pp. 8524–8533. Available at: https://doi.org/10.1016/j.biortech.2011.04.055

Si, B. et al. (2016) 'Towards biohythane production from biomass: Influence of operational stage on anaerobic fermentation and microbial community', *International Journal of Hydrogen Energy*, 41(7), pp. 4429–4438. Available at: https://doi.org/10.1016/j.ijhydene.2015.06.045

Stamatelatou, K., Antonopoulou, G., and Michailides, P. (2014). 15 - Biomethane and biohydrogen production via anaerobic digestion/fermentation. In K. Waldron (Ed.), *Advances in Biorefineries* (pp. 476–524). Cambridge, UK: Woodhead Publishing. Available at: https://doi.org/10.1533/9780857097385.2.476

Stams, A.J.M. and Plugge, C.M. (2009) 'Electron transfer in syntrophic communities of anaerobic bacteria and archaea', *Nature Publishing Group*, 7(auguST). Available at: https://doi.org/10.1038/nrmicro2166

Sun, Y. et al. (2019) 'A review of the enhancement of bio-hydrogen generation by chemicals addition', *Catalysts*, 9(4). Available at: https://doi.org/10.3390/catal9040353

Tiwari, M. K., Guha, S., Harendranath, C. S., & Tripathi, S. (2006). Influence of extrinsic factors on granulation in UASB reactor. Applied Microbiology and Biotechnology, 71(2), 145–154. Available at: https://doi.org/10.1007/s00253-006-0397-3

Ueno, Y. et al. (2007) 'Production of hydrogen and methane from organic solid wastes by phase-separation of anaerobic process', *Bioresource Technology*, 98(9), pp. 1861–1865. Available at: https://doi.org/10.1016/j.biortech.2006.06.017

Velásquez Piñas, J. A., Venturini, O. J., Silva Lora, E. E., and Calle Roalcaba, O. D. (2018). Technical assessment of mono-digestion and co-digestion systems for the production of biogas from anaerobic digestion in Brazil. *Renewable Energy*, 117, 447–458. Available at: https://doi.org/10.1016/j.renene.2017.10.085

Vögeli, Y., Lohri, C.R., Gallardo, A., Diener, S. and Zurbrügg, C. (2014). Anaerobic Digestion of Biowaste in Developing Countries: Practical Information and Case Studies. Dübendorf, Switzerland: Swiss Federal Institute of Aquatic Science and Technology (Eawag). Available at www.eawag.ch/fileadmin/Domain1/Abteilungen/sandec/publikationen/SWM/Anaerobic_Digestion/biowaste.pdf

Wang, W. et al. (2010) 'Performance and stability improvements in anaerobic digestion of thermally hydrolyzed municipal biowaste by a biofilm system', *Bioresource Technology*, 101(6), pp. 1715–1721. Available at: https://doi.org/10.1016/j.biortech.2009.10.010

Watts, N. et al. (2018) 'The Lancet Countdown on health and climate change: from 25 years of inaction to a global transformation for public health', *The Lancet*, 391(10120), pp. 581–630. Available at: https://doi.org/10.1016/S0140-6736(17)32464-9

Willquist, K. et al. (2012) 'Design of a novel biohythane process with high H 2 and CH 4 production rates', *International Journal of Hydrogen Energy*, 37(23), pp. 17749–17762. Available at: https://doi.org/10.1016/j.ijhydene.2012.08.092

Wu, H., Li, J., Wang, C., Liao, Q., Fu, Q., and Liu, Z. (2020). Sequent production of proteins and biogas from Chlorella sp. via CO2 assisted hydrothermal treatment and anaerobic digestion. *Journal of Cleaner Production*, 277, 123563. Available at: https://doi.org/10.1016/j.jclepro.2020.123563

Wu, L.J. et al. (2015) 'Upgrading of anaerobic digestion of waste activated sludge by temperature-phased process with recycle', *Energy*, 87(34), pp. 381–389. Available at: https://doi.org/10.1016/j.energy.2015.04.110

Xie, S. et al. (2016) 'Anaerobic co-digestion: A critical review of mathematical modelling for performance optimization', *Bioresource Technology*, 222, pp. 498–512. Available at: https://doi.org/10.1016/j.biortech.2016.10.015

Xu, S. et al. (2016) 'Effects of high-pressure extruding pretreatment on MSW upgrading and hydrolysis enhancement', *Waste Management*, 58, pp. 81–89. Available at: https://doi.org/10.1016/j.wasman.2016.07.012

Yang, L. et al. (2015) 'Challenges and strategies for solid-state anaerobic digestion of lignocellulosic biomass', *Renewable and Sustainable Energy Reviews*, 44, pp. 824–834. Available at: https://doi.org/10.1016/j.rser.2015.01.002

Yang, Z. et al. (2011) 'Hydrogen and methane production from lipid-extracted microalgal biomass residues', *International Journal of Hydrogen Energy*, 36(5), pp. 3465–3470. Available at: https://doi.org/10.1016/j.ijhydene.2010.12.018

Ye, Z. and Berson, R.E. (2014) 'Factors affecting cellulose hydrolysis based on inactivation of adsorbed enzymes', *Bioresource Technology,* 167, pp. 582–586. Available at: https://doi.org/10.1016/j.biortech.2014.06.070

Zainal, B.S. et al. (2018) 'Effects of process, operational and environmental variables on biohydrogen production using palm oil mill effluent (POME)', *International Journal of Hydrogen Energy*, 43(23), pp. 10637–10644. Available at: https://doi.org/10.1016/j.ijhydene.2017.10.167

Zhao, C. et al. (2012) 'Desulfobaculum xiamenensis gen. nov., sp. nov., a member of the family Desulfovibrionaceae isolated from marine mangrove sediment', *International Journal of Systematic and Evolutionary Microbiology*, 62(7), pp. 1570–1575. Available at: https://doi.org/10.1099/ijs.0.036632-0

Zhong, J., Stevens, D.K. and Hansen, C.L. (2015) 'Optimization of anaerobic hydrogen and methane production from dairy processing waste using a two-stage digestion in induced bed reactors (IBR)', *International Journal of Hydrogen Energy*, 40(45), pp. 15470–15476. Available at: https://doi.org/10.1016/j.ijhydene.2015.09.085

10 Sustainable Biomethanation Process for Energy Recovery from Faecal Sludge

A Promising Solution for India's Sanitation Challenges

Atun Roy Choudhury, Neha Singh, Kasturi Bidkar, Subhasmita Sahoo, S. Hemapriya, Jitesh Lalwani, and Sankar Ganesh Palani

10.1 INTRODUCTION

Faecal sludge (FS) refers to a mixture of human excreta, water, and other solid wastes such as toilet paper or other anal cleansing materials and menstrual hygiene products that are disposed of in onsite sanitation systems (OSS). Septage is a form of FS that refers to a partially digested slurry that results from the collection, storage, or treatment of combinations of excreta and blackwater, with or without greywater that builds up in septic tanks. FS and septage are far more concentrated than sewage waste. The BOD of septage ranges from 1,000 to 20,000 mg/l. This material has a foul odour, an unpleasant appearance, and high quantities of grease, grit, hair, debris, and several pathogens and causes surface and groundwater pollution (Samal et al., 2022).

Anaerobic digestion improves sludge stability, reduces pathogens and odour emissions, and reduces sludge dry matter, resulting in a considerable reduction in final sludge volume. Anaerobic treatment is a biological process that uses microorganisms to break down organic pollutants found in wastewater in the absence of oxygen. During anaerobic digestion, the organic matter is transformed into biogas and the residual sludge is known as slurry or digestate. Biogas is mainly composed of methane and carbon dioxide, and the digestate is biologically stable and can be utilized as a soil conditioner (Strande et al., 2018).

Anaerobic digestion occurs in a sealed vessel, to ensure anaerobic conditions, known as a reactor, which is designed and built in a variety of forms and sizes based on the site and feedstock conditions. Anaerobic digestion has been widely used in centralized wastewater treatment facilities to digest primary sludge and waste-activated

DOI: 10.1201/9781003364467-10

sludge, often using plug flow (PFR) or continuously stirred reactors (CSTRs). Up-flow anaerobic sludge blanket (UASB) reactors, anaerobic baffled reactors (ABRs), and anaerobic filters are more examples of anaerobic treatment systems. Anaerobic treatment is also widely used for industrial wastes and high-load wastewater treatment plants (e.g., agro-industries) (Arthur and Brew-Hammond, 2010). The onsite anaerobic digestion of animal manure, with or without the addition of FS, is routinely used throughout Asia (Koottatep et al., 2005). However, the potential for semi-centralized to centralized FS treatment in urban areas remains unexplored. The anaerobic digestion of FS has enormous potential for future growth.

The following operating conditions are critical in the design and operation of anaerobic digesters:

- Solids retention time (SRT)
- Hydraulic retention time (HRT)
- Temperature
- Alkalinity
- pH
- Toxic/inhibiting substances
- Bioavailability of nutrients and trace elements; and
- Loading pattern

While constructing an anaerobic reactor, it is crucial to know the estimated organic load in order to provide a long enough HRT for degradation to begin. For systems like plug flow reactors which do not include recycling, the SRT is equal to the HRT. The length of the HRT has a direct relationship with the anaerobic reaction: an increase or decrease in the HRT leads to an increase or decrease in the degree of hydrolysis, acidification, fermentation, and methanogenesis (Diaz-Elsayed et al., 2019). It is therefore essential to monitor the HRT in order to prevent reactor failure. At the same time, temperature influences both physical and chemical characteristics in reactors such as gas exchange, salt solubility, and pathogen inactivation. The degree and rate of hydrolysis and methane production are also affected by temperature.

Sewerage systems have been the preferred option for FSM in developing-country cities and urban areas. Sewerage networks, on the other hand, have substantial capital, operation and maintenance (O&M) expenditures (Shukla et al., 2022). The FSM technique adopted in India for census towns and urban areas focuses on septage management. For this, we need to create a new approach that has the following characteristics:

- Low capital and operational costs;
- Low-energy, low-skill, and simple-to-operate plant;
- Generates compost and treated water;
- Management techniques that integrate extreme climate technology and performance-based compensation;
- Makes use of information and communications technology (ICT) to track vehicles using GPS and mobile applications (apps) to centrally coordinate service providers.

10.2 CHARACTERISTICS OF FS

The characteristics of the FS vary drastically depending upon the source of generation and food preference of the habitants. Biodigesters, pit latrines, and septic tanks are a few widely used onsite containment systems and potential sources of FS generation. These containment systems tangentially impact pollutant concentration, pathogenicity, dry solid content, and type of digestion. However, FS tends to contain an excess concentration of nutrients regardless of its source. So, it is a mandate to treat FS to fulfil the discharge/reuse standard. In general, it has a tremendous concentration of suspended solids and pathogenicity which is roughly 10–20 times more than sewage.

One of the foremost factors that influences the overall characteristics is the duration of onsite digestion. Depending on this, FS is broadly categorized into two primary categories: fresh and digested sludge. Prolonged storage ensures better digestion, resulting in a decrease of dry solids and thereby pollution concentration. The representative parameters to ascertain the quality of the sludge are as follows: pH, Total Solids, Total Dissolved Solids, Total Suspended Solids, Total Volatile Solids, Chemical Oxygen Demand, Biological Oxygen Demand, Total Phosphorus, Total Kjeldahl Nitrogen, Total, and Faecal Coliform (Prasad et al., 2021).

The attributed parameters and their highest, lowest, and mean values related to FS collected from three different onsite containment sources are compiled in Table 10.1. Further, the comprehended parametric variability data from the existing literature is depicted in Table 10.2.

10.3 EXISTING PRACTICES OF FS TREATMENT

Since Indians are predominantly washers, septage coming for the treatment processes typically contains 96% of water and only 4% solid. However, the solid possesses major treatment challenges irrespective of its trivial count. Therefore, achieving optimal solid-liquid separation by dewatering remains the foremost treatment objective for any FS treatment technology. Further, unit operations that require careful attention are: the treatment of separated effluent and pasteurization of slurry for safe reuse or disposal. The technologies being reviewed here are therefore primarily concerned with how well they perform solid-liquid separation, how well they purify the effluent for reuse, and how well they ensure the stabilization and pasteurization of the biosolid for agricultural/horticultural applications. The assessment of the most popular FS treatment technologies is delineated below.

Pyrolysis: The pyrolysis process involves the heat breakdown of organic molecules to produce char, a carbon-rich by-product. It is known that biomass pyrolysis also produces energy-rich solid and liquid products (tars and oils) that can be utilized directly as fuels, as well as non-condensable gases, some of which release energy when burned. This process involves solid-liquid separation with a mechanical press followed by secondary and tertiary treatment of liquid, and pyrolysis of the segregated slurry. Additionally, it is reported that biochar may be able to clean water of impurities (Yacob et al., 2018).

Screw press: This process involves the physical separation of liquid and solid after coagulation and flocculation. It comprises a rotating screw that is set inside

TABLE 10.1
Physicochemical and Biological Characteristics of FS Collected from Three Different Onsite Containment Systems

			Septic Tank			Pit Latrines			Biodigesters			
Sl. No.	Parameter	UOM	Min	Max	Average	Min	Max	Average	Min	Max	Average	Reference
1.	pH @ 25^0C	-	6.50	8.82	7.66	5.70	8.22	6.96	7.4	7.7	7.55	Bassan etal.,2013; Niwagaba etal.,2014; Ahmed etal.,2019; Eliyan et al., 2022
2.	Total Solids	mg/L	2100	99000	50550	3200	158400	80,800	1300	59000	30150	
3.	Total Dissolved Solids	mg/L	950	2500	1725	1520	4250	2885	550	1700	1125	
4.	Total Suspended Solids	mg/L	870	97500	49,185	1480	156000	78,740	520	58500	29510	
5.	Total Volatile Solids	mg/L	116	1550	833	185	2500	1342	70	1080	575	
6.	Chemical Oxygen Demand	mg/L	4180	100800	52490	6270	151200	78735	2500	50500	26500	
7.	Biological Oxygen Demand	mg/L	1400	43260	22330	2380	69200	35800	840	30280	31120	
8.	Total Phosphorus	mg/L	145	875	510	261	1400	830	85	525	300	
9.	Total Kjeldahl Nitrogen	mg/L	628	2855	1741	1130	4850	2990	314	1715	1014	
10	Total coliform	MPN/100 ml	$280x10^5$	$540x10^7$	$27.14x10^6$	$43x10^6$	$70x10^8$	$352.15x10^5$	$13x10^5$	$25x10^7$	$125.65x10^4$	
11.	Faecal coliform	MPN/100 ml	$27x10^3$	$39x10^4$	$20.85x10^2$	$43x10^3$	$58x10^5$	$290.715x10^2$	$17x10^2$	$13x10^4$	$65.85x10^1$	

TABLE 10.2
Common Degree of Variation for Physicochemical and Biological Characteristics of FS Collected from Three Different Onsite Containment Systems

Sl. No.	Parameter	UOM	Septic Tank	Pit Latrines	Biodigesters	Reference
1.	pH @ 25°C	pH @ 25°C	7.10–7.50	6.10–6.98	7.35–7.60	Amoah et al., 2017; Krueger et al., 2021; Prasad et al., 2021
2.	Total Solids	mg/L	3600–8000	5500–13100	2000–4500	
3.	Total Dissolved Solids	mg/L	1500–2000	2450–3200	800–1200	
4.	Total Suspended Solids	mg/L	2053–5890	3380–9000	1100–3000	
5.	Total Volatile Solids	mg/L	240–450	360–700	150–225	
6.	Chemical Oxygen Demand	mg/L	4250–9000	6375–13500	2150–4500	
7.	Biological Oxygen Demand	mg/L	1200–3200	2100–5000	750–1750	
8.	Total Phosphorus	mg/L	60–80	95–150	25–40	
9.	Total Kjeldahl Nitrogen	mg/L	250–350	400–550	125–175	
10.	Total coliform	MPN/100 ml	34–40 x 10^7	240–540 x 10^7	110–220 x 10^6	
11.	Faecal coliform	MPN/100 ml	33–46 x 10^5	58–84 x 10^5	17–22 x 10^4	

a cylinder with holes in it. As the distance between the screw and the cylinder gets less, the liquid is forced out through the pores of the cylinder. The dewatered sludge is released at the opposite end. Screw presses offer dewatering at a comparatively cheaper price owing to its low capital and operational costs. The machine also requires very basic maintenance skills with negligible replacement of spares (Basamykina et al., 2020).

Belt-filter press: A belt filter press (BFP) enables sludge dewatering by squeezing the sludge to push the water through a permeable material. In the case of primary flocculated sludge, the procedure results in a cake (the dewatered product) with a dry solids percentage of 30% or more. The system typically consists of two to three recirculating belts, with two belts joining at a specific position to compress the sludge and squeeze out the water. A three-belt system is the foundation of major BFP systems, with the gravity belt operating independently from the two-pressurizing belt. This enables the thickening. However, the recirculation rate for each cycle requires manual intervention separately from the dewatering process. The source and qualities of the feed sludge, as well as the dosage parameters, affect the BFP's overall performance (McConville et al., 2020)

MBBR: The aerobic MBBR process is based on attached growth biodegradation using a moving media. The main goal of the MBBR development is to create an uninterrupted, non-clogging biofilm reactor that doesn't require backwashing. The system provides a high biofilm surface area with lower head loss. Efficiency is increased significantly by growing the biomass on tiny carrier elements that move with the water in the reactor. The aeration, mechanical, or hydraulic mechanisms in aerobic reactors produce movement within the reactor. A combination of anoxic and aerobic treatment mechanisms is used for the complete treatment of influent septage. The microbial treatment needs bioagents that are particularly flexible to combat high pollution loads (Osmani et al., 2021).

DEWATS: It is an organic process that stabilizes waste by increasing the time it spends in contact with active biomass. The leftover sludge and effluent are then treated and the dried sludge is used as compost. The entire process is a chemical-free biological treatment procedure. The most common methods for processing solids are stabilization reactors, sludge drying beds, and anaerobic baffled reactors with filter chambers, and planted gravel filters for treating liquids. This mechanism makes extensive use of the local topography and is largely designed as a gravity-based system with minimal electromechanical interventions. Liquid, biosolids, compost, and biogas are processed as end products to create an eco-friendly fuel (Varma et al., 2022).

Planted drying bed (PDB): Sludge is treated in PDBs using a combination of physical and biological processes. PDB is a biological system where sludge treatment and liquid treatment have naturally combined. An improvement over unplanted drying beds, planted drying beds have the added benefit of better sludge treatment and transpiration. The filters do not require desludging following each feeding or drying cycle. In a plated gravel filter, the liquid is largely treated. A gravity-based system that relies on natural treatment without the use of chemicals or energy makes up the entire end-to-end system. Planted drying beds are used to separate solids from liquids, while planted gravel filters and polishing ponds are used to treat liquids. There is the presence of multi-grade media beneath the soil layers that ensures efficient filtration and minimum percolation (Septiariva et al., 2022).

Soil Biotech: Soil Biotechnology (SBT) is an earthly system for wastewater treatment based on the principle of trickling filters. Three basic natural processes are involved in SBT: respiration, mineral weathering, and photosynthesis. The system's essential components include bioindicator plants, native microflora in the culture, and an appropriate mineral constitution. It includes a raw water tank, a bioreactor containment system, a processed water tank, pipelines, and pumps. First, the water passes through the additive layer and subsequently passes through the media. According to the desired water quality, the operation can be conducted in a single stage or multiple stages. Recirculation is offered for further polishing if needed (Kamble et al., 2017).

Phytorid: The Phytorid technology uses artificial wetland systems based on particular plants, such as elephant grass and others. The phytorid technology treatment system is a subsurface flow-type built wetland system with a cell that has baffles. Wastewater is applied to the cell/system filled with porous media such as crushed bricks, gravel, and stones, and the hydraulics is handled in such a way that wastewater does not rise to the surface, maintaining a free board at the top of the filled media. The procedure relies on the concept of bacterial action, filtration, adsorption, precipitation, and breakdown (Kumar et al., 2020).

Geotube: The Geo bag or tube (basically a non-mechanical membrane-based technology) technology is used to dry the FS from the on-site sanitation sector and is suitable for decentralized FS treatment. The distinctive geo bags are created from permeable geotextiles that may be able to dewater sludge. FS is delivered to these geo tubes/bags from tankers after being separated from undesirable items (such as plastics, pebbles, glass, and metal) and utilizing an organic flocculent to optimize floc formation, enabling the filtration and subsequently dewatering. Through underground conduits, the effluent from the geo bags is collected and transferred to a packaged sewage treatment plant for treatment (Banka et al., 2020).

Each of the above technologies has its own benefits and limitations in terms of the requirement of space, capital and operational expenditure, the requirement of consumables such as power, water, and lubricants, treatment period, ease of operation, treatment efficacy, quality and market of the by-products, etc. Based on the above criteria Roy Choudhury et al. (2021) have portrayed the hierarchy of the treatment schemes with the help of the Analytic Hierarchy Process. The comprehensive version of the finding is showcased in Figure 10.1.

Roy Choudhury et al. (2021) have reported the ascertained priority value for each of the above technologies as follows: Geotube – 39.1%, Screwpress – 24.5%, MBBR – 15.2%, Belt Filter Press – 9.4%, DEWATS – 5.8%, Phytorid – 3.6%, PDB –2.3%.

10.4 BIO-METHANE POTENTIAL OF FS

FS treatment provides an effective solution for improved biodegradability of FS and ensures optimum recovery in the form of methane gas. This treatment stabilizes the organic contents and can be disposed of with minimum impact on the environment. The biochemical methane potential (BMP) test has been widely used to assess the

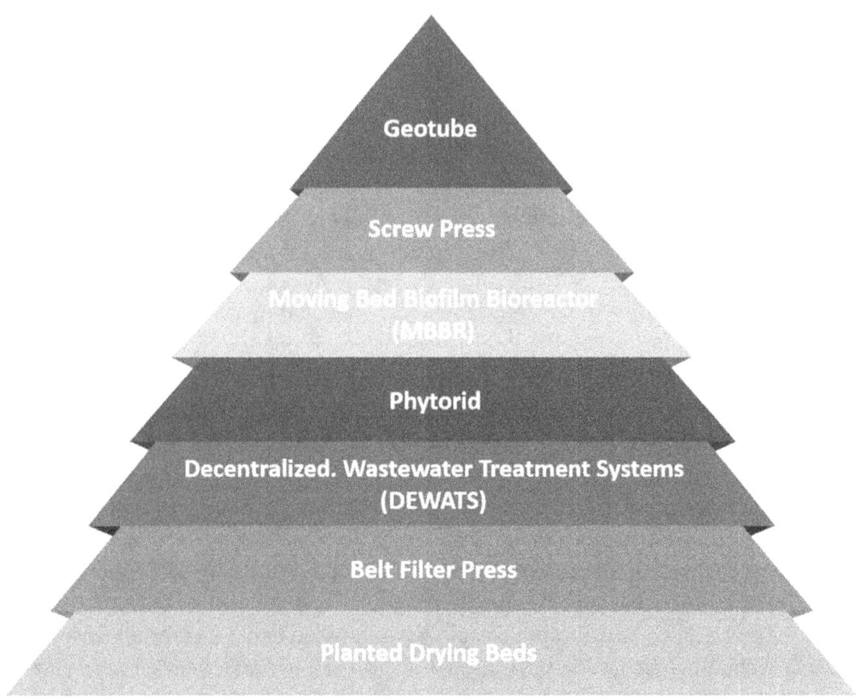

FIGURE 10.1 Hierarchy of treatment schemes for FS treatment.

biodegradability of various wastes. However, the findings of the BMP test are affected by several parameters, such as the substrate, inoculum, experimental settings, and operational conditions. It is challenging to compare values despite the larger number of research examining the BMP from various solid wastes due to the methodology utilized, results presentation, and discussion with different units. The maximum amount of methane produced per gram of available substrate is thus provided by the BMP test. The BMP is measured in NmL CH_4/gram of VS (Volatile Solids) or NmL/COD (Chemical Oxygen Demand). Despite the substantial work done over the past years, the BMP techniques used with solid organic substrates differ significantly (An et al., 2017). This feature is reflected in the biogas output from the anaerobic-digested substrates, and as a result, the outcomes vary.

10.4.1 Factors Affecting BMP

Temperature, mixing rate, ratio of different substrates, availability for carbon and nitrogen content in the substrate, which serves as the energy source for microorganisms, substrate pre-treatment used, headspace pressure, substrate particle size, substrate source, methane production, reactor volume, pH, inoculum source, measurement of biogas, and test duration are some of the factors that impact BMP.

Characterizing the substrate and the inoculum is crucial before beginning the BMP experiments. The following parameters are used for substrate characterization: TS,

VS, component composition (lipids, proteins, carbohydrates, and lignin), elemental composition, and COD. The initial concentration or reactor load is also a crucial parameter for BMP tests. A small load produces a meagre biogas volume, so the metabolic activity of microorganisms is low. More volatile fatty acids accumulate if the load is too large, which may hinder anaerobic digestion.

BMP test is greatly influenced by the inoculum used. The characteristics of the inoculum are determined by volatile solids or volatile suspended solids concentration. Although the inoculum source may not be uniform, it is feasible to use waste-activated sludge from municipal wastewater, soil extracts, rumens, and animal dung. The inoculum with a short lag phase is efficient for the BMP test. From the literature, several different concentrations were reported. Inoculum concentrations of 2.1 g VS/L were presented by El-Mashad and Zhang (2010). Whereas, Rincón-Pérez et al. (2021) presented a different value of VS = 21 g/L. The VDI 4630 (2006) indicates a range of 15 to 20 gVS/L. Angelidaki et al. (2009) suggested using fresh sludge if possible.

FS has a high level of excreted microorganisms that cause gastrointestinal infection and is rich in organic contents. It sets environmental security in danger. Additionally, the pathogen level in FS has to be analysed. It can be estimated by examining the concentration of *E.coli* in the FS as the pathogen indicator organism. The viability of using anaerobic digestion (AD) to treat FS and produce biogas as renewable energy is considered an efficient process. Organic matter is important for evaluating the level of stabilization of FS, biodegradation potential for treatment, and impact on receiving environment. The moisture content will, directly and indirectly, affect the biodegradability and viscosity of FS (Velkushanova et al., 2021).

10.4.2 Sample Storage

The collected FS sample should be stored in the refrigerator for no longer than one week as per the method of FS analysis (Velkushanova et al., 2021). Before utilizing the sample in the BMP test, the frozen sample should be thawed and slowly warmed to the ambient temperature

10.4.3 BMP Analytical Methods

Several methods are available to measure BMP. The manometric and volumetric approaches are currently the most widely used ones. A gravimetric technique has also recently and effectively been proposed. Using AMPTS devices, more sophisticated automated BMP measurement can be accomplished. The FS is loaded in batch mode under anaerobic conditions with a continuous stirring mechanism. The CO_2 is scrubbed using a sodium hydroxide solution. Using a flow cell array and the techniques of liquid displacement and buoyancy provides the methane generated throughout the experiment. The results are automatically loaded into the software.

BMP is defined as the volume of methane produced per amount of organic substrate material. Consequently, the accumulated volume of biogas is divided by the amount of organic substrate material. However, in the reactors, there will also be organic residues present in the added inoculum, which also contributes to the biogas produced during anaerobic digestion. Hence its volume fraction has to be subtracted from the total accumulated biomethane volume to determine the true biochemical methane potential of the substrate (Amodeo et al., 2020).

$$\text{BMP}(\text{NmL/G of VS}) = \frac{V_S - V_I}{M_{VS}}$$

Where:

V_S is the accumulated volume of biomethane from the reactor-containing sample (substrate + inoculum) (NmL)

V_I is the volume of biomethane originating from the inoculum present in the sample (NmL)

M_{VS} is the amount of organic substrate material contained in the sample (g).

10.4.4 Methane Yield

FS anaerobic digestion can produce methane but in low quantity due to the high nitrogen content. Since the biodegradability of FS is low compared to other organic substrates, a low specific methane yield is obtained. The origin of the substrate from a confined system, such as a septic tank or septage, where the condition for anaerobic digestion is naturally maintained, could be the reason for this low biodegradability. Also, stabilization of the organic components is achieved in the sedimentation process, and loss of volatile solids in the waste effluent is also a possible cause for low COD in the FS. Additionally, the high nitrogen content of FS contributes to a low C/N ratio. A low C/N ratio affects methanogens because it causes ammonia to build up when converting acetate, carbon dioxide, and hydrogen to methane. Therefore, the focus should be on using fresh sludge with short hydraulic retention time (HRT) to reduce the loss of volatile matter.

The anaerobic digestion of FS produces excess hydrogen sulphide, methane, and other gases. Also, more sulfuric acid production is reported, creating an unfavourable environment for the microorganisms, ultimately affecting the efficiency of the treatment process to produce methane. Studies have shown that increased methane yield and the digestion process are obtained by adding electron donors such as iron (Agani et al., 2016). Certain metals and heavy metals also contribute to the growth of methanogens. Studies proved that the addition of external electron donors enriches the treatment environment and promotes the growth of the microorganisms. Another method to improve the nutrient balance of FS is mixing it with other organic substrates, making it readily available for microorganism growth and development and enhancing biogas production (Afifah and Priadi, 2017). The comparative biogas yield reported from various organic substrates is summarized in Figure 10.2 (Rajendran et al., 2012; Ware and Power, 2016; Afifah and Priadi, 2017; Karne et al., 2018; Kilucha et al., 2022).

The assessment depicts that the biogas yield from FS enhances with temperature, addition of electron donors and carbon-rich co-substrates. The utmost reported biogas

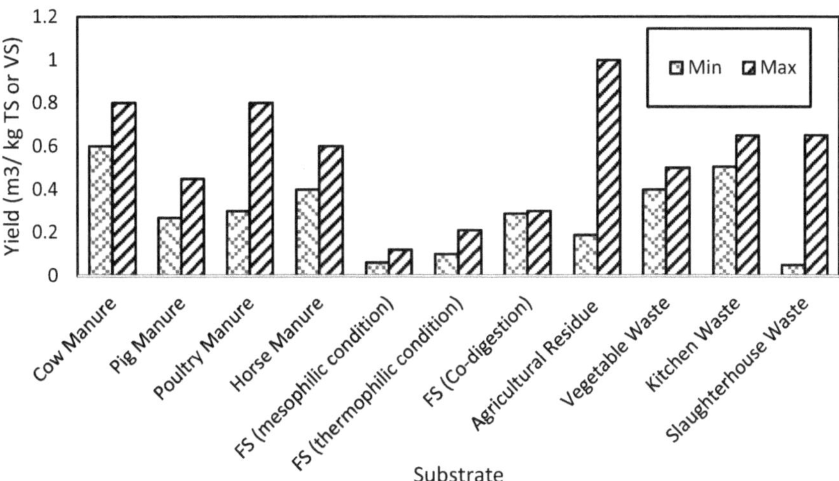

FIGURE 10.2 Comparative biogas yield reported from different substrates.

yield of 0.3 m^3/ kg VS is comparable with the production from other cattle manures, vegetable, and kitchen waste.

10.5 TECHNO-COMMERCIAL VIABILITY OF THE PROCESS

FS can be microbiologically transformed into biogas under anaerobic condition through the process of bio-methanation with ease. Fermenting and organic acid oxidizing bacteria, and methanogenic archaea are the three primary physiological groups which are responsible to facilitate the anaerobic digestion process (Angelidaki et al., 2011). Typically, bio-methanation is a less explored technology when it comes to FS treatment. However, citing the large biomethane potential of the substrate, researchers across the globe have recommended anaerobic digestion as a potential and profitable treatment solution to address the decentralized FS treatment issues in developing and under developed nations (Forbis-Stokes et al., 2016; Kilucha et al., 2022). In fact, biomethanation has been approved by Ministry of New and Renewable Energy (MNRE) as one of the most suited waste to energy (WtE) technologies for Indian conditions lately.

Owing to its global adaptation and success, especially in African nations (Soyingbe et al., 2019; Semiyaga et al., 2022), provoked the authors to ascertain the capital and operational expenditure to set up a decentralized anaerobic digestion-based faecal sludge treatment plant (FSTP) of 20 KLD capacity at the Indian outset. Table 10.3 and 10.4 delineates the list of electro-mechanical items involved and basic civil requirement to set up the FSTP of the desired capacity.

TABLE 10.3
List of Mechanical Components Required to Establish a 20 KLD Bio-methanation Plant

Material	Specifications	Unit	Quantity
Pre-processing unit	Screening, blending, and homogenization	No.	1
Feed & recirculation pump	60 m^3/h, 15 m head open impeller submersible cutter pump	No.	1
Recirculation pump	60 m^3/h, 15 m head open impeller submersible cutter pump	No.	1
Recirculation tank	5 m^3 PP/HDPE	No.	1
Gas holder dome for digester	9.67 m Dia., 300 m^3 volume	No.	1
Scrubber (gas purification)	100 m^3/h	No.	1
Biogas flare	100 m^3/h	No.	1
Power generation			
Engine	125 kVA	No.	1
Dewatering			
Screw Press	3.2 m^3/h	No.	1

Basic land requirement (development area) for establishing the FSTP of 20 KLD capacity is 600 m^2 (40 m x 15 m). Unlike many other processes of FS treatment, biomethanation is non-influenced by the climatic variance, especially monsoon. The process is not moisture sensitive as compared to thermal technologies. Therefore, pre-engineered structures are adequate to take up the operation. Further infrastructure details are depicted in Table 10.4.

Inside the digester, the volatile solids introduced are converted into biogas. The remaining digested organic residue and trace elements (if any) are removed in the form of slurry/digestate from the bottom of the reactor. Typically, biogas has a methane content of over 55% which is a greenhouse gas. Therefore, as a safety precaution, the availability of a flaring system is a mandate (Soyingbe et al., 2019). The gas yield is purified using a chemical scrubber to remove excess CO_2 and H_2S. The presence of H_2S even in the smallest capacity magnifies the threat of corrosion for gas engines (Mohammadpour et al., 2022). The general process flow of anaerobic digestion-based FSTP is showcased in Figure 10.3.

The major part of the operational expenditure for a biomethanation plant gets consumed to compensate for manpower charges other than the power requirement. The cost of power is neglected here as these plants are generally self-sustaining. The calculations were made based on the following production rates; biogas: 90 m^3/ ton, compost: 2000 kg/day (Yin et al., 2016). Further, a detailed cost break-up of the operational expenditure is presented in Table 10.5.

Table 10.5 depicts more than 13 years as the tenure for return on investment (ROI). The value of ROI in this case is extremely extended. Therefore, the biomethanation-based FSTP model will only be sustainable with the addition of treatment fees or support in the form of capital expenditure. Typically, the treatment fee for FS varies

TABLE 10.4
List of Civil/ Infrastructure Required for the Erection and Commissioning of a 20 KLD Bio-methanation Plant

Sl. No.	Description	Quantity (No.)	Dimensions (meter)		
			Length	Width	Height
A	Primary Work				
	Screening chamber	1	2.0	1.20	1.0
	Sludge holding and homogenization tank	1	4.0	3.0	1.6
1	Receiving shed civil work	1	10.0	6.0	6.0
2	Receiving shed pre-engineered building structure	1	10.0	6.0	6.0
3	Maturation Shed civil work	1	25.1	5.2	6.0
4	Maturation Shed pre-engineered building structure	1	25.1	5.2	6.0
5	Digester	1	12.6 m Diameter		6.5
6	Flare Stack Foundation	1	1.0	1.0	0.5
7	Engine foundation	1	2.0	4.0	
8	Feed recirculation tank	1	2.25	2.25	2.3
9	Gas Scrubbing foundation	1	1.0	1.0	
10	Recirculation tank foundation	1	2.0	2.0	
B	Allied and Development Works				
1	Compound wall		55.0 running meter		
2	Storm water drain	1	50.0 running meter		
3	Administrative building	1	10 m x 7 m		
4	Security cabin	1	1.5 m x 1.5 m		
5	Toilet	2	1.2 m x 1.2 m		
6	Bore well	1	Based on local conditions		
7	Internal road	1	Minimum 5m wide road		

between Rs. 250–300/ KL (NIUA, 2019). In this scenario, the ROI will come down to a little less than 7 years which is investable from a commercial perspective. However, up to 50% of the capital expenditure can be claimed from MNRE for waste-to-energy projects in India. This would slash the ROI to less than 3.5 years. Therefore, the present study highly recommends the adaptation of the biomethanation process for treating FS in suburban and remote areas.

10.6 COMPARATIVE ASSESSMENT AND COMMERCIALIZATION OPPORTUNITIES

The section presents a compilation of information on the various FS treatment technologies accessible in India, as well as their evaluation of techno-commercial effectiveness in relation to the existing treatment facilities. The attributes such as CAPEX, OPEX, area requirement, treatment efficacy, activation period, capability of handling shock load, and payback period are compared and delineated in detail (Table 10.6).

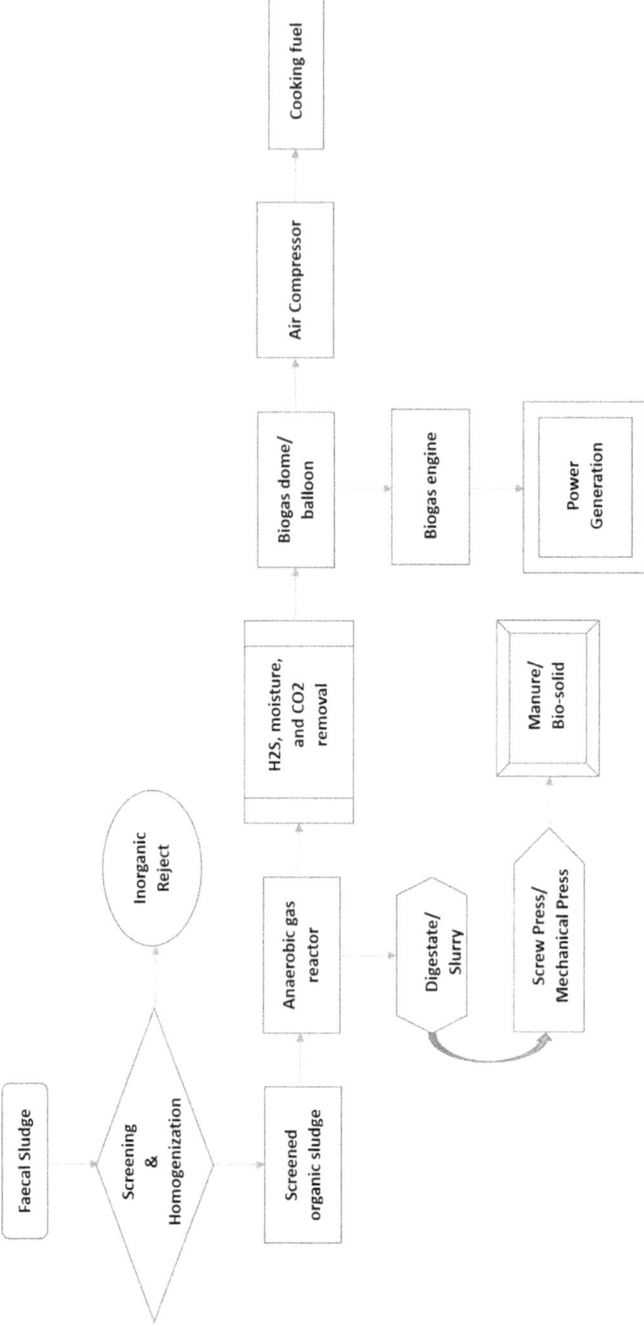

FIGURE 10.3 Process flow diagram of proposed 20 KLD anaerobic digestion-based FSTP.

TABLE 10.5
OPEX Break-up for 20 KLD Anaerobic Digestion-based FSTP

Sl. No.	Particulars	Quantity	Cost/ day (INR)	Cost/ month (INR)	Cost/Unit produced/ day (INR)	Revenue/ day (INR)	Profit/ day (INR)	CAPEX (INR)	Payback period (Years)
Manpower					5.05 (Production capacity of 1800 kWh/day)	13500 (@ Rs. 7.5/ unit rate; and Rs. 0.65/ kg of compost)	5695	2,80,00,000 (Comprehensive cost with Civil and installation (inclusive of taxes @ 18% on base price))	13.65
1	Operator	1	1000	30000					
2	Plant In-charge	1	1500	45000					
3	Labours	5	2500	75000					
4	Security	2	1000	30000					
Consumables									
5	Power	400 units	2325	69750					
6	Chemicals	1	200	6000					
7	PPEs	1	80	2400					
8	Water	20 KL	150	4500					
9	Grease and lubricants	1	200	6000					
10	Miscellaneous for Electromechanical repair and insurance	1	150	4500					
Total			9105	273150					

TABLE 10.6
Techno-commercial Assessment of Various FS Treatment Technologies

Sl. No.	Technology	CAPEX (INR/ KL)	OPEX (INR/ KL)	Footprint (m2)	Treatment Efficacy	Activation Period (weeks)	Capability of Handling Shock Load	Pay Back Period (years)	Location of FSTP	Reference
1	DEWATS	4,00,000–7,20,000	~295	~6000	> 80%	2–4	Somehow vulnerable	10	Baripada, Odisha	NIUA, 2017
2	Pyrolysis-based Operation	17,00,000	~535	~4000	> 98%	N.A.	Robust	9.5	Sanitation Resource Park Waralgal, Telangana	TTPL, 2018
3	Screw press-based Operation with HeliantisTM	10,00,000	~365	~2000	≤ 90%	2–4	Slightly venerable	5.6	Unnao, Uttar Pradesh	Vijayan et al., 2020
4	Mechanized De-watering and MBBR	4,13,000–8,83,000	~380	~2000	> 93%	Up to 5	Slightly venerable	13.7	Tenali, Andhra Pradesh	
5	Planted Drying Beds	1,50,000	~180	~27000	40–70%	5–10	Highly venerable	3–5	Jhansi, Uttar Pradesh	
6	Phytorid Wastewater Treatment Technology	< 1,00,000	~210	~12000	> 80%	8–10	Moderately venerable	3	Mumbai, Maharashtra	CPHEEO, 2020
7	Geotube-based Technology	5,00,000–5,50,000	~280	~2000	> 95%	2	Robust	10	Sanitation Resource Park Waralgal, Telangana	Banka et al., 2020
8	Anaerobic Digestion	15,00,000	~455	~2000	90–95%	30–40 days	Slightly venerable	5–16	Kalpetta, Kerala	Vijayan et., 2020

It's explicit from Table 10.6 that pyrolysis-based FS treatment incurs the highest cost both in terms of capital and operational expenditure. Whereas, PDBs are the most inexpensive treatment alternatives. However, the treatment scheme is susceptible to shock load and offers limited efficacy. Therefore, the reuse of treated effluent often gets difficult due to the excessive concentration of nutrients and may lead to eutrophication. So, treatment schemes such as mechanized dewatering and MBBR, Geotube-based technology, and screw press-based operation with Heliantis™ with affordable commercials and a high degree of treatment efficacy are gaining popularity. Anyhow, the commercial valorisation of the primary by-products (approx. 75% of the influent sludge volume) of the above processes, i.e., treated wastewater is far from reality (Banka et al., 2020). While power produced through WtE technologies such as biomethanation has a huge market demand and yields a revenue of Rs. 7–7.5 per kWh. This helps lower the operational expenditure and carbon footprint of the treatment facility. The higher initial investment can be taken care of with available government subsidies.

10.7 CONCLUSION

FSTPs based on aerobic treatment schemes generate nil to trivial revenue irrespective of the lower initial investment. Contrarily, biological WtE technologies such as biomethanation yield high-value by-products though being expensive. The capital and operational expenditures are ascertained as Rs. 280 lakhs and Rs. 455/ton respectively to set up a facility of 20 KLD capacity. The common yield-related hypothesis considered for the financial evaluation: biogas–90 m^3/ton, compost–2 tons/day. This can be further enhanced with the addition of electron donors or with the co-digestion of carbon-rich substrates. The revenue is calculated based on the following by-product values: power–Rs. 7.5/ unit; and compost–Rs. 0.65/ kg. A comprehensive revenue of Rs. 675/ KL is conveniently achievable for a 20 KLD biomethanation-based FSTP. However, the ascertained payback period of 13.65 years with an estimated profit of Rs. 285/ KL, makes it a difficult choice for private investors. So, the process can only be scaled up with financial aid from government agencies owning to its high capital investment.

REFERENCES

Afifah, U., & Priadi, C. R. (2017, March). Biogas potential from anaerobic co-digestion of faecal sludge with food waste and garden waste. In AIP Conference Proceedings (Vol. 1826, No. 1, p. 020032). AIP Publishing LLC. https://doi.org/10.1063/1.4979248

Agani, I. C., Suanon, F., Dimon, B., Ifon, E. B., Yovo, F., Wotto, V. D., Abass, O.K., & Kumwimba, M. N. (2016). Enhancement of fecal sludge conversion into biogas using iron powder during anaerobic digestion process. *American Journal of Environmental Protection*, *5*(6), 179–186. https://doi.org/10.11648/j.ajep.20160506.15

Ahmed, I., Ofori-Amanfo, D., Awuah, E., & Cobbold, F. (2019). A comprehensive study on the physicochemical characteristics of faecal sludge in greater Accra region and analysis of its potential use as feedstock for green energy. *Journal of Renewable Energy, 2019*, Article ID 8696058, 11 pages. https://doi.org/10.1155/2019/8696058

Amoah, P., Adamtey, N., & Cofie, O. (2017). Effect of urine, poultry manure, and dewatered faecal sludge on agronomic characteristics of cabbage in Accra, Ghana. *Resources*, *6*(2), 19. https://doi.org/10.3390/resources6020019

Amodeo, C., Hafner, S. D., Teixeira Franco, R., Benbelkacem, H., Moretti, P., Bayard, R., & Buffière, P. (2020). How different are manometric, gravimetric, and automated volumetric BMP results?. *Water*, *12*(6), 1839. https://doi.org/10.3390/w12061839

An, V. T. H., Thanh, V. T. M., & Anh, N. V. (2017). Bio-methane potential test for anaerobic co-digestion of faecal sludge and sewage sludge. *Vietnam Journal of Science and Technology*, *55*(4C), 27–32.

Angelidaki, I., Alves, M., Bolzonella, D., Borzacconi, L., Campos, J. L., Guwy, A. J., Kalyuzhnyi, S., Jenicek, P., & van Lier, J. B. (2009). Defining the biomethane potential (BMP) of solid organic wastes and energy crops: a proposed protocol for batch assays. *Water Science and Technology*, *59*(5), 927–934. https://doi.org/10.2166/wst.2009.040

Angelidaki, I., Karakashev, D., Batstone, D. J., Plugge, C. M., & Stams, A. J. (2011). Biomethanation and its potential. *Methods in Enzymology, 494*, 327–351. https://doi.org/10.1016/B978-0-12-385112-3.00016-0

Arthur, R., & Brew-Hammond, A. (2010). Potential biogas production from sewage sludge: A case study of the sewage treatment plant at Kwame Nkrumah university of science and technology, Ghana. *International Journal of Energy & Environment, 6*, 1009–1016.

Banka, N., Lanke, M., Roy Choudhury, A., Sarkar, J., Anandapu, S.C. and Banerjee, R. (2020). Evaluation of the functionality of GeoTube® based physicochemical faecal sludge treatment: A cursory alternate. *International Journal of Plant and Environment, 6*(2), 122–128. https://doi.org/10.18811/ijpen.v6i02.05

Basamykina, A., Kharlamova, M., & Mada, S. Y. (2020). Dewatering as a primary treatment of faecal sludge in individual residential sector (a technologies review). In *E3S Web of Conferences* (Vol. 169, p. 02008). EDP Sciences.

Bassan, M., Tchonda, T., Yiougo, L., Zoellig, H., Mahamane, I., Mbéguéré, M., & Strande, L. (2013). Characterization of faecal sludge during dry and rainy seasons in Ouagadougou, Burkina Faso. In *Proceedings of the 36th WEDC International Conference*, Nakuru, Kenya, 1–5 July 2013.

Central Public Health and Environmental Engineering Organisation (CPHEEO) (2020, July 20). *On-site and off-site sewage management practices*. Ministry of Housing and Urban Affairs. www.cseindia.org/static/mount/recommended_readings_mount/Advisory-On-Site-and-ffsite-Sewage-Management-Practices-MoHUA.pdf

Diaz-Elsayed, N., Rezaei, N., Guo, T., Mohebbi, S., & Zhang, Q. (2019). Wastewater-based resource recovery technologies across scale: a review. *Resources, Conservation and Recycling, 145*, 94–112. https://doi.org/10.1016/j.resconrec.2018.12.035

Eliyan, C., Vinnerås, B., Zurbrügg, C., Koottatep, T., Sothea, K., & McConville, J. (2022). Factors influencing physicochemical characteristics of faecal sludge in Phnom Penh, Cambodia. *Journal of Water, Sanitation and Hygiene for Development*, *12*(1), 129–140. https://doi.org/10.2166/washdev.2021.193

El-Mashad, H. M., & Zhang, R. (2010). Biogas production from co-digestion of dairy manure and food waste. *Bioresource Technology*, *101*(11), 4021–4028. https://doi.org/10.1016/j.biortech.2010.01.027

Forbis-Stokes, A. A., O'Meara, P. F., Mugo, W., Simiyu, G. M., & Deshusses, M. A. (2016). On-site fecal sludge treatment with the anaerobic digestion pasteurization latrine. *Environmental Engineering Science*, *33*(11), 898–906. https://doi.org/10.1089/ees.2016.0148

Kamble, S. J., Chakravarthy, Y., Singh, A., Chubilleau, C., Starkl, M., & Bawa, I. (2017). A soil biotechnology system for wastewater treatment: technical, hygiene, environmental LCA and economic aspects. *Environmental Science and Pollution Research, 24*, 13315–13334. https://doi.org/10.1007/s11356-017-8819-6

Karne, H. U., Bhatkhande, D., & Jabade, S. (2018). Mesophilic and thermophilic anaerobic digestion of faecal sludge in a pilot plant digester. *International Journal of Environmental Studies*, *75*(3), 484–495. https://doi.org/10.1080/00207233.2017.1406729

Kilucha, M., Cheng, S., Minza, S., Nasiruddin, S. M., Velempini, K., Li, X., Wang, X., Doroth, K. M., & Li, Z. (2022). Insights into the anaerobic digestion of fecal sludge and food waste in Tanzania. *Frontiers in Environmental Science*, *10*, 911348. https://doi.org/10.3389/fenvs.2022.911348

Koottatep, T., Surinkul, N., Polprasert, C., Kamal, A. S. M., Koné, D., Montangero, A., Heinss, U., & Strauss, M. (2005). Treatment of septage in constructed wetlands in tropical climate: lessons learnt from seven years of operation. *Water Science and Technology*, *51*(9), 119–126. https://doi.org/10.2166/wst.2005.0301

Krueger, B. C., Fowler, G. D., & Templeton, M. R. (2021). Critical analytical parameters for faecal sludge characterisation informing the application of thermal treatment processes. *Journal of Environmental Management,* 280, 111658. https://doi.org/10.1016/j.jenvman.2020.111658

Kumar, M. D., Tortajada, C., Kumar, M. D., & Tortajada, C. (2020). Wastewater treatment technologies and costs. In *Assessing Wastewater Management in India* (pp. 35–42). SpringerBriefs in Water Science and Technology. Springer, Singapore. https://doi.org/10.1007/978-981-15-2396-0_7

McConville, J. R., Kvarnström, E., Nordin, A. C., Jönsson, H., & Niwagaba, C. B. (2020). Structured approach for comparison of treatment options for nutrient-recovery from fecal sludge. *Frontiers in Environmental Science, 8*, 36. https://doi.org/10.3389/fenvs.2020.00036

Mohammadpour, H., Cord-Ruwisch, R., Pivrikas, A., & Ho, G. (2022). Simple energy-efficient electrochemically-driven CO2 scrubbing for biogas upgrading. *Renewable Energy, 195*, 274–282. https://doi.org/10.1016/j.renene.2022.05.155

National Institute of Urban Affairs (NIUA) (2017, January 11). *City sanitation plan for Baripada*. Government of Odisha. https://scbp.niua.org/sites/default/files/City%20Sanitation%20Plan-BARIPADA_Draft%20Report%20_1.pdf

National Institute of Urban Affairs (NIUA). (2019). Cost Analysis of Faecal Sludge Treatment Plants in India. https://scbp.niua.org/sites/default/files/FSTP_Cost_Analysis_1_0.pdf

Niwagaba, C. B., Mbéguéré, M., & Strande, L. (2014). Faecal sludge quantification, characterisation and treatment objectives. In *Faecal Sludge Management: Systems approach for implementation and operation* (pp. 19–44). London: IWA Publishing. www.eawag.ch/forschung/sandec/publikationen/ewm/dl/fsm_book.pdf Accessed Feb 2023.

Osmani, S. A., Rajpal, A., & Kazmi, A. A. (2021). Upgradation of conventional MBBR into Aerobic/Anoxic/Aerobic configuration: A case study of carbon and nitrogen removal based sewage treatment plant. *Journal of Water Process Engineering,* 40, 101921. https://doi.org/10.1016/j.jwpe.2021.101921

Prasad, P., Andriessen, N., Moorthy, A., Das, A., Coppens, K., Pradeep, R., & Strande, L. (2021). Methods for estimating quantities and qualities (Q&Q) of faecal sludge: field evaluation in Sircilla, India. *Journal of Water, Sanitation and Hygiene for Development*, *11*(3), 494–504. https://doi.org/10.2166/washdev.2021.269

Rajendran, K., Aslanzadeh, S., & Taherzadeh, M. J. (2012). Household biogas digesters – A review. *Energies*, *5*(8), 2911–2942. https://doi.org/10.3390/en5082911

Rincón-Pérez, J., Celis, L. B., Morales, M., Alatriste-Mondragón, F., Tapia-Rodríguez, A., & Razo-Flores, E. (2021). Improvement of methane production at alkaline and neutral pH from anaerobic co-digestion of microalgal biomass and cheese whey. *Biochemical Engineering Journal, 169*, 107972. https://doi.org/10.1016/j.bej.2021.107972

Roy Choudhury, A., Singh, N., Banka, N., Lanke, M., Mavuduru, M. (2021, July-December). Ablest Faecal Sludge Management: Challenging a Dormant Opportunity. *Energy Future, 10*(1), 60–69. https://bookstore.teri.res.in/magazines/2278-7186

Samal, K., Moulick, S., Mohapatra, B. G., Samanta, S., Sasidharan, S., Prakash, B., & Sarangi, S. (2022). Design of faecal sludge treatment plant (FSTP) and availability of its treatment technologies. *Energy Nexus, 7*, 100091. https://doi.org/10.1016/j.nexus.2022.100091

Semiyaga, S., Nakagiri, A., Niwagaba, C.B., Manga, M. (2022). Application of Anaerobic Digestion in Decentralized Faecal Sludge Treatment Plants. In: Meghvansi, M.K., Goel, A.K. (eds) *Anaerobic Biodigesters for Human Waste Treatment. Environmental and Microbial Biotechnology* (pp. 263–281). Springer. https://doi.org/10.1007/978-981-19-4921-0_14

Septiariva, I. Y., Suryawan, I. W. K., Zahra, N. L., Putri, Y. N. K., Sarwono, A., Qonitan, F. D., & Lim, J. W. (2022). Characterization sludge from drying area and sludge drying bed in sludge treatment plant Surabaya city for waste to energy approach. *Journal of Ecological Engineering*, *23*(7), 268–275. https://doi.org/10.12911/22998993/150061

Shukla, A., Patwa, A., Parde, D., & Vijay, R. (2022). A review on generation, characterization, containment, transport and treatment of fecal sludge and septage with resource recovery-oriented sanitation. *Environmental Research*, *216*, 114389. https://doi.org/10.1016/j.envres.2022.114389

Soyingbe, A. A., Olayinka, O., Bamgbose, O., & Adetunji, M. T. (2019). Effective management of faecal sludge through co-digestion for biogas generation. *Journal of Applied Sciences and Environmental Management*, *23*(6), 1159–1168. https://dx.doi.org/10.4314/jasem.v23i6.25

Strande, L., Schoebitz, L., Bischoff, F., Ddiba, D., Okello, F., Englund, M., Ward, B. J., & Niwagaba, C. B. (2018). Methods to reliably estimate faecal sludge quantities and qualities for the design of treatment technologies and management solutions. *Journal of Environmental Management, 223*, 898–907. https://doi.org/10.1016/j.jenvman.2018.06.100

Tide Technocrats Private Limited (TTPL) (2018, January 08). *Faecal Sludge and Septage Treatment Plant–Thermal Processes by Tide Technocrats Private Limited.* Atal Mission for Rejuvenation and Urban Transformation. https://cdn.cseindia.org/attachments/0.02276200_1542097169_Thermal-technologies-in-FSTP-Shriram.pdf

Varma, V. G., Jha, S., Raju, L. H. K., Kishore, R. L., & Ranjith, V. (2022). A review on decentralized wastewater treatment systems in India. *Chemosphere*, *300*, 134462. https://doi.org/10.1016/j.chemosphere.2022.134462

VDI-4630 (2006) Fermentation of Organic Materials: Characterisation of the Substrate, Sampling, Collection of Material Data, Fermentation Tests. Verlag des Vereins Deutscher Ingenieure, Düsseldorf, 92.

Velkushanova, K., Strande, L., Ronteltap, M., Koottatep, T., Brdjanovic, D., & Buckley, C. (2021). *Methods for faecal sludge analysis.* IWA publishing.

Vijayan, V., Mallik, M., & Chakravarthy, S. K. (2020). *Performance Evaluation: How Faecal Sludge Treatment Plants Are Performing.* Centre for Science and Environment, New Delhi. www.cseindia.org/performance-evaluation-how-faecal-sludge-treatment-plants-are-performing-10048

Ware, A., & Power, N. (2016). Biogas from cattle slaughterhouse waste: Energy recovery towards an energy self-sufficient industry in Ireland. *Renewable Energy, 97*, 541–549. https://doi.org/10.1016/j.renene.2016.05.068

Yacob, T. W., Linden, K. G., & Weimer, A. W. (2018). Pyrolysis of human feces: gas yield analysis and kinetic modeling. *Waste Management, 79*, 214–222. https://doi.org/10.1016/j.wasman.2018.07.020

Yin, F., Li, Z., Wang, D., Ohlsen, T., & Dong, H. (2016). Performance of thermal pretreatment and mesophilic fermentation system on pathogen inactivation and biogas production of faecal sludge: Initial laboratory results. *Biosystems Engineering, 151*, 171–177. https://doi.org/10.1016/j.biosystemseng.2016.08.019

11 Regulating Total Soluble Products During Food Waste Biomethanation for Material and Energy Recovery

S. Hemapriya and Sankar Ganesh Palani

11.1 INTRODUCTION

Food waste contributes significantly to municipal solid waste. Nowadays, with an increase in population and urbanization, food wastage has increased sharply. According to The Food and Agricultural Organisational ((FAO, 2013) report around 1.3 billion tons of food is wasted globally. Food wastage includes the food that is lost right from the production (agricultural stage) to the consumption cycle. Certain types of food waste is unavoidable. The food which is produced but never consumed represents an inappropriate use of valuable natural resources. Mostly, food waste as municipal solid waste ends up in landfill sites or undergoes incineration treatment. These conventional techniques of handling food waste are inefficient, as they contribute to increasing greenhouse gas emissions, high chances of polluting the groundwater, air pollution, leachate generation, and space constraints. Problems associated with improper handling of food waste have resulted in massive resource waste and prompted environmental hazards. Therefore, food waste management is crucial because it protects the environment and generates bioenergy.

Anaerobic digestion is a significant replacement for conventional and less significant food waste treatment, which is considered a cost-effective technology that enables the production of methane-rich biogas via microbial digestion. It results in a sustainable and promising solution for the problem of untreated food waste disposal in the environment. Additionally, providing a useful remaining residue in the form of sludge from anaerobic digestion called digestate, which can be bioprocessed to produce biofertilizer.

During anaerobic digestion, a number of intermediates are produced, some of them will inhibit the process. Such intermediates are collectively called total soluble products (TSPs), among them, the most dominant is the volatile fatty acid (VFA). The VFA accumulates in excess and hinders anaerobic digestion due to acidification. It is

DOI: 10.1201/9781003364467-11

important to implement a precise technique like adsorption using ion exchange resin (IER) to counteract the product inhibition imposed by the undissociated acids that predominate at low pH. Excess VFA can be adsorbed during adsorption to maintain the optimal concentration for enhanced bioenergy production.

The extracted VFAs can also be desorbed from the resins and they can be downstream processed and purified. They support the circular bioeconomy and have a significant economic value. The regenerated resin can be again used in the adsorption-desorption cycle.

11.2 COMPOSITION AND CHARACTERISTICS OF FOOD WASTE

Food waste composition is significantly associated with the food type utilized, as rice and vegetables contain more carbohydrates whereas, fish, egg, and meat contain more concentration of protein (Katarzyna et al., 2022). However, the universal characteristic of food waste includes high nutrient and moisture levels, and high volatile solids concentration that makes it suitable for the fermentation process. The design and operation of bioconversion methods are adopted according to the physical and chemical attributes of food waste. Significant amounts of organic materials like proteins, lipids, and carbohydrates can be identified in food waste. Depending upon the source, food waste also contains varying amounts of suspended solids due to this it has high biological oxygen demand (BOD). In addition, the microbes present in food wastes require nitrogen, carbon compounds, phosphorous, potassium, and other elements to thrive.

11.2.1 Solids and Moisture Content of Food Waste

Generally, food waste has a high moisture content, which makes them unsuitable for conventional food waste treatments like incineration and composting treatment. Additionally, the demand for green energy has increased significantly necessitating the incorporation of energy recovery in waste management treatment.

To analyze the potential and actual utilization of an organic amendment, the dry matter (total solids) must also be taken into consideration. Regarding digestates, the dry matter composition is greatly influenced by the digestion method used or by any potential post-treatment that the digestion residues may undergo.

Volatile solids act as the energy content for microorganisms making the food waste an efficient substrate for anaerobic digestion. Figure 11.1 represents the amount of moisture and solid content present in food waste from select regions (Amirhossein et al., 2017, Y. Zhang et al., 2015, Zhang et al., 2011, Mirzaman et al., 2016).

11.2.2 Carbohydrates, Proteins, and Lipid Content

Organic matter constitutes the following three primary macromolecules: carbohydrates, lipids, and proteins. When compared to the other two macromolecules, some prepared food and food waste had a considerably high amount of carbohydrates. Typically, food waste contains about 50% carbohydrates, whereas protein constitutes nearly

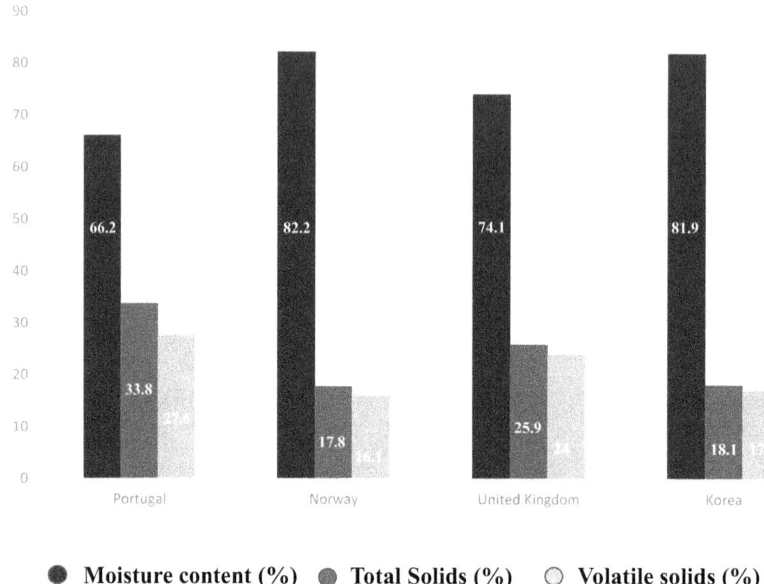

FIGURE 11.1 Major characteristics of food waste in select countries.

about 15%. Carbohydrates are quickly broken down by anaerobic digestion to yield subsequent sugar monomers, which acidogens then break down into volatile fatty acids (VFA). Then, acetogens transform VFA to yield acetate, hydrogen, and carbon dioxide. The accumulation of VFA by acidogens as a result of the quick breakdown of carbohydrates causes a sharp decrease in pH. The proteins present undergo hydrolysis to produce ammonium which aids to some extent in buffering the acids produced. Lipids are not easily degradable by microbes because of the presence of cellulose and hemicellulose contents. However, lipids are more capable of producing methane than proteins and carbohydrates. More hydrogen is produced in the case of the carbohydrate-rich substrate.

11.2.3 Elemental Composition

Many macronutrients and micronutrients present in food waste are significant for the growth of micro-organisms making it suitable for anaerobic treatment. In fact, a feedstock that produces products that are high in organic carbon and nitrogen will be added to the soil as soil amendments to enhance its biological characteristics.

Nitrogen is an essential element for better soil microbial activity and vegetation growth. It is the common growth-limiting nutrient that is required for plant growth in large amounts. Figure 11.2 depicts the quantity of carbon, nitrogen, oxygen, and hydrogen present in food waste from select countries (Wanqin et al., 2015, Adebayo Opatokun et al., 2015, James and Murphy., 2012, Zhang et al., 2011, Najam et al., 2019, Dennehy et al., 2018).

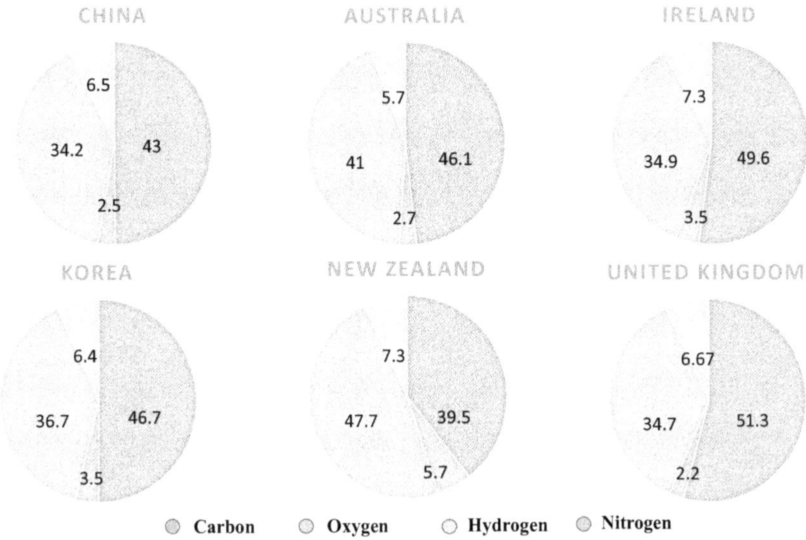

FIGURE 11.2 Elemental composition of food waste in select countries.

11.3 SOURCE AND TREATMENT OF FOOD WASTE

Food loss happens throughout the food cycle at various points. It includes the following stages starting from agricultural production, harvesting, handling and storage, packaging, and distribution and ends with the consumer consumption stage. Among this, about 55% of food wastage occurs during the initial food production and storage cycle and about 35% of wastage occurs at the consumer level (Ammaiyappan et al., 2021)When compared to low-income countries, in developed and urbanized countries there is a high prevalence of food wastage at the consumer level. In low-income countries or underdeveloped countries, wastage occurs due to a lack of advanced techniques for harvesting, improper storage facilities, and marketing strategies.

It is more straightforward to quantify the concentrated and consistent waste generated at the industrial level during the processing stage of food waste than the wastage that occurs at the consumer level. The possibility of underestimating the food wastage that occurs at the consumer level is high.

Among the generated food waste more than 78% is landfilled and the rest is incinerated and composted (EPA, 2018)). The aforementioned food waste generation and treatment statistics are shown in Figure 11.3. Food waste treatment by a conventional method like landfill represents the simplest method to manage food waste as the cost incurred is low and does not require highly qualified personnel. While they have various negative impacts as they release Green House Gases which are more potent and 25% more harmful than carbon dioxide which contributes to global warming and climate change. Landfill emissions have the greatest impact on ozone depletion and requires huge space. Additionally, leachate generated from it will also have a high possibility of polluting the environment. Incineration is not suitable for

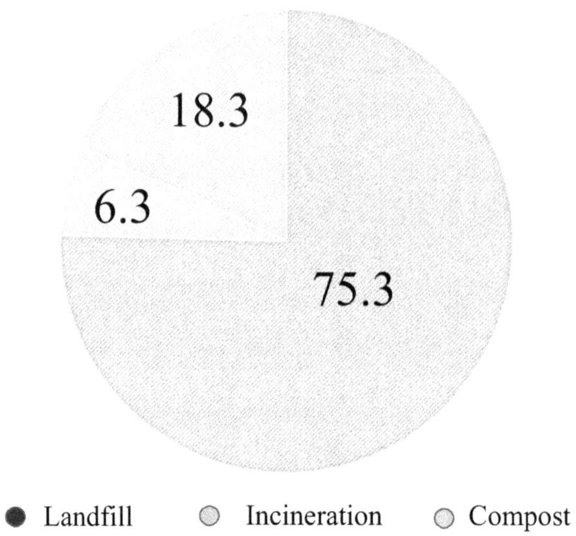

FIGURE 11.3 Food waste generation and conventional treatment statistics.

waste with high moisture content and it leads to air pollution causing long-term negative health effects such as cancer, neurological problems, reproductive abnormalities, and other health effects that occur due to exposure to hazardous pollutants and it does not contribute to waste reduction. These conventional techniques are non-efficient and negatively impact the environment.

11.3.1 Food Waste Treatment by Anaerobic Digestion

Anaerobic digestion of food waste has numerous advantageous over the traditional waste treatment techniques, including a reduction in the formation of biomass sludge (Lalak et al., 2015), minimal odor emission and generation of bioenergy in form of methane-rich biogas (Ward et al., 2008). Furthermore, there is a high level of compliance with various waste regulations established to lower the quantity of biodegradable waste going to landfills. Organic substances including polysaccharides, proteins, and lipids are hydrolyzed or broken down into various intermediate products during anaerobic digestion. These compounds are then oxido-reduced to generate methane-rich biogas. It is indeed suitable owing to less sludge production and energy requirement. Anaerobic digestion involves four biochemical processes, as shown in Figure 11.4. Different types of microbial populations govern these four reaction stages in anaerobic digestion. In the first step, the high molecular weight and granular complex organic polymers such as proteins, lipids, and carbohydrates are hydrolyzed by hydrolytic bacteria to simple soluble monomers such as glucose and amino acids. This initial step is regarded as the rate-determining step since synergistic relationship persists between different microorganisms. In acidogenesis, the soluble monomers are degraded into acetic acid, propionic acid, butyric acid, and valeric acid which

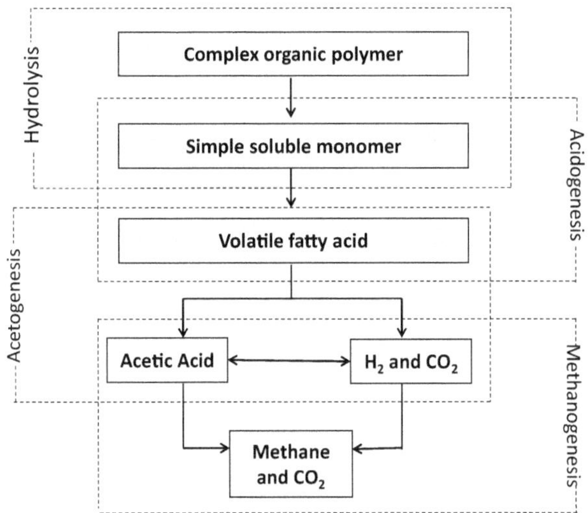

FIGURE 11.4 Biochemical conversion pathway in anaerobic digestion.

are collectively called volatile fatty acids. In the further step, long-chain VFAs are oxidized further to acetate along with carbon dioxide and hydrogen for further reaction, and this process of conversion to acetate is called acetogenesis. In methanogenesis, methane is generated by two groups of bacteria, acetoclastic methanogens utilize acetate to generate methane, while the other group, called hydrogenotrophic methanogens, utilizes hydrogen and carbon dioxide to form methane.

11.3.2 Intermediate Products

Anaerobic digestion has been essential in converting food waste to renewable fuels and electricity. Various products are produced during the process and some are already present in the reactor that includes water, biomass, organic waste, total soluble products (TSP), particulates, salt, and dissolved gases. Alcohols (ethanol, propanol, butanol), solvents (acetone), lactate, and fatty acids (acetate, propionate, iso-butyrate, valerate, and caproate) are all formed as intermediate products and are collectively called total soluble products (TSP), which has an impact on methane production (Chakraborty et al., 2022).

Among various intermediates produced during the process of anaerobic digestion, some will function as enhancers and some will function as inhibitors. The stability of the process necessitates having control over the generated intermediates. Due to the syntropic correlation existing between the microbes of anaerobic digestion, the product of one microbe will act as the substrate for another. If the inhibitory intermediate has high economic value then it can be recovered as it can act as a source for generating value-added products, which will contribute to the circular bioeconomy and concomitantly maintain the optimal concentration for normal functioning of anaerobic digesters.

VFAs will be formed as one of the major intermediates during the anaerobic process of converting food waste to methane-rich biogas. Easily hydrolyzable food waste results in increased production of VFAs by acidogens and acetogens. However, the ability of the methanogens to consume the produced VFAs occurs at a lower rate when compared to the production rate, leading to the accumulation of VFA and concomitant acidification of the reactor. The lipophilic VFA can easily cross the cell barrier and dissociate within cells, decreasing intracellular pH, negatively impacting microbial growth and metabolic pathways, and ultimately leading to digester failure. Co-digesting easily degradable wastes with other substrates that are difficult to degrade to maintain the effective C/N (carbon to nitrogen) ratio, using commercially available alkaline chemicals as a buffering agent, elutriation of the methanogenic phase, and recirculation of digestate that functions as a self-buffering agent are some specific ways to overcome acidification. However, all these methods increase the cost involved in biogas production and just delay the acidification without preventing it.

11.4 VOLATILE FATTY ACID: AN OVERVIEW

Organic acids like formic, acetic, propionic, butyric, valeric, caproic, lactic, succinic, fumaric, citric, gluconic, ascorbic, etc. are among the group of aliphatic mono–and dicarboxylic acids known as "carboxylic acids". Carboxylic acids have significant applications in the chemical, pharmaceutical, food, and fuel industries. Interestingly, some carboxylic acids have already met biochemical production at an industrial scale. When compared to other carboxylic acids, volatile fatty acids are monocarboxylic acids with 1 to 6 carbons and have comparatively high volatility (Bhatia and Yang, 2017).

Commercial VFAs production is highly dependent on chemical synthesis utilizing non-renewable petroleum as a raw material, however, these petroleum-based methods are not sustainable. A better alternative for VFA production would be microbial fermentation. The simplest methods rely on using sugars like glucose or xylose, which allow for higher production and fewer by-products, but there are also drawbacks such as a higher cost for the raw materials. In contrast, waste biomass such as food waste can be used as a substrate, as they contain highly available and abundant carbon sources that can be fermented to VFAs.

VFA is having larger market value and it is the fastest-growing biochemical. VFAs are demanded by chemical industries for the production of alcohol, ketones, esters, olefins, or aldehydes. Furthermore, VFA can be used in textiles and cosmetics and also in biopolymer plastic applications. VFA has the potential to replace petrochemicals addressing the concern of plastic accumulation in the ecosystem. This would also offset the requirement for petrochemicals to produce biodegradable PHA (polyhydroxyalkanoates) and PHB (polyhydroxy butyrate). The cost of production of PHA and PHB is strongly correlated with substrate cost. If VFAs could be produced from waste then production costs could be more than halved. Thus, VFA production by microbial fermentation is considered an effective alternative to chemical routes. Figure 11.5 shows the diverse application of VFA in industries.

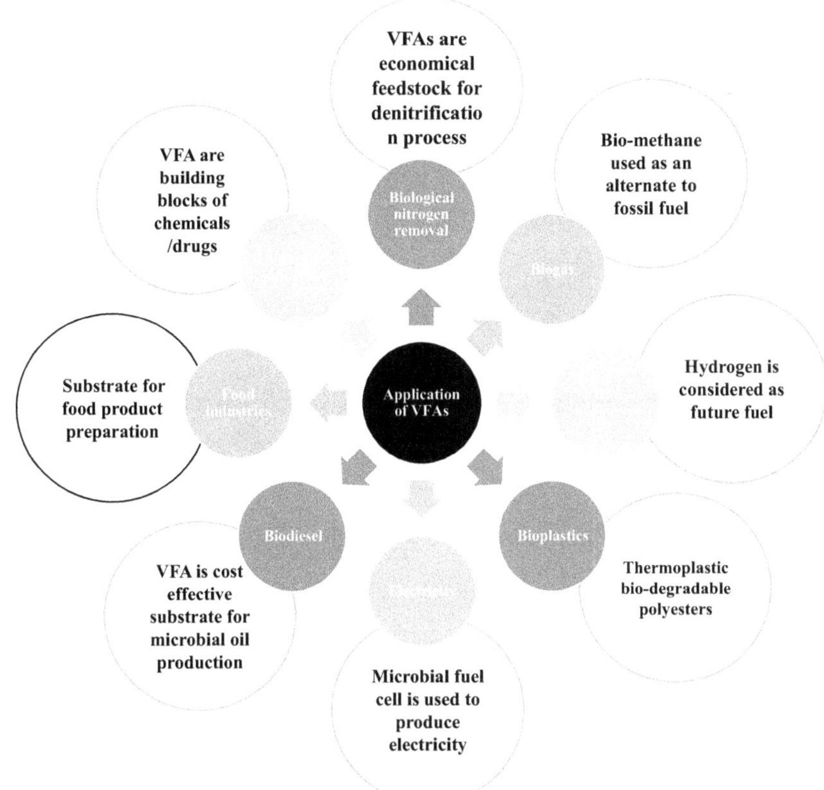

FIGURE 11.5 Diverse application of volatile fatty acid in industries.

11.5 VFA RECOVERY PROCESS

The intermediate metabolites such as volatile fatty acids can be valuable and compete economically with the methane and carbon dioxide end-products (Silva et al., 2013). The wide range of organic acids and other molecules that can be obtained during anaerobic digestion will depend on the raw material composition and operational conditions, which will control the equilibrium between different organisms and determine the final microbial community.

Regardless of the financial potential of VFA production from the anaerobic process, their extraction and recovery from the heterogeneous reactor mixture is still a significant challenge. Existing conventional processes for the recovery of organic acids include stripping, electrodialysis, direct distillation, solvent extraction, evaporation, chromatographic methods, precipitation, liquid-liquid extraction ultrafiltration, and drying. Although some of the conventional techniques are simple to set up and have a high extraction yield. The usage of chemical solvents further leads to increased generation of solid wastes instead of mineralizing the wastes.

All these qualities make them unsuitable and non environmentally friendly. Thus, a precise solution would be the removal of the excess VFAs, and maintaining the ideal concentration inside the reactor which leads to continuous reactor operation. Techniques such as membrane-based separation and ion exchange resin adsorption will assist with the selective recovery of the desired VFA. In this study, we compare the aforementioned methods to identify the most effective one for enriching VFA recovery. The different techniques used to recover carboxylic acids are discussed in detail below.

11.6 LIQUID-LIQUID EXTRACTION

The selection of the solvent is crucial in liquid-liquid extraction. Also, it operates as one of the affinity separation procedures, and the characteristics of the extractant have a substantial impact on the function. Three main groups of solvents are identified for the extraction of carboxylic acids:

- carbon-bonded and oxygen containing extractants like alcohols, ethers, and esters;
- organophosphorus extractants like trioctylphosphine oxide and tributyl phosphate; and
- aliphatic amine extractants like trioctylamine (Keshav et al., 2008; (Chand et al., 2009).

Standard extractants such as alcohols and ketones have a weak ability to extract carboxylic acids because of their low distribution coefficients (Eda et al., 2017). Thus, extractants with higher distribution coefficients and greater efficiency such as amines or organo-phosphoric compounds are used. Particularly, trioctylamine (TOA) and tributylphosphate (TBP) are found to be the best amine and phosphate group extractants respectively (Mungma et al., 2019). The physical extraction technique uses a variety of diluents and offers efficient extraction outcomes. To reach equilibrium, the sample is kept at the appropriate temperatures in a mechanical stirring incubator for a period of time. Following incubation, the samples were provided at certain times to reach equilibrium. Then, using a separating funnel the organic and aqueous layers are separated (Sprakel et al., 2019). The residual VFA concentration in the aqueous layer was then analyzed, and the mass balance calculation was used to determine how much VFA had been transported to the organic layer. Another type of liquid-liquid extraction is reactive extraction, where the VFA can sometimes be changed to different compounds with a better distribution co-efficient for simple extraction. The most used extractants are phosphorous solvents and aliphatic amines (Sameena Begum et al., 2020). Nowadays, ionic liquids are also employed for liquid-liquid extraction process. The trihexyl(tetradecyl)phosphonium cation and mixed chloride and bis(trifluoromethylsulfonyl)imide anion in the ionic liquid show good esterification and extraction capabilities.

11.7 MEMBRANE BASED SEPARATION

Recent investigations have demonstrated that membrane-based separation methods show better yield and selectivity of VFAs with less energy input. Membranes enable continuous VFA extraction which controls the pH independent solid retention time. Continuous VFA extraction is accomplished by membranes, which concomitantly adjusts the pH independent of solid retention period, which reduces the fermentation inhibition.

Many membrane-based technologies using various transport mechanisms are available to separate VFAs from the fermented waste stream, including reverse osmosis (RO) and nanofiltration (NF). Both NF and RO are well-established pressure-driven membrane technologies for downstream biotechnology as well as the chemical and water treatment industries (Bona et al., 2020). They are considered feasible techniques because of the VFAs' less molecular weight and chemical characteristics of VFAs. The RO membrane consists of relatively minimal pore size for the process and benefits from high ion selectivity, which enables it to reject other ionic species. NF has larger pore size and surface charge, which makes it more advantageous than RO and anticipated to enhance the selectivity (Zhu et al., 2021). Additionally, NF has a low molecular weight cut-off ranging from 100–1000 Da. Despite all of its merits, the membrane fouling that occurs in industrial-scale reactors is a drawback of the membrane-based bioreactors.

11.8 ELECTRODIALYSIS

Electrodialysis is a crucial technology that enables carboxylic acid product recovery by application of electrical field assisted ion transport toward electrodes for the optimum acid removal. It can result in high acid purity and a study refers to the possibility of removing up to 99% of acetic acid and butyric acid (Tartakovsky et al., 2011). But one of the most limiting factors for using electrodialysis at an industrial scale is electricity consumption as it can result in high processing costs.

A new processing method aiming to recover VFA and nitrogen merges the bio-electrochemical system with anaerobic digestion (Cerrillo et al., 2016). Bio-electrochemical system such as microbial electrolysis cells, that use micro-organisms attached to one or both bioelectrodes in order to catalyze oxidation/reduction reactions, coupled to anaerobic digestion in order to improve its performance.

11.9 ADSORPTION–ION EXCHANGE RESIN

Adsorption is a surface process that leads to transfer of molecule/ions (adsorbate) from gas or liquid phase to solid surface (adsorbent). The adsorption process can occur through physical forces or chemical bond formation between the adsorption medium and the adsorbate. This procedure is generally reversible, and desorption is the term used to describe the release of adsorbate from an adsorbent. Chemical compounds can be easily separated from either diluted or complicated solutions using the technique of adsorption. One of the most efficient and significant ways of recovering volatile fatty acid produced during fermentation with the adsorption and desorption cycle is by using ion exchange resins.

The adsorption occurs between the oppositely charged carboxylic acid group and the functional group of solid matrixes. It is a simple method to carry out and grants high selectivity (Reyhanitash et al., 2017). The aforementioned benefits of ion exchange adsorption make it an efficient technique to implement in industrial-level VFA extraction. The adsorbents relevant to the recovery of carboxylic acids and carboxylates can be classified according to their electronic properties in ionic and non-ionic materials and more thoroughly by their functionality and support structure. Several adsorbents have been explored for the recovery of carboxylic acid. The common functional groups that operate as reactive sites in these adsorbents are primary, secondary, and tertiary amines and quaternary ammonium (López-Garzón et al., 2014) . IERs are mostly generated from the polymerization reaction and consist of a uniform matrix. Anion exchange resins are widely used to extract carboxylic acids. Weak anion exchangers become charged over a limited pH range and otherwise are not able to exchange anions whereas strong anion exchangers exchange anions over a broad pH range. Quaternary ammonium-based adsorbents are also referred to as strong basic adsorbents. Under normal fermentation conditions the only nitrogen-based adsorbents capable of recovering carboxylate anions through anion exchange (Cerrillo et al., 2016). Even though, a quaternary ammonium-based adsorbent seems advantageous for recovering carboxylate from fermented waste, the replacement anion of the ammonium was exchanged with the mineral acid. The VFA obtained after adsorbent regeneration must go through an additional chemical and processing stage, and it is not pure and is present in an aqueous solution with a sizable amount of mineral impurities. Polystyrene-divinylbenzene (PS-DVB) resins are used in recent days for an efficient process, which was either nonfunctionalized or functionalized with a primary, secondary, or tertiary amine. Characteristics of certain PS-DVBs like Lewatit, Amberlyst, Amberlite, and Purolite were shown in Table 11.1 (Rizzioli et al., 2021).

To encourage VFA consumption during anaerobic digestion, carbon-based conductive materials are currently exploited. These materials enhance the syntropic interactions between microorganisms by encouraging the growth of biofilms on their surface (Capson-Tojo et al., 2018). Activated carbon (granular and powdered) has been extensively developed in favor of reactor performance by utilizing VFA at a faster rate which leads to quicker methane production. The potent benefit of activated carbon is the high-surface-area and porous structure. Additionally, activated carbon has high electrical conductivity and enhances direct interspecies electron transfer (DIET) as shown in Figure 11.6. In DIET there will be a free flow of electrons without being carried by reduced molecules unlike MIET (mediated interspecies electron transfer). The electrons are transferred with the aid of conductive materials between electron-donating and accepting partners (Capson-Tojo et al., 2018). DIET is anticipated to be a more rapid and efficient electron transfer mechanism than MIET. Thus, the application of activated carbon materials favors DIET and avoids VFA accumulation during the anaerobic digestion process.

TABLE 11.1
Characterization of Ion Exchange Resins and Carbon-based Adsorbents

Description	Lewatit	Amberlyst	Amberlite	Purolite	GAC	PAC
Chemical composition	Styrene-divinyl benzene primary amine (benzyl amine)	Styrene-divinyl benzene tertiary amine (benzyl amine)	Styrene-divinyl benzene (Weak base anion exchange resin)	Macroporus polystyrene crosslinked with divinyl benzene (Tertiary amine functionalized resin)	carbon	carbon
Particle size (µm)	0.47–0.57	0.49–0.69	0.3–1.18	0.42–1.2	0.4–1.2	0.001–0.15
Approx. pore volume (cm3/g)	0.27	0.10	0.10	0.6	0.97	0.65
Approx., surface area (m2/g)	0.005	0.0035	35	450	950–2000	0.000012

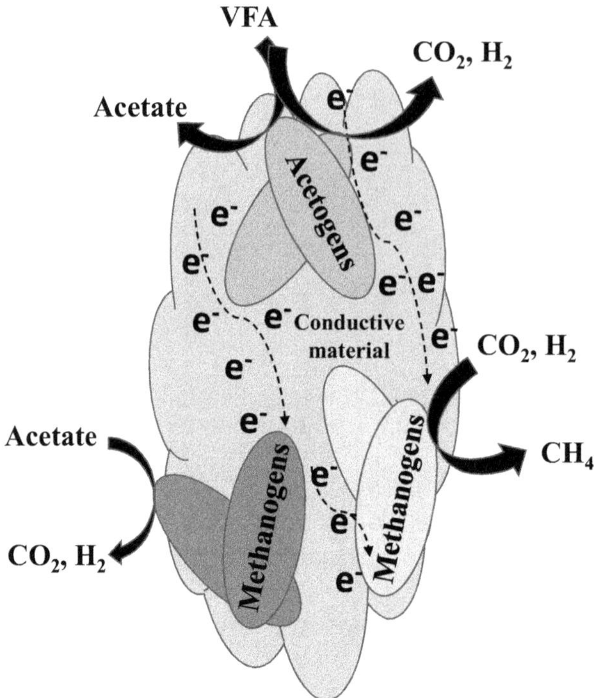

FIGURE 11.6 Direct interspecies electron transfer.

11.9.1 Types of Resin

Ion active sites are dispersed evenly throughout the cross-linked polymer matrix that makes up an ion exchange structure. The ion exchange resin has a spherical morphology with specific size and uniformity.

11.9.1.1 Weak Base Resins

Weak base resins serve as acid adsorbents because they lack exchangeable ionic sites. These resins are easily regenerable with caustic soda and possess a high capacity for adsorbing strong acids (López-Garzón et al., 2014). The pH has a significant impact on this reaction cycle's adsorption capacity because a lower pH makes more protons accessible to permit ion pairing between the soluble acid and the amine.

Recovery of a more concentrated carboxylate solution results from the use of a smaller volume of a more concentrated desorbant. However, recovering carboxylic acid from them requires extra processes. As a result, mineral acids may be employed to desorb carboxylic acid rather than carboxylate. It is anticipated that mineral acids with lower pKa values than carboxylic acids will interact with resin more strongly.

11.9.1.2 Strong Base Resins

Strong base resins are functionalized with a quaternary amine, a permanently charged functional group, and contain negative counterion to maintain the electroneutrality

(López-Garzón et al., 2014). When carboxylates are absorbed by ionic paring, the counterion (OH-) is released. When compared to other amine groups, the interaction between the quaternary amine and the carboxylate is stronger.

The quaternary ammonium is the most fundamental strong base resin commonly available and has a stronger affinity for weak acids. It is more stable and can tolerate applications involving high temperatures. Another type of resin is obtained by the reaction of the styrene-DVB copolymer with dimethylethanolamine (Wheaton et al., 2018). The regeneration efficiency is more in this type of resin.

11.9.2 Structure of Resin

The structure of the host polymer must facilitate hydrated ion diffusion to confer ion exchange capabilities. Modern organic ion exchange resins are synthesized by polymerization mechanisms. Monomers with vinyl double bonds are used to create polymers; one of the monomers has two or more vinyl double bonds to enable crosslinking. Without crosslinking, the polymerization would result in linear polymers, and the ion exchange material would be soluble (Gerin et al., 2018)

There are two distinct structure types for resins, the first kind is a continuous polymeric phase-based gel with a uniform structure and non-porous. The second kind consists of a bead-like structure and is referred to as microporous beads. These beads are heterogeneous; they are made up of interconnected macropores surrounded by gel like microbeads.

11.10 ADSORPTION ISOTHERMS AND KINETICS

The distribution of VFAs on the surface of the adsorbent can be described and characterized through isotherm models. The VFA adsorption process can be elucidated using Langmuir and Freundlich models. The Langmuir adsorption isotherm is based on the assumption that each discrete adsorption site on an adsorbent surface can only retain one adsorbate. The adsorption sites available on the adsorbent surface have equal energy and adsorbate forms a monolayer on the adsorbent surface. The Freundlich adsorption isotherm model describes the empirical relationship between pressure at a given temperature and the amount of adsorbate that is absorbed by a unit mass of adsorbent. Additionally, it provides an exponential distribution of adsorption sites and energy and explains the surface heterogeneity of adsorbents. This isotherm model assumes that multiple molecular layers of adsorbate will form on the adsorbent surface (Eregowda et al., 2019).

An important factor in any adsorption process is the impact of contact time and concentration of resins. The total number of sites that are available for exchange makes up the total capacity of the resin. Kinetics also determines the speed at which the ion exchange takes place. This speed is related to the diffusion through the film of solution in close contact with the beads called the Nernst film, and diffusion in the porosity of the resin. The intra-particle mass transfer diffusion model is commonly used to describe diffusion-controlled processes.

11.11 DESORPTION AND PURIFICATION TECHNIQUES

Regeneration of the loaded adsorbents is necessary for product recovery (Reyhanitash et al., 2017). Certain techniques involved in desorption techniques are solvent wash. It is required to recover the organic acid that has been adsorbed in the IER. The desorbants can be water or alkali solution. Using such desorbents has the disadvantage of increasing the water content of the resultant stream. Thermal operation is a significant alternative method. In order to condense and recover the VFAs in the vapor phase in a concentrated form, thermal desorption employs the application of heat. The condensate samples were analyzed by using High-performance liquid chromatography (HPLC) or gas chromatography (GC), which represents evaporated species that were condensed in the condenser. Desorption can also be done using organic solvents like methanol, and ethanol directly, as the hydroxide ion replaces the carboxylic acid in the ion active site of the resin.

Either distillation or crystallization, which often include the evaporation of water, are required for final VFA purification. Distillation is a method of separating a mixture based on differences in volatility of the components (Gerbaud et al., 2019). Even if there are other methods for separating organic acids from fermentation broths, distillation is still a significant alternative technology, particularly in the last stages of refinement. Water will evaporate first during distillation of a VFA-water mixture because all VFAs have boiling points greater than water (Yalkowsky et al., 2010). Water has a high latent heat of vaporization, so distillation of VFAs is only economically attractive as a final step of purification, after the VFAs in aqueous phase have already been concentrated (Gangadwala et al., 2003). Figure 11.7 depicts the adsorption-desorption process to regulate the VFA concentration using IER.

11.12 CONCLUSION

According to the Environmental Performance Index, 2022, India gets the lowest ranking among 180 countries. Greater attention is needed to achieve the spectrum of sustainable living by maintaining air and water quality, preserving natural resources, and reducing greenhouse gas emissions. Management of food waste is vital since it continues to increase as population and industrialization increase. Since food waste is highly concentrated and contains readily degradable components it is used for the biological treatment process. In recent decades, the generation of biogas using anaerobic digestion with food waste as a substrate is becoming popular practice worldwide.

Anaerobic digestion is a complex biological process. Thus, when dealing with highly concentrated and rapidly hydrolyzable organic waste that is susceptible to failure if not properly regulated. The reactor can readily get overloaded due to the easily digestible carbohydrates that make up most of food waste, leading to excessive build up of the inhibitory intermediate product VFA. Methanogens, being the most delicate organism, are unable to tolerate the sudden drop in pH and will ultimately fail to produce biogas. Recovering the VFA enables precise pH regulation and maintains the stability of the process, subsequently enhancing the methanogenesis which leads to high biomethane production. The IER and carbon-based adsorbents are used for the recovery of carboxylic acids. Since, the resins made from polystyrene and

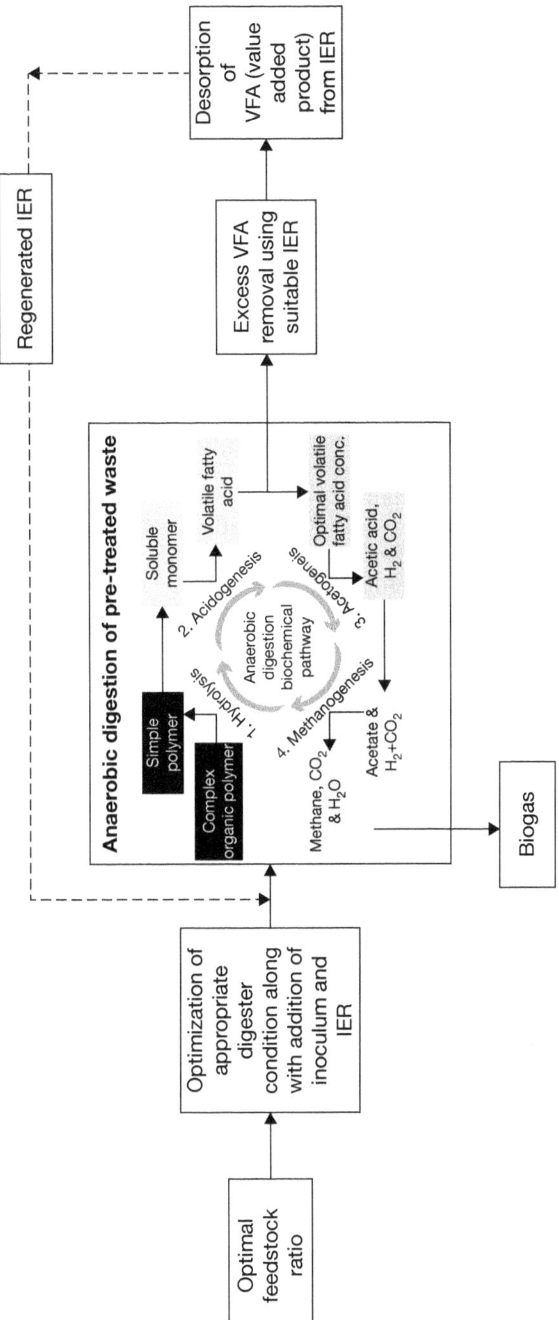

FIGURE 11.7 The regulation process of VFA concentration using IER.

divinylbenzene exhibit very high selectivity for VFAs, regeneration of these resins showed that they are remarkably robust during adsorption and desorption cycles. Ion exchange resins can produce high-purity biobased VFA in general. The adsorbent can be used in the continuous cycle of the adsorption-desorption process without losing its capacity for a longer period.

REFERENCES

Alan Henrique da Silva and Everson Alves Miranda (2013). Adsorption/Desorption of Organic Acids onto Different Adsorbents for Their Recovery from Fermentation Broths. *Journal of Chemical Engineering and data*. American Chemical Society Publication 2013, 58, 1454–1463. dx.doi.org/10.1021/je3008759

Alastair J. Ward, Phil J. Hobbs, Peter J. Holliman, David L. Jones (2008). Optimisation of the anaerobic digestion of agricultural resources. *Bioresource Technology*. Volume 99, Issue 17, November 2008, Pages 7928-7940

Amirhossein Malakahmad, Motasem S. Abualqumboz, S.R.M. Kutty, Taher Abunama (2017). Assessment of carbon footprint emissions and environmental concerns of solid waste treatment and disposal techniques; case study of Malaysia. *Waste Management*. Volume 70, December 2017, Pages 282–292.

Amit Keshav, Kailas L. Wasewar, Shri Chand (2008). Extraction of propionic acid with tri-n-octyl amine in different diluents. Separation and Purification Technology. Volume 63, Issue 1, 1 October 2008, Pages 179–183. DOI:10.1016/j.seppur.2008.04.012

Amit Keshav, Shri Chand, Kailas L. Wasewar (2009). Recovery of propionic acid from aqueous phase by reactive extraction using quarternary amine (Aliquat 336) in various diluents. *Chemical Engineering Journal* Volume 152, Issue 1, 1 October 2009, Pages 95–102. https://doi.org/10.1016/j.cej.2009.03.037

Ammaiyappan Selvam, Petchi Muthu K. Ilamathi, Muthulingam Udayakumar, Kumarasamy Murugesan (2021). Food Waste Properties. In book: *Current Developments in Biotechnology and Bioengineering* (pp.11–41). DOI:10.1016/B978-0-12-819148-4.00002-6

Aron Bona, Peter Baknoyi, Ildiko Galambos, Katalin Bélafi-Bakó (2020). Separation of volatile fatty acids from model anaerobic effluents using various membrane technologies. *Membranes* 2020, Volume 10, page 252.

B Tartakovsky, P Mehta, J-S Bourque, S R Guiot (2011). Electrolysis-enhanced anaerobic digestion of wastewater. *Bioresource Technology*. Volume 102, Issue 10, May 2011, Pages 5685–5691. DOI: 10.1016/j.biortech.2011.02.097

Camilo S. López-Garzón, Adrie J. J. Straathof (2014). Recovery of carboxylic acids produced by fermentation. *Biotechnology Advances*. Volume 32, Issue 5, September–October 2014, Pages 873–904. DOI: 10.1016/j.biotechadv.2014.04.002

Conor Dennehy, Peadar G. Lawlor, Zhenhu Hu, Matthew McCabe, Paul Cormican, Xinmin Zhan (2018). Inhibition of volatile fatty acids on methane production kinetics during dry co-digestion of food waste and pig manure. *Waste Management*. Volume 79, September 2018, Pages 302–311. https://doi.org/10.1016/j.wasman.2018.07.049

Debkumar Chakraborty, Sankar Ganesh Palani, Makarand M. Ghangrekar, N. Anand, Pankaj Pathak (2022). Dual role of grass clippings as buffering agent and biomass during anaerobic co-digestion with food waste. *Clean Technologies and Environmental Policy*. Volume 24, pages 2787–2799, (2022)https://doi.org/10.1016/j.biortech.2021.126396

Esan Reyhanitash, Sascha R. A. Kersten, Boelo Schuur (2017). Recovery of volatile fatty acids from fermented wastewater by adsorption. *ACS Sustainable Chemistry and Engineering* 2017, 5, 10, 9176–9184. DOI: 10.1021/acssuschemeng.7b02095

Fabia Rizzioli, Federico Battista, David Bolzonella, Nicola Frison (2021). Volatile fatty acid recovery from anaerobic fermentate: focussing on adsorption and desorption performances article. Industrial and Engineering Chemistry Research. https://doi.org/10.1021/acs.iecr.1c03280

Food and Agricultural Organization (FAO) (2013). Food wastage footprint Impact on natural resources: Summary report (2013). www.fao.org/3/i3347e/i3347e.pdf

Food: Material-Specific Data (2018). 2018 wasted food report. EPA United States Environmental protection agencies. Website: 19january2021snapshot.epa.gov

Gabriel Capson-Tojo, Roman Moscoviz, Diane Ruiz, Gaëlle Santa-Catalina, Eric Trably, Maxime Rouez, Marion Crest, Jean-Philippe Steyer, Nicolas Bernet, Jean-Philippe Delgenès, Renaud Escudié(2018). Addition of granular activated carbon and trace elements to favour volatile fatty acid consumption during anaerobic digestion of food waste. *Bioresource Technology*. Volume 260, July 2018, Pages 157–168. doi: 10.1016/j.biortech.2018.03.097.

Gerin, Patrick A., Castel, Guillaume (2018). Extraction and purification of volatile fatty acids. Faculté des bioingénieurs, Université catholique de Louvain, 2018. URL: http://hdl.handle.net/2078.1/thesis:14913

James Denis Browne, Jerry D Murphy (2012). Assessment of the resource associated with biomethane from food waste. *Applied Energy* 104:170–177. DOI:10.1016/j.apenergy.2012.11.017

Jignesh Gangadwala, Surendra Mankar, Sanjay Madhusudan Mahajani (2003). Esterification of Acetic Acid with Butanol in the Presence of Ion-Exchange Resins as Catalysts. *Industrial & Engineering Chemistry Research*. DOI: 10.1021/ie0204989

Justyna Lalak, Agnieszka Kasprzycka, Ewelina M. Paprota, Jerzy Tys, and Aleksandra Murat (2015). Development of optimum substrate compositions in the methane fermentation process. International Agrophysics.

Katarzyna Slopiecka, Federica Liberti, Sara Massoli, Pietro Bartocci, Francesco Fantozzi (2022). Chemical and physical characterization of food waste to improve its use in anaerobic digestion plants. *Energy Nexus*. Volume 5, 16 March 2022, 100049. https://doi.org/10.1016/j.nexus.2022.100049

Lei Zhang, Yong-Woo Lee, Deokjin Jahng (2011). Anaerobic co-digestion of food waste and piggery wastewater: focusing on the role of trace elements. *Bioresource Technology*. Volume 102, Issue 8, April 2011, Pages 5048–5059. DOI: 10.1016/j.biortech.2011.01.082

Lisette Sprakel, Boelo Schuur (2019). Solvent developments for liquid-liquid extraction of carboxylic acids in perspective. *Separation and Purification Technology*. 211 (2019) 935–957. https://doi.org/10.1016/j.seppur.2018.10.023

Miriam Cerrillo, Marc Vinas, Agust Bonmati (2016). Removal of volatile fatty acids and ammonia recovery from unstable anaerobic digesters with a microbial electrolysis cell. *Bioresource Technology*. Volume 219, November 2016, Pages 348–356. DOI: 10.1016/j.biortech.2016.07.103

Mirzaman Zamanzadeh, Live Heldal Hagen, Kine Svensson, Roar Linjordet (2016). Anaerobic digestion of food waste – Effect of recirculation and temperature on performance and microbiology. *Water Research*. DOI:10.1016/j.watres.2016.03.058

Najam Ul Saqib, Hari Bhakta Sharma, Saeid Baroutian, Brajesh Dubey, Ajit K. Sarmah (2019). Valorisation of food waste via hydrothermal carbonisation and techno-economic feasibility assessment. *Science of The Total Environment*. Volume 690, 10 November 2019, Pages 261–276. https://doi.org/10.1016/j.scitotenv.2019.06.484

Nuttakul Mungma, Marlene Kienberger, Matthäus Siebenhofer (2019). Reactive Extraction of Lactic Acid, Formic Acid and Acetic Acid from Aqueous Solutions with Tri-n-octylamine/1-Octanol/n-Undecane. *ChemEngineering* 2019, 3(2), 43. DOI:10.3390/chemengineering3020043

Robert M Wheaton, L. J. Lefevre (2018). "Fundamentals of Ion Exchange." Dow Liquid Seperations. DOW Chemical USA, June 2000.

Sameena Begum, Vijayalakshmi Arelli, Gangagni Rao Anupoju, Sridhar S., Suresh K. Bhargava, Nicky Eshtiaghi (2020). Optimization of feed and extractant concentration for the liquid–liquid extraction of volatile fatty acids from synthetic solution and landfill leachate. *Journal of Industrial and Engineering Chemistry*. Volume 90, 25 October 2020, Pages 190–202. https://doi.org/10.1016/j.jiec.2020.07.011

Samuel H. Yalkowsky, Yan He, Parijat Jain (2010). Handbook of Aqueous Solubility Data, Second Edition. CRC Press.

Shashi Kant Bhatia, Yung-Hun Yang (2017). Microbial production of volatile fatty acids: current status and future perspectives. *Reviews in Environmental Science and Bio/Technology*. Volume 16, pages 327–345, (2017) https://doi.org/10.1007/s11157-017-9431-4

Sumalatha Eda, Alka Kumari, Prathap Kumar Thella, Bankupalli Satyavathi, Parthasarathy Rajarathinam (2017). Recovery of volatile fatty acids by reactive extraction using tri-n-octylamine and tri-butyl phosphate in different solvents: Equilibrium studies, pH and temperature effect, and optimization using multivariate taguchi approach. *Canadian Journal of Chemical Engineering*. Volume95, Issue7, July 2017 Pages 1373–1387 https://doi.org/10.1002/cjce.22803

Suraj Adebayo Opatokun, Vladimir Strezov, Tao Kan (2015). Product-based evaluation of pyrolysis of food waste and its digestate. *Energy*. Volume 92, Part 3, 1 December 2015, Pages 349–354. http://dx.doi.org/10.1016/j.energy.2015.02.098

Tejaswini Eregowda, Eldon R. Rene, Jukka Rintala, Piet N. L. Lens (2019). Volatile fatty acid adsorption on anion exchange resins: kinetics and selective recovery of acetic acid. *Separation Science and Technology*. 55:8, 1449–1461. https://doi.org/10.1080/01496395.2019.1600553

Vincent Gerbaud, Ivonne Rodriguez-Donis, Laszlo Hegely, Peter Lang, Ferenc Denes, XinQiang You (2019). Review of extractive distillation. Process design, operation, optimization and control. *Chemical Engineering Research and Design*. Volume 141, January 2019, Pages 229–271 https://doi.org/10.1016/j.cherd.2018.09.020

Wanqin Zhang, Shubiao Wu, Jianbin Guo, Jie Zhou, Renjie Dong (2015). Performance and kinetic evaluation of semi-continuously fed anaerobic digesters treating food waste: Role of trace elements. *Bioresource Technology*. Volume 178, February 2015, Pages 297–305. http://dx.doi.org/10.1016/j.biortech.2014.08.046

Xiaobo Zhu, Aaron Leininger, David Jassby, Nicolas Tsesmetzis, and Zhiyong Jason Ren (2021). Will membranes break barriers on volatile fatty acid recovery from anaerobic digestion?. *ACS ES&T Engineering*. 2021, 1, 1, 141–153 https://dx.doi.org/10.1021/acsestengg.0c00081

12 Comparative Life Cycle Assessment and Carbon Footprint Analysis of Waste Treatment Facilities

Azam Akhbari and Shaliza Ibrahim

12.1 INTRODUCTION

Growing concern about global warming and its impact on the environment as well as energy security promotes research and development into green fuels, mostly from renewable sources (Tepari et al. 2020). As a reliable and eco-friendly alternative source of energy, it is essential to develop bioenergy. Biohydrogen and biomethane are the most appealing forms of bioenergy, because of their high energy densities of 141 MJ kg^{-1} and 50 MJ kg^{-1}, respectively (Mishra et al. 2017). In addition, biohydrogen can be used as a catalyst to enhance other gaseous biofuels' combustion properties. Biohydrogen and biomethane can be sequentially produced in a two-stage process that enhances energy recovery by approximately 10–43%. A further benefit is that they complement each other by converting to hythane, which greatly improves the fuel's economy and efficiency (Prawit et al. 2018). Biohythane could prove to be a promising fuel similar to biohydrogen due to its higher energy content and low primary and secondary pollution. In the anaerobic digestion of organic materials, hythane is produced. There are two processes involved in the anaerobic digestion of substrate (Chontich et al. 2019). The first phase is the formation of hydrogen, and the second phase is the generation of methane. Biohythane is formed during the following stages after the production of hydrogen through dark fermentation and the digestion of effluents to generate methane (Chontich et al. 2019). Thus, the two-stage process has been modified to cogenerate biohydrogen and biomethane in existing anaerobic digesters. To successfully produce hythane, the choice of substrate is crucial. It is possible to produce biogas from biomass if key components are present (such as carbohydrates, proteins, fats, cellulose, and hemicellulose) (Hawkes et al. 2002). Various types of organic substrates include organic solid waste, food waste, POME, etc. A significant amount of attention has been paid to POME for biogas production in recent years (Akhbari et al. 2021a). The production of biohythane from

DOI: 10.1201/9781003364467-12

POME was achieved with H_2 and CH_4 yields of 201 mL H_2/g-COD and 315 mL CH_4/g-COD, respectively (Mamimin et al. 2015). Optimizing different parameters for biohythane production is another important aspect. There are several aspects of the biochemical process that can be modified during fermentation to enhance biohythane productivity and purity. Throughout this chapter, the palm oil milling process, the potential of POME in sustainable approach toward biogas/biohythane production. By using life cycle assessment (LCA), this study compares the environmental impacts and identifies biogas production from palm oil mill effluent to reduce environmental impacts and to demonstrate the advantages and prospects of this innovative practice of bioenergy.

12.2 PALM OIL MILL EFFLUENT (POME) GENERATION

Extraction of crude palm oil (CPO) and palm kernel from fresh fruit bunch (FFB) is typically achieved by a wet milling process (Akhbari et al. 2020a). The milling process comprises four major steps, namely sterilization, pressing, oil clarification, and kernel recovery. The sterilization of FFB is followed by the separation of the fruits and stalks from the sterilized bunches. On the press station, fruits are mashed and pressed using a digester and screw press, respectively (Akhbari et al. 2020a). Through this process, crude oil is extracted and nuts are separated from mesocarp fibers. CPO is produced by screening, clarifying, purifying, and drying crude oil in oil rooms. To recover kernels from pressed cakes, nuts are separated from pressed cakes, cracked, and separated from kernel shells. In palm oil mills, water consumption typically ranges from 1–12 m^3/t of FFB processed, which is a water-intensive process (Akhbari et al. 2020a). For mills that use dilution, water consumption per t CPO was 7.5 m^3, compared to 3.5 m^3 for mills that do not use dilution. In palm oil mills, dilution is commonly practiced, resulting in high water consumption and POME production. In the milling process, about 50% of the water consumed is discharged as POME. The remaining 50% is used for processing, boiler steam blowdowns, evaporation, and leaks from washing machines. In the total volume of POME, 60%, 36%, and 4% are wastewaters produced from clarification processes, sterilization, and kernel recovery stations (Akhbari et al. 2020a).

12.2.1 Characteristics of POME

Palm oil mill effluent discharged from palm oil mills physically appears as a thick viscous brownish-dark liquid with a pH of about 4.5, a temperature ranges of 80–90 °C, high in suspended solids content (Akhbari et al. 2019). There is 95% water in it, followed by 4% organic solids and 1% oil and grease. In POME, organic carbon is abundant, and biochemical oxygen demand is high, along with O&G (6,000 mg/L), and suspended solid (SS) (18,000 mg/L) (Akhbari et al. 2020b). POME is classified as highly polluting wastewater with nutrient-rich characteristics based on these values. Due to the high oxygen depletion, untreated discharge of this product is highly harmful to the environment despite the lack of chemicals used during milling (Akhbari et al. 2020b). According to Table 12.1 on the POME characterization, the Malaysian Department of Environment regulates the parameters and limits

TABLE 12.1
Palm Oil Mill Effluent Characteristics

(A) General properties		(B) Amino acid contents		(C) Fatty acid contents		(D) Mineral contents	
Parameter	**Value[a]**	**Parameter**	**Value[a]**	**Parameter**	**Value[a]**	**Parameter**	**Value[a]**
pH	3.4–5.5	Arginine	1384–1688	Caprylic acid	835–878	B	0.25–2.00
Temperature	44–90	Aspartic acid	3437–3546	Capric acid	1518–1583	Na	3.19–87.92
Density	876–985	Cystine	1192–1243	Lauric acid	1142–1185	Mg	29.51–1144
Viscosity	4.17–4.69	Glutamic acid	3886–3980	Myristic acid	4533–4619	Al	6.30–344
Turbidity	65,590–69,410	Glycine	3343–3473	Pentadecanoic acid	784–813	Si	55–125
Oil & grease	2234–27,166	Histidine	466–567	Palmitic acid	7435–8795	P	74.50–2370
BOD_5	10,250–76,457	Isoleucine	1518–1756	Heptadecanoic acid	491–513	S	60–400
COD	15,000–100,000	Leucine	1677–1836	10-heptadecanoic acid	394–415	K	314.32–5533
Total solids	11,500–79,000	Lysine	910–1012	Stearic acid	3567–3958	Ca	53.85–607
Total suspended solids	5000–71,275	Methionine	2288–2685	Oleic acid	4804–5708	V	0–0.01
Total dissolved solids	20,554–41,098	Phenylalanine	1102–1211	Linoleic acid	3011–3878	Cr	0.13–0.16
Total volatile solids	9000–72,000	Proline	1612–1691	Linolenic acid	1511–1901	Mn	1.27–1.53
Total protein	460–1580	Serine	2429–2530	Arachidic acid	1265–1308	Fe	0.32–0.48
Total organic carbon	9584–38,616	Threonine	842–1023	Eicosatrienoic acid	708–766	Co	0.07–0.10
Total nitrogen	204–1708	Tryptophan	441–469	Eicosatetraenoic acid	394–415	Ni	0.04–0.06
Total Kjeldahl nitrogen	330–1530	Tyrosine	1073–1211	Eicosapentaenoic acid.	108–151	Cu	0.35–0.43
NH_3–N	17–254	Valine	1200–1373			Zn	0.56–0.71
Carotene	6.50–7.23					Se	0.40–0.49
Hemicellulose	2530–8433					As	0.31–0.35
Cellulose	3976–14,300					Mo	0.22–0.25
Lignin	9044–15,182					Cd	0.01–0.02
Extractive	4370–14,459					Sn	0.07–0.09
Ash	1178–5867					Pb	0.17–0.21

of discharged POME at much stricter levels than those imposed on raw POME. It is a resource that can be biotechnologically processed due to its high carbohydrate and nitrogen content, including protein and amino acids, lipids, organic nutrients, and minerals. It also provides an environment and resources that are conducive to the growth of microorganisms for the production of biogas through AD (Kongjan et al. 2011).

12.2.2 Potential of POME for Producing Biogas

Biomass can be converted into biogas when it contains key components (including carbohydrates, proteins, fats, cellulose, and hemicellulose) (Akhbari et al. 2020b). POME is considered a suitable biomass for biogas production; since it contains a high level of carbohydrates, proteins, and lipids (Akhbari et al. 2020a). The unique properties of POME can also be used in compost production as well as converting it into biodiesel, biobutanol, and biohydrogen. POME wastes contain significant organic solid fractions and micronutrients. The production of 1 tonne of crude palm oil generates $2.5*10^3$ kg of POME, resulting in approximately 70 m^3 of biogas. Thus, one tonne of POME produces 28.13 m^3 of biogas (Akhbari et al. 2021b). A complex mechanism, involving the interactions of microorganisms, is needed for the production of biogas from POME (Akhbari et al. 2020b). This process is called anaerobic digestion. This process has four stages: hydrolysis, acidogenesis, acetogenesis, and methanogenesis. As a result of hydrolysis, carbohydrates, lipids, and proteins are broken down into smaller sugar, fatty acid, and amino acid molecules, respectively (Li et al. 2015). The process is carried out by a group of bacteria called hydrolytic bacteria, such as *Clostridium* and *Bacillus*. The process of hydrolysis allows all raw materials to be converted into amenable forms for further microorganism activity. As a result of the hydrolysis of compounds, acidogenic bacteria, produce intermediary compounds (Li et al. 2015). Acetogenic bacteria, degrade the intermediaries during acetogenesis to form acetate, carbon dioxide, and hydrogen. During methanogenesis, two types of bacteria are involved, *hydrogenotrophic* and *acetotrophic*. The acetotrophic methanogen splits acetate into methane and carbon dioxide, while the hydrogenotrophic methanogen uses hydrogen to form methane (Liu and Whitman 2008). This process produces 70% of its methane through the acetotrophic methanogen pathway and 30% through the hydrogenotrophic methanogen pathway (Lee et al. 2013). To complete the anaerobic digestion process, 30–40 days at a mesophilic temperature are obligatory. Meanwhile, the entire process takes only 7–14 days at thermophilic temperatures. During digestion's acidogenic and methanogenic stages, pH also shows a significant role. An acidogenic environment requires a pH of 5, while methanogenic environments require a pH of 6.5–7.2 (Liu et al. 2008). In the anaerobic digester, pH and alkalinity are related, because alkalinity regulates pH. Alkalinity is a measure of how well a substance neutralizes hydrogen ions, while pH measures hydrogen ion concentration. Biogas production is increased when alkalinity is increased at the beginning of digestion. In order to produce methane from carbohydrate waste, an alkalinity-to-COD ratio of 1.2 to 1.6 (w/w) is necessary (Singer et al. 2018). Calcium carbonate at a rate of 5500 mg/L is beneficial for methanogenic bacteria in coping with sensible shock loads of VFAs. POME production of biogas is also influenced by environmental factors, such as pretreatment and bioreactor configuration.

12.2.3 A Feasible Approach to Biogas Production from POME

Two distinct metabolic processes in AD convert organic material into biogas: mesophilic (35 °C) and thermophilic (55 °C) (Akhbari et al. 2019). In these conditions, anaerobic methanogenic archaea produce biogas by dismutating acetic acid and reducing H_2. During AD, four metabolic reactions occur: hydrolysis, acidogenesis, acetogenesis, and methanogenesis. In the first step, organic compounds are hydrolyzed into smaller units, such as sugars and acid amino acids. Acidogenic bacteria then break down the hydrolysis products into organic acids, primarily hydrogen and carbon dioxide. As a result of the acetogenesis reaction, acid-phase products become acetate, hydrogen, and carbon dioxide. The methanogenesis process involves the conversion of acetate and hydrogen to CH_4 and CO_2 by bacteria that are azeotropic and hydrogenotrophic. Several factors influence the production of biogas, including the environment and internal factors. Temperature and pH are the main environmental factors that determine how well the biogas technology works internally (Akhbari et al. 2020b). Several internal factors can affect the production of biogas in anaerobic environments, including mixing, nutrients, organic loading rates (OLR), hydraulic retention time (HRT), microbial populations and activities, inhibitory materials, pressure, and chemical equilibrium (Liu et al. 2008). If these factors are not addressed carefully, scaling up to an industrial scale will be a major challenge. It is commonly reported that 28 m^3 biogas per m^3 POME and 0.25 kg CH_4 per kg COD removed are generated by the anaerobic digestion of POME. The composition of biogas ranges from 60–70% CH_4, 30–40% CO_2, and 800–1500 ppm H_2S. In biogas, only CH_4 is combustible, so a higher CH_4 indicates better quality biogas. As a result of methane emissions from the ponds, 6.54 kg/t FFB are emitted, equivalent to 137.4 kg of carbon dioxide. It is not possible to produce higher biogas yields using conventional anaerobic-open ponding or tanks, which do not include capturing and using the generated CH_4 as an energy source. Using a closed anaerobic digester for POME treatment and capturing biogas can be both cost-effective and environmentally friendly (Yeshanew et al. 2016).

12.2.4 Palm Oil Mill Biogas Industrial Plant Technology

Since palm oil is Malaysia's largest biomass producer, it produces many by-products that can be used as fuel. Several benefits are provided to the palm oil industry to improve renewable energy development. In Malaysia, palm oil mills produce large amounts of POME, which is used as an energy source. POME could potentially become Malaysia's sustainable biogas source as its anaerobic digester technology continues to advance. In most palm oil mills, biogas is captured using a closed digester tank system and a covered lagoon system. As mature technology, closed digester tanks are preferred due to their, developed biogas yields, reduced footprints, and shorter HRT, although their low capital outflow and effective costs are significantly higher than those covered lagoons (Loh et al. 2017). According to a study, closed digesters produce 0.23 kg of CH_4 per kilogram of COD treated, compared to only 0.16 kg from covered lagoons. Because of a deficiency of process mechanism and an extended hydraulic retention time, covered lagoon systems are generally less efficient. The covered lagoon strategy has been upgraded pointedly, mainly with several feeding and desludging systems and involvement mechanisms using biogas and hydraulics (Lok et al. 2020). In contrast to conventional systems, these systems produced higher biogas yields and reduced HRT. A covered lagoon system is an excellent choice for mills with larger capacities and a large land area. Developing local commercial biogas technologies is advancing toward cost-effective technology that captures and conserves biogas. As technology providers provided the most information about these technologies, they must be assessed from the viewpoint of long-term commercialization. The addition of biogas capture facilities to POME treatment systems gains a great deal of attention to complement conventional POME treatment systems that are not financially viable. POME treatment and biogas production with high-rate bioreactors are summarized in Table 12.2. As a result of improved control and process systems, high-rate AD bioreactors have been industrialized over the years, improving POME treatment efficiency and biogas yields, as well as catching productivities in shorter HRTs and lesser footprints in the future for large-scale applications. In previous years, Zainal et al. (2019) investigated UASFF and found that it was the best technology for COD removal efficiencies, as well as high CH_4 yield at minimum HRT, compared to other technologies. Commercial applications of membrane technology have been hindered by fouling and high costs. There are only two technologies most commonly deployed for commercial operation: CSTR and UASB. CSTRs are recognized for their straightforwardness in design, ease of set-up, and enhanced mixing rates. It offers similar advantages to UASB technology, and significantly increases CH_4 yields.

12.3 LIFE CYCLE ASSESSMENT OF BIOGAS WITH A GLOBAL OVERVIEW OF PREVIOUS STUDIES' PERSPECTIVES

Biogas is becoming more popular as a green and renewable energy source, and LCAs are increasingly used to evaluate its environmental performance (Akhbari et al. 2021b). According to ISO 14040 and 14044, LCA methods were used to conduct the studies. Increasing research on biogas generation from organic wastes has been driven by environmental crises such as global warming, fossil fuel depletion, and greenhouse gas emissions (Akhbari et al. 2021b). There is a growing interest in LCA for the assessment of biogas production and utilization, as well as for assessing the sustainability of energy pathways. A list of some articles on the LCA of biogas is shown in Table 12.3, along with their scope, methodology, and categories of biogas and digestate utilization. Further enhancing biogas' potential as a sustainable renewable energy source may require a comprehensive life cycle assessment. Biogas energy systems are analyzed using a variety of specific LCA goals according to the reviewed studies. Biogas systems were primarily evaluated in terms of their environmental impacts in the majority of the reviewed studies. Biogas energy sustainability can be assured through environmental performance evaluation of biogas systems according to prior studies. An investigation conducted by (Vu et al. 2015) examined whether

TABLE 12.2
POME Treatment and Biogas Production with High-rate Bioreactors

Type of reactor	Operational conditions	CH_4 production rate	H_2 production rate	COD removal efficiency, %	Ref.
Up-flow anaerobic sludge-fixed film (UASFF)	Mesophilic-38 °C, HRT-1.5 days	0.34 L CH_4/gCOD	-	97	(Najafpour et al. 2006)
UASFF (Two-stage)	Thermophilic 54 °C, HRT-1 day	15.63 L CH_4/d	-	93	(Bidattul et al. 2020)
UASFF (Two-stage)	Mesophilic 44 °C, HRT-1 day	9.60 L CH_4/d	5.29 L H_2/d	83.70	(Zainal et al. 2019)
UASFF Pilot-scale (One-stage)	Mesophilic-37 °C, HRT-1 day	-	36 L H_2/d	48	(Akhbari et al. 2020b)
UASFF (Single stage)	Mesophilic-37 °C, HRT 8h	-	4.1 L H_2/cycle	33	(Akhbari et al. 2021a)
UASFF (Single stage)	Mesophilic-37 °C, HRT 10.5h	-	4800 ml/cycle	24.33	(Akhbari et al. 2021a)
Continuous stirred tank reactor (CSTR)	Mesophilic 37 °C, HRT-7 days	0.3	-	71	(Choorit and Wisarnwan 2007)
CSTR	Thermophilic 55 °C, HRT-6 days	0.8	-	82	(Irvan et al. 2018)
High-rate CSTR	Thermophilic 54 °C, HRT-3 days	0.27	-	82	(Khemkhao et al. 2015)
CSTR	Thermophilic 55°C, HRT-2 days	0.26	-	85	(Mahmod et al. 2020)

TABLE 12.3
The Overview of Some Studies on the LCA of Bioenergy Production

		Methodology						Impact assessment		
Scope of study	Location	Functional unit	Reference system	System boundaries	Method/ software	Feedstock	Biogas utilization	Midpoint	Endpoint	Ref.
Bioenergy and Environmental Performance	Malaysia	1 kg Biohydrogen production from POME	Biogas	Cradle-to-grave	SimaPro LCA software and ReCiPe 2016 impact assessment method	POME	-	*	-	(Akhbari et al. 2021b)
Environmental Performance	Germany	1 ton of feedstock mixture	German condition	Biogas production and utilization	SimaPro7.2, Ecoinvent v2.1	Straw	Biomethane	*	*	(Poeschl et al. 2012)
Energy and Environmental Performance	China	1 MJ produced by the combustion of biogas	Traditional agrosystem	Biogas and digested system	SimaPro 7.1	Straw	Electricity, heat, biofuel	*	-	(Chen et al. 2012)
Energy and Environmental Performance	Italy	100 kWh of electricity produced	Biogas plant	Cradle-to-grave	Ecoinvent database	MS,TS	CHP	*	-	(Lijó et al. 2014)
Environmental Performance	United Kingdom	1 m^3 of biogas	-	Cradle-to-grave	SimaPro, Ecoinvent 99	-	Heat	*	*	(Mezzullo et al. 2013)

biogas digesters have positive or negative effects on the environment in Vietnam. Biogas digesters in pig farms were compared with other manure management systems to assess their environmental impacts. As a result of the study, the biogas digester was shown to reduce global warming impacts. However, biogas systems must be improved in order to minimize biogas losses from digesters and manure storage emissions. Energy and environmental performance were also considered in other studies, taking into account all flows of energy and materials and emissions. In biogas production, energy balance has been assessed through energy performance evaluations. As a result, several studies have examined the explicit factors that affect the performance of biogas systems in order to stun these boundaries. According to (Claus et al. 2014), diverse biogas substrate cropping methods have different emissions of GHG and nitrate leaching potentials. Based on the results of the study, it appears that energy crops produce biogas which contributes to the mitigation of greenhouse gas emissions. In addition, cropping systems and nitrogen fertilization should be adapted to site conditions if sustainable biogas production is to be achieved. Björnsson and Prade (2014) evaluated greenhouse gas emissions using grass crops as a biogas feedstock. In addition to improving the soil's organic carbon content, the cultivation system would also result in a reduction in greenhouse gas emissions. Dressler et al. (2012) investigated the factors that influence biogas production from maize and biogas conversion to electricity based on regional parameters. An assessment of the life cycle costs for POME-based energy generation, focusing covered lagoon biodigesters and continuous stirred tank reactors (CSTRs) was investigated by Dressler et al. (2012). Compared to covered lagoon biodigesters, CSTR has a global warming potential (GWP) of 4.48 kg CO_2 eq/kWh and an acidification potential (AP) of 2.21 kg SO_2 eq/kWh, while CLB has 4.09 kg CO_2 eq/kWh and 0.15 kg SO_2 eq/kWh.

12.3.1 Life Cycle Assessment of Biogas Production in Malaysia: Potential and Prospects

Towards industrial development and environmental protection, the Ministry of Energy, Green Technology, and Water focuses on green technologies. The Malaysian government also participated in the Kyoto Protocol's Clean Development Mechanism (Chin et al. 2013). The clean development mechanism allows developing countries to profit from trading certified emission reductions with developed countries. In addition to reducing greenhouse gas emissions, the program allows palm oil millers to also trade certified emission reductions through the capture of methane from POME (Chin et al. 2013). Green and sustainable technologies have been the subject of a variety of research and development activities by both public and private institutions. In the current plan, the energy supply is to be made sustainable and emission reductions are to be achieved, along with increasing community awareness about the significance of sustainable and green technology. Policies and decision-makers need to conduct and initiate more research in order to achieve these goals and face the existing challenges. Implementing LCA in biogas production could help optimize the whole biogas system. Performing LCAs on bioenergy performance could deliver innovative visions into the system operation, both in terms of energy use and environmental impact.

Biogas production involves the consumption and release of materials and energy. In order to improve overall environmental sustainability, a holistic approach to assessing biogas deployment is needed. Stakeholders could use LCA studies to assess the environmental profiles of their products. POME contributes up to 50% of GHG emissions. Hence, as an effort to reduce the carbon footprint, mills are beginning to install methane capture facilities. A number of life cycle assessments have been conducted by the Malaysian Palm Oil Board, ranging from seedlings to palm biodiesel production and consumption. By studying oil palm seedlings from gate-to-gate, greenhouse gas emissions from the oil palm supply chain have also been measured. In order to maintain a sustainable environment and resource management, the government has provided policies, schemes, and funding in support of green policies and schemes. Consequently, it may influence the LCA studies involving biogas systems. It has been observed that Malaysian palm oil mills are increasingly developing biogas plants to reduce their operational emissions. Although only about half of palm oil mills have installed biogas plants, the launch of these plants displays a commitment to clean advance and promoting low-carbon systems (Lansche et al. 2017).

12.4 BIOHYTHANE

Using renewable resources to generate biohydrogen and biomethane has many environmental benefits, such as reducing carbon dioxide emissions and nitrogen oxide emissions. The mixture of biohydrogen and biomethane is called biohythane (Veluswamy et al. 2019). Biohythane has a great deal of potential, which opens up the possibility of active research in this field that requires a growing investment in intellectual, technical, and logistical investments (Pasupuleti and Mohan 2015). Through two-stage anaerobic digestion, biohydrogen and biomethane can be co-produced from organic substrates and can increase energy recovery by 8–43% (Liu et al. 2013). The production of biohythane could be enhanced by combining biohydrogen and biomethane, which in turn could enhance the combustion process efficiency. In two-stage anaerobic fermentation processes, biohythane is produced over a 13–18-day fermentation period. Biohythane produced by two-stage fermentation processes is an efficient energy recovery energy through the degradation of organic components of biomass (about 67.70%). Hydrolysis, acidogensis, and acetogenesis are involved in the first stage (dark fermentation), while methanogenesis is involved in the second stage (anaerobic digestion) (Friedrich et al. 2005). Using hydrolytic microorganisms, complex organic matter is converted into monomeric units in the first stage. The monomers are then converted into hydrogen, carbon dioxide, alcohols, intermediate metabolites, and lactic acid by acidogenic bacteria. A methanogen converts alcohols, intermediate metabolites, and lactic acid into methane and carbon dioxide in the second stage. Further, methanogens produce methane from the volatile fatty acids produced during the previous stage (Zainal et al. 2019). In addition, it boosts the hydrogen and methane yields by switching intermediate metabolites from the first to second stages. Two-stage processes have several advantages, including being flexible in adjusting the hydrogen/methane ratio, reducing fermentation time, increasing degradation efficiency, generating numerous products, and increasing energy recovery (Akhbari et al. 2021a). The integration of biohydrogen and biomethane into biohythane could be worth exploring for industrial applications. There are a number of advantages to using biohythane as a fuel over currently available fuels, including cleaner, eco-friendly, and more energy-efficient. With the two-stage process, fermentation time is also reduced with increased organic loading rates, and methane and hydrogen production are optimized separately. As a result, the two-stage process provides increased mechanical reliability, increased shock resistance when heavy loads are applied, improved pH regulation, reduced chances of failure due to the accumulation of volatile fatty acids and high efficiency in converting organic substrates into energy. Hence, the most efficient way to synthesize biohythane is by producing biohydrogen and biomethane in two steps. A circular economy can be achieved by utilizing POME as a bioenergy feedstock. Additionally, some researchers have investigated the possibility of generating biohythane using single-step or single-stage processes (David et al. 2019). The biokinetic performance of biohythane production in two-stage processes is ostensibly superior to those of one-stage processes. The rate-limiting step in the one-stage biohythane production is the production of H_2, which ultimately determines the composition of biohythane. Acidogens produce VFAs, which in turn are consumed by methanogens, resulting in the simultaneous production of H_2 and CH_4. In one-stage biosystems, both the production and consumption of VFAs are crucial for acidogens and methanogens function. H_2-producing acidogens are favored by pH 6.0. It is possible to observe a steady rise in pH after 8 hours of operation, due to increased hydrolytic utilization of the fatty acids produced from the substrate. The most efficient biohythane production technology involves the production of biohydrogen, followed by the production of biomethane. A disadvantage of this two-stage process is the keeping of an impeccable hydrogen and methane ratio, resulting in very low and unstable gas yields. Several factors play a role in biohythane production, including pretreatment techniques, substrate types, inoculum characteristics, bioreactor configurations, and etc. In a one-stage process, biohythane can be generated without the need for a secondary bioreactor, making it feasible to implement a sustainable process that supports both production increases and user demands. It is more economically viable to use this system than two-stage systems with individual acidogenic and methanogenic reactors. Biohythane can therefore be produced using a one-stage system that is innovative.

12.4.1 Palm Oil Mill Effluent as Substrate in Biohythane Production

A large amount of organic matter and micronutrients are present in POME wastes. Its low pH at higher temperatures benefits biohythane production (Akhbari et al. 2020b). Two-stage solid anaerobic digesters were used by (Suksong et al. 2015) to produce biohythane by co-digesting POME waste with empty palm fruit bunches. Anaerobic digesters with two stages have been shown to enhance the variolization of lignocellulosic biomass (cellulose is 57–69%, hemicellulose is 35–40%, and lignin is 16–27%) (Suksong et al. 2015). The solid waste residues could also be co-digested with POME wastes to produce biohythane. Biohythane production was improved significantly and cost-effectively by combining oil industry solid waste and POME. The two-stage thermophilic reactor developed by (Seengenyoung et al. 2019) is designed to produce biohythane from POME waste substrate.

12.4.2 Microbial Community in Biohythane Production

The two-stage fermentation process can generate biohythane if the correct microbes are identified that produce both hydrogen and methane in the desired ratio. In biohythane production, two separate groups of enzymes, namely acetogens and methanogens are enriched because their physiology, growth rates, nutritional requirements, and environmental sensitivity differ (Akhbari et al. 2021a). Thus, under various reaction conditions, these microorganisms work synergistically to decompose the feedstock and provide H_2 and CH_4. In the first stage of biohythane fermentation, different bacterial strains are involved in the DF reaction to generate H_2. A mesophilic and thermophilic bacterium (chemoautotroph) was found to be one of the most efficient H_2 producers among the numerous microbes that contribute to hydrogen production (Akhbari et al. 2021). High hydrogen production by obligate anaerobic *Clostridium sp.* is recognized under mesophilic conditions. Based on their theoretical yield, they can produce 2.1–2.2 mol H_2/mol sugar consumed. Under mesophilic conditions, *Enterobacter sp.* produced H_2 at a yield of 1.0 mol H_2/mol sugar (Knappe and Sawers 1990). The major metabolic end product of *Bacillus sp* is lactic acid, which was produced at a rate of 0.5 mol H_2/mol glucose. Furthermore, dark fermentation in thermophilic conditions shows a very favorable kinetic and stoichiometric profile for hydrogen production, consequently, the inoculum is less likely to contain contaminants or partial hydrogen pressure. During the dark fermentation of hydrogen, thermophilic bacteria were identified as moderately thermophilic (45 °C-55 °C), thermophilic (55 °C-75 °C), or extremophilic (above 75 °C) based on their optimal growth temperature. Under thermophilic conditions, some thermophilic *Clostridium spp* and *Thermoanaerobacterium spp* play a role in DF. *Thermoanaerobacterium sp* (Seengenyoung et al. 2019) are moderate thermophiles that ferment butyrate and acetate to produce hydrogen. Whereas *C thermocellum*, an obligatory anaerobe, produces H_2 from hexoses with a yield of 1.2–1.6 mol/mol hexose/H_2. Various hexoses derived from lignocellulosic waste were fermented by *Thermotoga sp.* and *Caldicellulosiruptor sp.* under extreme thermophilic conditions, yielding 3.3 mol H_2/mol hexose, respectively (Seengenyoung et al. 2019). The second stage of biohythane production in anaerobic digesters involves bacteria and archaea-producing methane. Several classes of Archaea are capable of generating CH_4 under anaerobic conditions. *Methanococcus sp.*, *Methanoculleus sp.*, and *Methanospirillum sp.* contribute significantly to methane production from carbon dioxide, hydrogen, and acetic acid under mesophilic conditions. Under thermophilic conditions, *Methanothermobacter sp.* and *Methanosarcina sp.* are active in converting CO_2 into CH_4, by using acetic acid or methyl-containing compounds and H_2. Methane is produced by *Methanothermococcus sp.* using carbon dioxide, hydrogen, under thermophilic conditions (85 °C) (Taubner and Rittmann 2016).

12.5 CONCLUSION

A good alternative waste management solution, especially in developing countries, is biogas production for waste disposal. Malaysia, like many other countries around the globe, has switched from fossil fuels to renewable energy sources. In an industry that is one of the largest producers and exporters of palm oil biomass, it is imperative that biogas production from palm oil be thoroughly assured to be sustainable. Several advantages taken by the Malaysian government, including biogas production from POME, will produce stimulating consequences for growing the country's environmental outline. As Malaysian palm oil companies take steps toward making environmental investments and improving their international competitiveness, it is vital to implement LCA. Exhaustive inventories may help open the eyes of many companies toward making environmental investments in the future.

REFERENCES

Akhbari A, Zinatizadeh AA, Mohammadi P, Syirat Z, Ibrahim S, 2019. Effect of operational variables on biological hydrogen production from palm oil mill effluent by dark fermentation using response surface methodology. *Desalination Water Treat* 137:101–13.

Akhbari A, Kutty PK, Chuen OC, Ibrahim S, 2020a. A study of palm oil mill processing and environmental assessment of palm oil mill effluent treatment. *Environ Eng Res* 25:212–21.

Akhbari A, Chuen OC, Zinatizadeh AA, Ibrahim S, 2020b. Start-up study on biohydrogen from palm oil mill effluent in a pilot-scale reactor. *Clean: Soil, Air, Water* https://doi.org/10.1002/clen.202000192.

Akhbari A, Chuen OC, Ibrahim S, 2021a. Start-up study of biohydrogen production from palm oil mill effluent in a lab-scale up-flow anaerobic sludge blanket fixed-film reactor. *Int J Hydrogen Energy* 46:10191–204.

Akhbari A, Chi CO, Ibrahim S, 2021b. Analysis of biohydrogen production from palm oil mill effluent using a pilot-scale up-flow anaerobic sludge blanket fixed-film reactor in life cycle perspective. *Int J Hydrogen Energy* 46:34059–72.

Akhbari A, Ibrahim S, Shakeel Ahmad M, 2023. Optimization of up-flow velocity and feed flow rate in up-flow anaerobic sludge blanket fixed-film reactor on biohydrogen production from palm oil mill effluent. *Energy* 266: 126435.

Bidattul Z, Danaee M, Syuhadaa N, Ibrahim S, 2020. Effects of temperature and dark fermentation effluent on biomethane production in a two-stage up-flow anaerobic sludge fixed-film (UASFF) bioreactor. *Fuel* 263: 116729.

Björnsson L, Prade T, 2014. Introduction of grass-clover crops as biogas feedstock in cereal-dominated crop rotations. Part II: effects on greenhouse gas emissions, Proceedings of the 9th International Conference on Life Cycle Assessment in the Agri-Food Sector. San Francisco, USA, pp. 8–10 ISBN:978-0-9882145-7-6.

Chen S, Chen B, Song D, 2012. Life-cycle energy production and emissions mitigation by comprehensive biogas–digestate utilization, Bioresour. *Technol.* 114; 6: 357–364.

Chin MJ, Poh PE, Tey BT, Chan ES, Chin KL, 2013. Biogas from palm oil mill effluent (POME): opportunities and challenges from Malaysia's perspective, Renew. *Sustain Energy Rev* 26; 6: 717–726.

Chontich M, Prawit K, Sompong O-T, Poonsuk P, 2019. Enhancement of biohythane production from solid waste by co-digestion with palm oil mill effluent in two-stage thermophilic fermentation. *Int J Hydrogen Energy.* 44; 32: 17224–37.

Choorit W, Wisarnwan P, 2007. Effect of temperature on the anaerobic digestion of palm oil mill effluent. *Electron J Biotechnol* 10: 376–385. https://doi.org/ 10.4067/S0717-34582007000300005.

Claus S, Taube F, Wienforth B, Svoboda N, Sieling K, Kage H, 2014. Life-cycle assessment of biogas production under the environmental conditions of northern Germany: greenhouse gas balance, *J Agric Sci* 152: 172–181.

David B, Federico B, Cristina C, Marco G, Federico M, Paolo P, 2019. Biohythane production from food wastes. *Biohyd. Elsevier* 347–68.

Dressler D, Loewen A, Nelles M, 2012. Life cycle assessment of the supply and use of bioenergy: impact of regional factors on biogas production, *Int J Life Cycle Assess* 17;9:1104–1115.

Friedrich MW, 2005. Methyl-coenzyme M reductase genes: unique functional markers for methanogenic and anaerobic methane-oxidizing Archaea. *Methods Enzymol.* 397:428–42.

Hawkes F, Dinsdale R, Hawkes D, Hussy I, 2002. Sustainable fermentative hydrogen production: challenges for process optimization. *Int J Hydrogen Energy* 27:1339–47.

Irvan I, Trisakti B, Maulina S, Daimon H, 2018. Production of biogas from palm oil mill effluent at pilot scale: effect of recycle sludge. *Orient J Chem* 34; 1: 161–8. https://doi.org/10.13005/ojc/340118.

Khemkhao M, Techkarnjanaruk S, Phalakornkule C, 2015. Simultaneous treatment of raw palm oil mill effluent and biodegradation of palm fiber in a high-rate CSTR. *Bioresour Technol* 177:17–27. https://doi.org/10.1016/j.biortech.2014.11.052.

Knappe J, Sawers G, 1990. A radical-chemical route to acetyl–CoA: the anaerobically induced pyruvate formate-lyase system of Escherichia coli. *FEMS Microbiol Rev* 6:383–98.

Kongjan P, Thong S, Angelidaki I, 2011. Performance and microbial community analysis of two-stage process with extreme thermophilic hydrogen and thermophilic methane production from hydrolysate in UASB reactors. *Bioresour Technol* 102:4028–35.

Lansche J, Müller J, 2017. Life cycle assessment (LCA) of biogas versus dung combustion household cooking systems in developing countries–a case study in Ethiopia. *J Clean Prod* 165:828–35.

Lee SH, Park JH, Kang HJ, Lee YH, Lee TJ, Park HD, 2013. Distribution and abundance of Spirochaetes in full-scale anaerobic digesters. *Bioresour Technol* 145:25–32.

Li D, Liu S, Mi L, Li Z, Yuan Y, Yan Z, Liu X, 2015. Effects of feedstock ratio and organic loading rate on the anaerobic mesophilic co-digestion of rice straw and cow manure. *Bioresour Technol* 189: 319–326 http://dx.doi.org/10.1016/j.biortech.2015.04.033.

Lijó L, González-García S, Bacenetti J, Fiala M, Feijoo G, Lema JM, 2014. Life cycle assessment of electricity production in Italy from anaerobic Co-digestion of pig slurry and energy crops. *Renew Energy* 68: 625–635.

Liu Y, Whitman WB, 2008. Metabolic, phylogenetic, and ecological diversity of the methanogenic archaea. *Ann N Y Acad Sci* 1125:171–89.

Liu Z, Zhang C, Lu Y, Wu X, Wang L, Wang L, 2013. States and challenges for high-value biohythane production from waste biomass by dark fermentation technology. *Bioresour Technol* 135:292–303.

Loh SK, Nasrin AB, Azri SM, Adela BN, Muzzammil N, Jay TD, Eleanor RA, Lim WS, Choo YM, Kaltschmitt M, 2017. First report on Malaysia's experiences and development in biogas capture and utilization from palm oil mill effluent under the economic transformation programme: current and future perspectives. *Renew Sustain Energy Rev* 74: 1257–1274. https://doi.org/10.1016/j.rser.2017.02.066.

Lok X, Chan YJ, Foo DC, 2020. Simulation and optimisation of full-scale palm oil mill effluent (POME) treatment plant with biogas production. *J Water Proc Eng* 38: 101558 https://doi.org/10.1016/j.jwpe.2020.101558.

Mahmod SS, Azahar AM, Luthfi, AAI, Abdul PM, Mastar MS, Anuar N, Takriff MS, Jahim JM, 2020. Potential utilisation of dark-fermented palm oil mill effluent in continuous production of biomethane by self-granulated mixed culture. *Sci Rep* 10: 1–12. https://doi.org/10.1038/s41598-020-65702-w.

Mamimin C, Singkhala A, Kongjan P, Suraraksa B, Prasertsan P, Imai T, 2015. Two-stage thermophilic fermentation and mesophilic methanogen process for biohythane production from palm oil mill effluent. *Int J Hydrogen Energy* 40;19:6319–28.

Mezzullo WG, Mcmanus MC, Hammond GP, 2013. Life cycle assessment of a small scale Anaerobic digestion plant from cattle waste. *Appl Energy* 102: 657–664.

Mishra P, Thakur S, Singh L, Krishnan S, Sakinah M, Wahid ZA, 2017. Fermentative hydrogen production from indigenous mesophilic strain Bacillus anthracis PUNAJAN 1 newly isolated from palm oil mill effluent. *Int J Hydrogen Energy* 42:16054–63.

Najafpour GD,. Zinatizadeh AAL, Mohamed R, Hasnain Isa M, Nasrollahzadeh H, 2006. High-rate anaerobic digestion of palm oil mill effluent in an upflow anaerobic sludge-fixed film bioreactor. *Proc Biochem* 41; 2: 370–379.

Pasupuleti SB, Mohan SV, 2015. Single-stage fermentation process for high-value biohythane production with the treatment of distillery spent-wash. *Bioresour Technol* 189:177–85.

Poeschl M, Ward S, Owende P, 2012. Environmental impacts of biogas deployment–Part II: life cycle assessment of multiple production and utilization pathways, *J Clean Prod* 24;10: 184–201.

Prawit K, Kulla Sama K, Rattana J, Alissara R, 2018. Feasibility of bio-hythane production by co–digesting skim latex serum (SLS) with palm oil mill effluent (POME) through two-phase anaerobic process. *Int J Hydrogen Energy* 9577–90.

Seengenyoung J, Mamimin C, Prasertsan P, Sompong O, 2019. Pilot-scale of biohythane production from palm oil mill effluent by two-stage thermophilic anaerobic fermentation. *Int J Hydrogen Energy* 44:3347–55.

Singer S, Magnusson L, Hou D, Lo J, Maness P-C, Ren ZJ, 2018. Anaerobic membrane gas extraction facilitates thermophilic hydrogen production from Clostridium thermocellum. *Environ Sci: Water Res Technol* 4:1771–82.

Suksong W, Kongjan P, Sompong O, 2015. Biohythane production from co-digestion of palm oil mill effluent with solid residues by two-stage solid state anaerobic digestion process. *Energy Procedia* 79:943–9.

Taubner RS, Rittmann SKM, 2016. Method for indirect quantification of CH4 production via H2O production using hydrogenotrophic methanogens. *Front Microbiol* 7:532.

Tepari EA, Nakhla G, Haroun BM, Hafez H, 2020. Co-fermentation of carbohydrates and proteins for biohydrogen production: statistical optimization using Response Surface Methodology. *Int J Hydrogen Energy* 45:2640–54.

Veluswamy G, Laycock C, Shah K, Ball A, Guwy A, Dinsdale R, 2019. Biohythane as an energy feedstock for solid oxide fuel cells. *Int J Hydrogen Energy* 44:27896–906.

Vu T, Vu D, Jensen LS, Sommer SG, Bruun S, 2015. Life cycle assessment of biogas production in small-scale household digesters in Vietnam, Asian-Australas. *J Anim Sci* 28; 5: 716–729.

Yeshanew MM, Frunzo L, Pirozzi F, Lens PN, Esposito G, 2016. Production of biohythane from food waste via an integrated system of continuously stirred tank and anaerobic fixed bed reactors. *Bioresour Technol* 220: 312–22.

Zainal BS, Akhbari A, Zinatizadeh AA, Mohammadi P, Danaee M, Mohd NS, 2019. UASFF start-up for biohydrogen and biomethane production from treatment of Palm Oil Mill Effluent. *Int J Hydrogen Energy* 44: 20725–37.

13 Evaluation of Social Acceptance and Market for Human Excreta-Derived Products

Sarah Lebu, Chimdi C. Muoghalu, Aaron Salzberg, Swaib Semiyaga, Charles B. Niwagaba, and Musa Manga

13.1 INTRODUCTION

A critical link in the management of the human excreta value chain is the reuse of by-products generated from treatment (Mallory et al., 2020). Effectively treated human excreta in the form of fecal sludge (FS) and sewage sludge (SS) has the potential to significantly reduce risks related to public health and direct disposal of untreated waste into the environment. At present, over 3.6 billion people lack adequate access to safely managed sanitation (World Health Organization, 2021). Of those, 494 million still practice open defecation, depositing untreated excreta directly into the environment. Median fecal wet mass production of fecal waste is at 128g/cap/day. This is equivalent to 290 billion kg of feces and 1.98 liters of urine each year, globally (Rose et al., 2015). A rapid rise in population growth is expected to exponentially increase volumes of excreta generated, therefore, calling for increased efforts to safely manage human waste.

Resources recovered from human excreta can offset current and growing pressures on nutrient, energy, and water systems (MacArthur, 2013; Trimmer et al., 2020). Elements and micronutrients found in excreta can be effectively treated to generate added value for economic gain while protecting human and environmental health. For example, the daily average energy value of per capita human feces is 4 115 kcal/kg for dry solids. An average person generates 4.5 kg of nitrogen, 0.5 kg of phosphorus, and 1.2 kg of potassium yearly (Mcconville et al., 2020). Useful by-products include a variety of energy forms, e.g., electricity, biogas, bioenergy, and products to recover nutrients in soil, e.g., fertilizers, biomass, soil conditioners, etc.

Establishing resource recovery opportunities within the sanitation value chain can be crucial in shifting from a traditional and linear model toward a more circular, sustainable economy (MacArthur, 2013). Human excreta contain carbon, water, and

DOI: 10.1201/9781003364467-13

nutrients—key resources that can be recovered and recirculated, helping to tackle a myriad of Sustainable Development Goals (SDG). SDG 6.2, to ensure universal access to safe and equitable sanitation, can be partly achieved by improving FS and SS management. Soil amendment products like fertilizers derived from excreta can help improve global food security by enhancing agricultural inputs and boosting yields (SDG 2.3). Energy products such as biogas and methane contribute to global renewable energy (SDG 7.2). Effective sludge management can offset up to 93% of greenhouse gas (GHG) emissions from sewage sludge (SDG 13) (Chen et al., 2022). Altogether, improving resource recovery from human excreta accentuates and provides opportunities to tackle complex societal challenges and contributes toward achieving multiple SDGs.

A circular economy is defined as material cycle where the key design is restoration or regeneration (MacArthur, 2013). In the sanitation value chain, a circular economy is achieved often by targeting a market segment that is interested in value-added excreta-derived products. An additional revenue flow through the sale of value-added products could generate more revenue from the sanitation value chain. Revenue from the bioresource could potentially offset some sanitation expenses that are currently being taken on by households, thus improving sanitation access and public health overall (Diener et al., 2014).

The full realization of the benefits of human excreta-derived by-products has been limited by constraints of social acceptance and market barriers that hinder their application and commercialization at scale (Mallory et al., 2020). For example, the application of fertilizers from human excreta is hindered by mistrust and unclear regulatory standards guiding their commercialization (MacArthur, 2013). Examples of successful resource recovery businesses in the sanitation sector remain uncommon and at a low scale (Ferguson et al., 2022; Moya et al., 2019b; Tarpeh et al., 2023). Consequently, the ability of organizations to recover value from excreta has a lot to do with social acceptance and ready markets.

The underlying premise guiding this body of work is that resource recovery from the sanitation value chain is widely underutilized and poorly established. Through the lens of varied behavior-change theoretical underpinnings, the drivers, and barriers to the successful uptake of resource recovery programs are discussed in detail. The body of work is written for policymakers, decision-makers, sanitation actors, and practitioners (including the private sector), with the hope of guiding the improvement of more feasible and socially accepted reuse programs.

13.2 THE CIRCULAR ECONOMY IN THE CONTEXT OF SANITATION

Sanitation systems can be transformed from linear value chains to closed loop, "circular" economies where nutrients, energy, and other resources are restored at the end of the chain. Human excreta poses both environmental difficulties and the potential for resource utilization (Mallory et al., 2020; Moya et al., 2019b). The relevance of a circular economy in the context of sanitation is of economic, environmental,

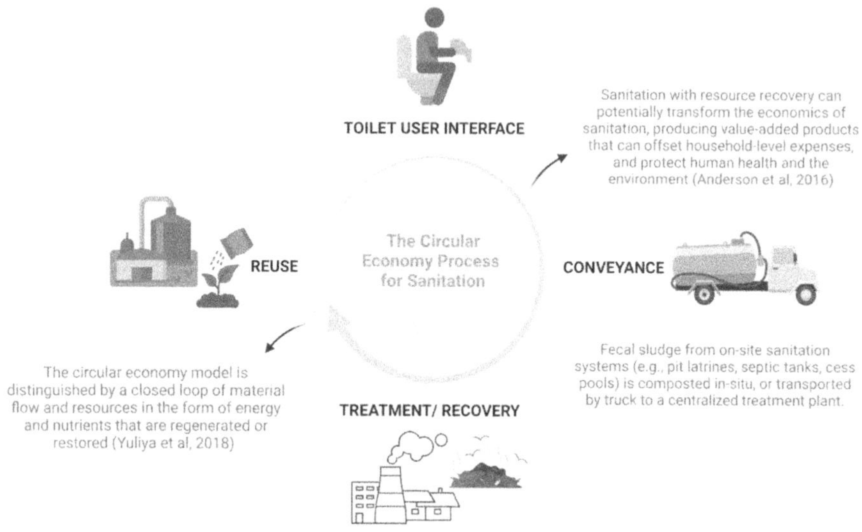

FIGURE 13.1 The circular economy cycle for sanitation.

and public health importance. Moreover, there is a direct linkage between sanitation resource recovery and the promotion of food security, as common by-products such as organic compost and urine fertilizer can be used in soil amendment to improve agricultural yield (Moya et al., 2019b). It is estimated that wastewater contains nutrients that can replace 50 million tons of organic fertilizer—an equivalent market value of US$15 billion which would otherwise be dumped in waterways globally (Mallory et al., 2020). A circular economy for sanitation can help recuperate these losses and turn them into positive streams of revenue that can help keep sanitation businesses afloat and offset initial investment into the front end of the sanitation value chain.

However, it is common knowledge that the notion of a closed-loop economy is hampered within the context of sanitation services. One factor that precludes the effective implementation of sanitation resource recovery is the issue of social acceptance and public trust. Prominent features of the sanitation circular economy are toilet use, collection, transport, treatment, and reuse of treated by-products, as illustrated in Figure 13.1 (Kalmykova et al., 2018). Literature supports that this model operates in the context of an enabling environment with governance, regulatory framework, business activities, markets, innovation, and end-user practices (Iacovidou et al., 2021). By turning human waste from the point of production, through the entire chain, to the point of treatment and reuse, bioresource and energy can be optimally conserved and regenerated. Understanding the dynamics that drive this circularity is a prerequisite for building the capabilities to support a flourishing sanitation resource recovery model (Iacovidou et al., 2021).

13.3 FECAL SLUDGE MANAGEMENT FOR SANITATION RESOURCE RECOVERY

The scope of human excreta discussed encompasses FS, defined as a partially treated slurry that accumulates in non-sewered sanitation technologies (e.g., pit latrine, septic systems and dry, urine-diverting toilets (Peal et al., 2014). It is a distinct type of sludge composed of excreta and other residuals found in on-site containment systems, e.g., flush water and greywater. It may contain solid waste. The reuse of products from SS, sometimes known as wastewater sludge is also covered. In contrast, SS comprises residual material, semi-solid in nature that is produced from a wastewater treatment plant.

Improper handling of untreated FS that contains significant levels of organic substances, nutrients, and disease-causing agents when released into the environment can lead to eutrophication and an increased likelihood of disease spread (Manga et al., 2022b; Semiyaga et al., 2015). Pit latrines often have more solid and organic matter in comparison to septage. In unlined pit latrines, liquid often infiltrates into the soil, resulting in a highly concentrated sludge in the pit. Septage, on the other hand, is a more stabilized form of FS due to longer detention time and dilution by greywater. Pathogen levels and organic content decay and decrease with the age of FS. In urban informal settlements where many users of pit latrines and septic systems live, it is usually not possible for FS to sit for long periods due to the shared nature and high user loads of latrines.

Composting is a common solution for treating FS and it involves pathogen die-off through organic matter decomposition at high temperatures achieved at the thermophilic stage (ranging from 3–15 consecutive days at 55^{0}C for windrow turning). The composted product is usually attractive as it effectively sanitizes FS and helps in recovering nutrients once applied to the soil (Manga et al., 2022a). Studies suggest testing different bulking agents and time-temperature configurations to identify the most viable conditions for pathogen die-offs (López-González et al., 2021; Manga et al., 2021, 2023). For example, the process of co-composting FS with sawdust and coffee husks, which are used as bulking agents, can effectively eliminate harmful pathogens. This is achieved by maintaining temperatures between 48 to 60 degrees Celsius during a windrow period of up to 38 days, while turning the compost every 7 days (Manga et al., 2021). Although composting of municipal solid waste has been extensively studied, the composting of FS remains insufficiently researched and understood.

Compared to FS generated from non-sewered sanitation systems, lower concentrations of nitrogen, phosphorus, and helminth eggs in sewage sludge can mainly be attributed to the dilution effect. Notably, composted SS comprises 50% organic matter and therefore is attractive to recovering soil nutrients. A conventional approach for the treatment of SS is through sludge thickening, anaerobic digestion, and dewatering. The process of sludge thickening involves the separation of excess water from the solid components of sludge, using a low-energy approach, prior to the onset of anaerobic digestion. The result is a reduced sludge volume and an increase in the solid content. Sludge contains 2–4% solids by weight before thickening and 16–18% solids by weight after thickening (Wiśniowska, 2019). Anaerobic digestion is a process where naturally occurring microorganisms in the sludge decompose organic matter without the presence of oxygen. This process takes place in large, sealed tanks, where the sludge is kept at an elevated temperature for around 15 to 30 days. The anaerobic digestion process can occur at either mesophilic temperatures, which are optimal between 30–38°C, or at thermophilic temperatures, which are optimal at 49–57°C. Although thermophilic digestion is less commonly used, it requires less time in the digestion tanks than mesophilic digestion. The product of anaerobic digestion, called biosolids, contains a high amount of water, with a dry solids content of about 6–12%.

13.4 HUMAN-EXCRETA-DERIVED BY-PRODUCTS

The literature identifies five classifications of excreta-derived by-products, namely (1) agricultural application; (2) fuel and energy for combustion; (3) biomass and protein; (4) construction materials; and (5) water. The most used application is fertilizers and soil conditioners in agriculture. An adaptation of a compendium of excreta-derived by-products and end uses presented elsewhere in the literature is shown in Table 13.1 (Mcconville et al., 2020). The table describes important elements of each product, including a description of real-life application. The low application indicates hypothetical examples or reuses at the pilot scale; medium application indicates emerging reuse with demonstrations in one or few contexts; high application indicates wide uptake with demonstrations in multiple contexts.

13.5 THEORETICAL FRAMEWORK AND FRAMING

This body of literature is guided by two theoretical frameworks that are commonly used to explain consumer behavior and social acceptance of recycled products (Mensah, 2020). This section describes two key theories namely, the Theory of Planned Behavior (TPB) and the Value-Belief-Norm Theory (VBN). These theories were preferred for understanding underlying motivations for social acceptance of excreta-derived products because they capture the spectrum of purchasing decisions and they have been used in the context of environmentally friendly market products. The TPB explains a user's psychological tendency to purchase recycled products. The model holds that an individual experiences subjective norm (the idea that the society surrounding them approves of their behavior), a level of environmental awareness and trust in the product as well as a willingness to pay (Ajzen, 1991). In essence Strydom (2018) suggests that a user's intention to use a recycled product is primarily influenced by their attitude, subjective norms, and perceived control over the technology process. These factors have the strongest impact on a user's willingness to engage in recycling activities. Ultimately, the individual makes a purchasing decision based on a set of values, shown in Table 13.2. Empirical demonstration of the TPB has been achieved elsewhere in the literature (Bigliardi et al., 2020; Strydom, 2018).

Secondly, a reference of the Value-Belief-Norm Theory (VBN) is used to examine the elements that stimulate the adoption of excreta-derived by-products by society and influence consumer behavior (Stern et al., 1999). The VBN suggests that one's

TABLE 13.1
Taxonomy and Characteristics of Identified Excreta-derived by-products

Use group	Product	Product description	Market factors	Application	References on implementation
Agricultural application –soil conditioners	Dewatered sludge	A concentrated sludge product stripped of moisture as a way of stabilization and treatment to remove pathogens.	Dewatered sludge produced at a local scale is relatively cheap, although transport costs to the field may accrue.	*High*	(Mantovi et al., 2005; Strande et al., 2014)
	Compost	A soil-like substance high in nutrients and organic matter and resulting from controlled aerobic degradation of organic material.	Capital costs are generally low.	*High*	(Tilley et al., 2014)
	Biochar	Biochar is a solid material obtained from pyrolysis, the thermochemical conversion of biomass in an oxygen-limited environment.	The majority of the global market share comes from the United States (65%), followed by Europe (25%), Asia (7%), and Africa (3%).	*High*	(Kambo and Dutta, 2015)
	Ash from sludge	Forms at the bottom of an incinerator from heavy components that are neither combustible nor volatile. Residues contain large proportions of phosphorous and potassium.	The use of ash in large infrastructure projects offer better delivery security for the supplier of bottom ash.	*High*	(Donatello and Cheeseman, 2013)
	Pit humus	Refers to material removed from double pit systems.	– capital costs for tools and equipment to apply pit humus are low	*High*	(Thuries et al., 2019)

(continued)

TABLE 13.1 (Continued)
Taxonomy and Characteristics of Identified Excreta-derived by-products

Use group	Product	Product description	Market factors	Application	References on implementation
	Dried feces	Feces stored in the absence of moisture.	– high market costs can be incurred in transport to the field, labor, application, and protective equipment – potentially cheaper than fertilizers and soil conditioners	*High*	(Buit and Jansen, 2016)
Agricultural application – solid fertilizers	Dry urine	A nutrient-rich fertilizer produced by dehydrating and concentrating human urine in an alkaline substrate.	– easy to package and transport to market – appropriate for household and large-scale application – large-scale application requires costly supply chain	*Low*	(Senecal et al., 2018; Senecal and Vinnerås, 2017; Simha et al., 2017)
	Struvite	A phosphate material that naturally occurs in sanitation systems. Composed of magnesium ammonium phosphate hexa-hydrate.	– as a solid, product is easy to package and transport to market – appropriate for large-scale agricultural application – restrictive legislation results into low commercialization – Restrictions on international trade as struvite may be registered as a product in one country and waste in another	*High*	(Etter et al., 2011)
Energy	Biogas	Biogas is produced through anaerobic digestion of sludge and other organic matter. If generated from FS, methane potential is low.	For commercialization of biogas, it is important to consider the calorific efficiency of biogas in different applications, which is 55% in stoves and 24% in engines, 3% in lamps and 88% in electricity generation.	*High*	(Vögeli et al., 2014)

	Energy briquettes	A calorific value of 17.3 MJ/kg dry solids of FS, which is comparable with that of other biomass fuel	A calorific value of 17.3 MJ/kg dry solids of FS is obtained from briquettes.	*Medium*	(Semiyaga et al., 2015)
Protein and biomass	Black Soldier Flies	Nutrient-rich insect larvae that grow from the treatment of organic waste using black soldier fly larvae (BSFL), used for animal feed.		*Medium*	(Surendra et al., 2020)
Construction materials	Nutrient-enriched filter material	Nutrient-enriched filter materials are mineral or organic materials that have been charged with nutrients through the process of adsorption.	May be sold directly as fertilizers/ or soil conditioners.	*High*	(Cucarella and Renman, 2009)
Water	Irrigation water	Wastewater that has undergone tertiary treatment (i.e., filtration and/or disinfection) can be used in agriculture.	Commercial-scale irrigation systems for industrial production are expensive, requiring pumps and an operator. Small-scale drip irrigation systems can be constructed out of locally available low-tech materials and are inexpensive.	*High*	(Bahri et al., 2009)
	Aquaculture	Fish, aquatic plants, and other organisms can be grown in ponds that receive sludge, where they can feed on algae and microorganisms that grow in nutrient-rich water.	Due to preservation, storage and transportation constraints, plant aquaculture is often restricted to local use, can be labor intensive.	*Medium*	(Castine et al., 2013)

TABLE 13.2
A Taxonomy Illustrating the Application of the Theory of Planned Behavior in the Social Acceptance of Excreta-derived by-products

TPB values	Hypothesis in the context of sanitation resource recovery
Attitude	Consumers' attitude toward eco-friendly products significantly influences their intention to purchase/use the product, e.g., when a farmer has a positive toward products that improve the environment, they are more likely to use excreta by-products.
Subjective norms	The more favorable an individual's subjective norm, the more they are likely to purchase a product. Subjective norm refers to approval of others i.e., in the case of sanitation, as the collective societal decision to use excreta by products.
Perceived behavior control	An individual is likely to face difficulties in the process of purchasing/using excreta derived by-products, e.g., lack of adequate information on product. When they perceive that these constraints are beyond their control, they are unlikely to proceed with the decision to purchase/use the product.
Perceived value	The sum benefits and perceived performance of a product can influence whether it's socially accepted or not. For example, in sanitation, farmers are more likely to buy organic fertilizer if they perceive that there will be a differential increase in crop yield after its application.
Environmental awareness	Having information about the consequences of a product on the environment can influence social acceptance.
Environmental trust	Having an expectation that a product will have a positive effect on the environment can influence social acceptance.
Willingness to pay	Users may be willing to pay premium for environmentally friendly products compared to alternative market products.

Source: Ajzen, 1991.

values impacts their belief, and beliefs impact personal norms, which finally affect behavior. This model has been commonly used to understand purchasing behavior for recycled and eco-friendly products. The premise of the model is that human behavior is facilitated through the development of beliefs and norms. Biospheric value refers to the innate concern for the environment while altruistic values create an innate understanding of others (Stern et al., 1999). Individual environmental perspectives increase consciousness about the outcomes of actions and the assignment of accountability for environmental damage. Finally, a set of personal norms and behaviors are formed, which take into consideration the perceived risk of the product and the intention to purchase, before a final decision is made. This theory has been used to empirically investigate the social acceptance of recycled products such as household waste, and hog waste (Al Mamun et al., 2022; Dursun et al., 2017; Ghazali et al., 2019). A taxonomy of hypothesized behaviors around social acceptance of excreta

TABLE 13.3
A Taxonomy Illustrating the Application of the Value-Belief-Norm Theory on the Social Acceptance of Excreta-derived by-products

VBN values	Hypothesis in the context of sanitation resource recovery
Biospheric values	Biospheric values (concern for the environment) positively influence environmental concern. Users may purchase by-products because of the perceived environmental benefit.
Altruistic values	Altruistic values (concern for others) positively influence environmental concern. Although less common, users may purchase by-products due to the environmental nature that may accrue to society.
Environmental concern	Environmental concern positively influences awareness of consequences.
Awareness of consequences	Awareness of consequences positively influences ascription of responsibility. Users who are more aware of the consequences that environmental problems entail will be more likely to purchase by-products.
Ascription of responsibility	Ascription of responsibility positively influences personal norms. This entails who is blamed for the consequences of certain actions.
Personal norms	Personal norms positively influence purchase intention.
Perceived risk	Perceived risk has a negative influence on purchase intention. Expectation of a loss (e.g., in quality or financial) that is associated with the purchase will likely deter the purchase.
Purchase intention	Perceived risk negatively moderates the relationship between personal norm and purchase intention.

Source: Hein, 2022.

by-products and how they map to the VBN theory is illustrated in Table 13.3. In the context of sanitation-related by-products, end users decide to purchase if they perceive that the product is safe, good for the environment, provides high financial gain, is convenient, and is of high quality.

A comprehensive examination of these fundamental theories is conducted to facilitate comprehension of the notion of social acceptance and markets for by-products derived from excreta. An examination of social acceptance dynamics and the feasibility of different sanitation business models toward achieving a circular economy is reviewed.

13.6 SOCIAL ACCEPTANCE

Based on the frameworks described above, social acceptance is considered to mean the overall level of approval for a product or innovation by individual persons (including attitudes, behavior, and tolerance), or simply the favorable reaction expressed by members of a particular social group (such as a country, region, town, household, or organization) toward a suggested technology or product (Wolsink, 2018). This pertains to the supportive conditions (such as decision-makers and socio-cultural acceptance), market participants (market acceptance), or the acceptance of the local community (community acceptance). (Figure 13.2). The stakeholders engaged in the sanitation system's value chain encompass a broad range of individuals and groups, such as sanitation users, policymakers, community planners, transport and logistics actors (like exhauster truck operators), private sector entities and companies, waste treatment plant operators, regulators, and end-users. The actors belong to multitiered structures and subsystems that may overlap (Trimmer et al., 2020). Figure 13.1 shows an illustration of the structure of actors and sub-system-level contribution to the concept of social acceptance.

Despite its benefits, the practice of resource recovery from excreta-derived by-products remains poorly implemented and at a low scale. Before significant progress can be achieved, one of the most significant obstacles to surmount is the acceptance of products. According to the literature, a reluctance to handle human excrement due to its foul odor and perceived risk to health is known to affect social acceptance. As a result, there has been a hesitancy to incorporate excreta resource recovery into national policies that promote circular nutrient systems. Therefore, it is crucial to comprehend the factors that influence acceptance and the specific performance of various by-products derived from excreta.

It is important to mention that the majority of studies examining social acceptance indicate that at least half of the participants express a willingness to consider the

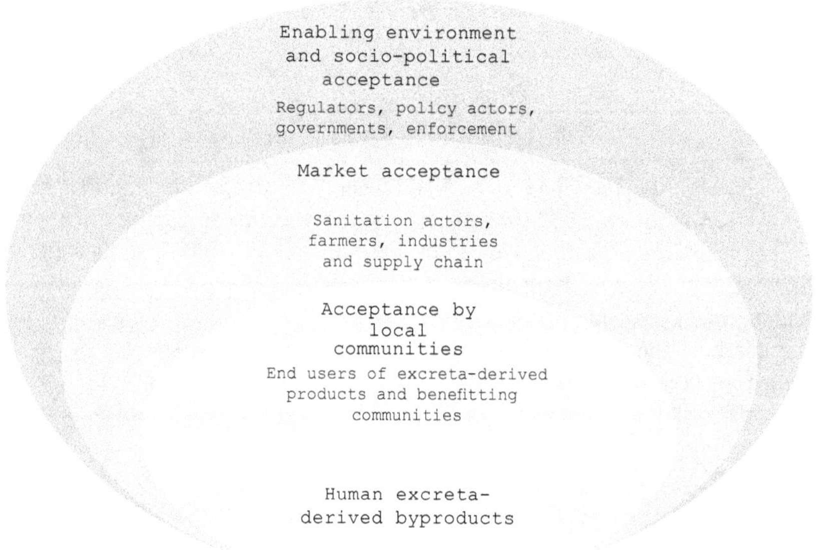

FIGURE 13.2 Illustration of stakeholders and actors who constitute social acceptance of a product or innovation.

concept of reuse (McConville et al., 2023). Earlier studies indicate that the public has limited knowledge regarding resource recovery from sanitation, but their beliefs and perspectives about the handling of human excreta are often firm. Moreover, new technologies and systems must be viewed as being as convenient, comfortable, clean, and safe as the present status quo (Mcconville et al., 2020).

The perception of risks to human and environmental health is strongly linked to social acceptance of excreta products. Nonetheless, it's important to note that perceived risk does not always correspond to actual risk. One illustration from McConville et al. (2020) is that people may perceive child excreta to be less risky than that of adults, but in actuality, the opposite is true. To sum up, the level of acceptance for excreta-derived products is diverse, with higher acceptance reported for end uses that are not related to food.

13.6.1 Drivers of Uptake of Human Excreta Products

To align with the earlier theoretical framework, this section is divided into four main groups of factors that affect the social acceptance of excreta-derived products. These groups include individual-level factors such as perceived value, willingness to pay, and perceived risk, socio-cultural and public trust, governance and regulatory landscape, and markets and innovation.

13.6.1.1 Individual-Level Factors

The attitudes toward the use of excreta-derived products can be influenced by individual-level factors such as age, income, education, and farm size. According to research, experience, income, farm size, and level of education are significantly associated with social acceptance (Gwara et al., 2021). Several studies have suggested that farmers who have received more formal education are more accepting of using human excreta products compared to those with less or no education (Eskelinen et al., 2022; Huang et al., 2022; Laborda et al., 2023). Other researchers who examined factors for social acceptance reported no difference in age, gender, education, or country of residence (Eskelinen et al., 2022). However, caution should be exercised when interpreting these findings due to differences in context and methodology across studies.

To determine whether to adopt a product or not, end users often consider perceived benefits, such as an increase in crop yield from organic fertilizers and compost. The lack of perceived benefits can have a negative impact on product uptake. To evaluate the expenses related to the implementation of sanitation technologies, it is important to consider both the upfront capital costs as well as ongoing expenses for operation and maintenance. Decentralized non-sewered systems like biogas-connected latrines typically put the responsibility of maintenance and repairs on the facility owner. As a result, the facility may be neglected until it starts malfunctioning, which can lead to high repair costs.

Providing users with information about excreta-derived products can be a key factor in promoting their use. Such knowledge can help dispel myths and reduce stigma, which can encourage more widespread adoption. In the past, this information has been disseminated through various means, such as public awareness campaigns or targeted materials focused on specific brands.

13.6.1.2 Socio-Cultural and Public Trust

Human excreta products should be safe to use in the environment without causing contamination. Sanitation system outputs should be well-treated so that the receiving environment is not compromised. This is pertinent in agricultural settings where excreta products are used as soil amendments to boost soil nutrients. The prospects of using feces for fertilizer are informed by possible cultural norms on the perception of handling feces; some communities are characterized as feces-friendly or fecophilic, while others are characterized as feces-fearing or fecophilic (Buit and Jansen, 2016). Often, feces may bear definitive cultural meaning, for example, the feces certain kin should not mix or that a person's feces can be used as a medium for witchcraft. Such perceptions are difficult to demystify in communities where social cohesion is high and people tend to have similar beliefs (Drangert, 1998a). Parallel to fecophobia is the concept of urine blindness which has been used in previous literature to describe negative attitudes toward urine in the context of organic fertilizer application (Drangert, 1998b). It is argued that when urine and feces are mixed, the mixture produces a foul odor, which decreases the desirability of reuse (Drangert, 1998b). This matters because one's action is driven by their perception as well as the perception of their friends and family. More positively, studies fortifying market acceptance of excreta products suggest that people prefer products with healthy properties and that they may be willing to pay a premium for such goods, for their interests, for altruism, and for the biospheric value that could be potentially incurred (Eskelinen et al., 2022). Products derived from human excreta are considered "green" or "environmentally friendly" in the eyes of end users, and so may be attractive to environmentally-conscious users (Polyportis et al., 2022).

13.6.1.3 Governance and Regulatory Landscape

An enabling regulatory environment should ensure that the product meets requirements for quality, safety, and labeling to be traded and used freely in the market. Quality covers the minimum nutrient content of the product and organic matter content, necessary for their performance. Safety depicts the maximum limits for contaminants such as heavy metals, organic contaminants, etc., specific to each product. Both the safety for human health and the environment should be considered. Labeling of products with nutrient content information, their proper use, and consequences of improper use or disuse should be required. Regulatory standards, guidelines, and legislation provide an opportunity to address all three dimensions of a product, which in turn boost reliability and public trust. A stamp of approval (i.e., a standardization mark in many instances) can signal if the product follows laid down standards and ensures consistency. In many countries, the application of products derived from human excreta is regulated by standards and regulations integrated into member countries' legislation. However, the literature reports that the legislative framework that governs sanitation resource recovery is not well developed. There is evidence that political recognition and certification of products can drive social acceptance. Some farmers may not be able to sell their excreta derive by-products if they lack regulatory permission to participate in the market (Mallory et al., 2020).

13.6.1.4 Markets and Innovation

The main concern relating to the social acceptance of by-products derived from human excreta is safety from pathogens (Moya et al., 2019a). Inaccurate information about a product can affect the public's perception of its safety. Quality assurance processes, standards, labels, and information systems have been used in more advanced economies to improve information about product quality (Laborda et al., 2023). The perception of a product can significantly influence its acceptance in the market, and a positive image can lead to long-term acceptance. Effective product labeling can aid market stimulation by communicating key information such as the product's source, nutrient content, and quality checks. To strengthen social acceptance and markets, it is beneficial to have a standardized process for measuring the attributes of a product and assessing compliance with quality standards. Governments can play a role in this process by developing and implementing standardization processes that conduct quality assessments of products, enhancing public trust, and incentivizing the market for excreta-derived products.

13.6.2 Summary of Case Studies on Product-Specific Drivers of Acceptability

13.6.2.1 Acceptability of Human-Derived Fertilizers

Struvite, which is one of the organic fertilizers obtained from treating human excreta, has been found to have the highest level of social acceptance compared to other types. The reason for this is that the techniques used to produce it involve the recovery of phosphorus from sewage ash and precipitation at different stages of the wastewater treatment process, which include the liquid fraction and sludge. According to a pilot study conducted in Sweden, farmers tended to be more accepting of fertilizers derived from human excreta when they were further removed from direct human consumption (McConville et al., 2023). Users expressed greater willingness to use struvite or dehydrated urine on non-food plants than on edible plants, with food safety being the most important factor influencing their decision to use the product. To reduce disgust and stigma, users suggested avoiding direct references to feces or urine when naming the product. Perceptions of risk were found to have a significant impact on acceptance, with product packaging containing information on possible pharmaceutical residues being the least desired. Lack of product-specific knowledge led to some reluctance to invest in the product, emphasizing the need for clear information about each product for users. The study revealed that limited knowledge of a product was closely linked to perceptions of food safety.

13.6.2.2 Acceptability of Biochar

Biochar is created by heating biomass, such as crop residues, in a process called pyrolysis with limited oxygen. When biochar is added to soil, it can enhance the soil's ability to retain water, adjust soil pH, and has potential applications in soil remediation

(Latawiec et al., 2017; Muoghalu et al., 2023). The application of biochar is well-established in the United Kingdom, Germany, and Finland. However, there is sparse knowledge on the adaptation of biochar in other countries. In Poland, farmers with acreage below 100 ha have shown reluctance toward implementing technological innovations—biochar included. Small-scale farmers tend to be more cautious about risks and prefer to minimize production costs. Farmers who had prior knowledge and were familiar with biochar were more likely to adopt it. Out of the 161 farmers interviewed, 20% showed interest in using biochar, with 55% of them having some knowledge and 27% having familiarity with biochar. Interest in adopting biochar was higher among farmers with more experience in agriculture, between 5–10 years. Less experienced farmers were found to be more resistant to change and innovation. Farmers who considered ecological factors were more interested in using biochar.

13.6.2.3 Acceptability of Biogas

Anaerobic digestion of organic materials produces biogas which can serve as a clean cooking alternative, especially in rural areas. Nepal's biogas program installed 350,000 domestic biogas units in 2015, and a majority (79%) of these units were connected to in-home toilets (Boyd Williams et al., 2022). The success of Nepal's biogas program challenges the idea that biogas and fecal sludge are culturally unacceptable, despite being associated with the dominant religion of Hinduism in the country. This contrasts with neighboring India, where there is lower acceptance of toilets connected to biodigesters.

The use of biogas as a fuel source was encouraged by several factors, such as improved cleanliness, reduced need for wood collection, and perceived health benefits. The fast-cooking process associated with biogas also provided more time for other activities, especially for women and girls. Additionally, the connection to a toilet provided an additional benefit by improving access to both sanitation and clean fuel for cooking. Having a biodigester connected to a toilet has several advantages such as reducing odors and the need for frequent emptying, as well as improving one's reputation and status within the community. The amount of biogas produced affects its usage, with some users resorting to other fuel sources like wood or liquefied petroleum gas when biogas supply is low due to factors such as loss of cattle, repairs, or cold temperatures. Success in using a biodigester depends on having operational knowledge and understanding alternative feedstocks like goat dung and food waste.

13.6.3 Barriers to Adoption

When attempting to make treated excreta resources commercially available, a primary concern is ensuring the safety of humans and animals, as FS can contain high concentrations of pathogens that may not be fully eliminated by treatment processes (Mcconville et al., 2020). The acceptance and uptake of reuse products are often negatively influenced by perceived risks to human and/or environmental health. These risks include the transmission of diseases by pathogens and the potential adverse effects of medicinal residues, such as pharmaceuticals, hormones, and antibiotics (McConville et al., 2023).

Feelings of disgust and shame are also significant factors that influence acceptance of reused products. In certain contexts, communities have coined the term fecophobia to describe the socio-cultural aversion to feces (Gwara et al., 2021). The cultural and religious values of a community are closely tied to their sanitation practices, making it challenging to introduce new sanitation methods (Moya et al., 2019b).

The absence of regulations and certification schemes can deny legitimacy to organic fertilizers from human excreta and thwart the potential for wide-scale commercialization. Existing policies are often disjointed, lacking consistency and specificity. The process of regulation is often influenced by industry and advocacy groups, who may prioritize different concerns compared to the public.

The economic feasibility of human excreta products has been shown to affect their adoption. For instance, compost is less cost-effective than chemical fertilizers due to its bulkiness, resulting in higher transportation costs relative to its value. Additionally, it is seen as less convenient to use than pelletized chemical fertilizers. The profitability of a product can also be affected by modifying its properties. Farmers are more likely to pay a higher price for a product they perceive as high quality (Moya et al., 2019b). To recover the full cost of production is a challenge when it comes to resource recovery in the sanitation sector (Segrè Cohen et al., 2020). Table 13.3 provides a summary of the obstacles that hinder the social acceptance and widespread adoption of excreta products.

13.7 DEMAND FOR EXCRETA-DERIVED BY-PRODUCTS

The U.N. has estimated that globally, the conversion of human waste to fuel is worth \$9.5 billion. The waste generated by about 1 billion people without access to safe sanitation is valued at up to \$376 million in methane production alone, which is enough to provide power to 10 to 18 million households. The successful commercialization of excreta-derived products highlights the importance of understanding market dynamics, market size, and growth, as well as the willingness of consumers to pay for specific excreta end products. Studies on willingness to pay for fecal sludge compost have shown that farmers are willing to pay \$0.4 more than the current market price for a certified product (Diener et al., 2014). The recycling of organic waste for agriculture and the production of FS compost can bring significant benefits to farmers, businesses, and the environment. This is evident from the demand for FS compost in the market. Table 13.4 provides an overview of the evidence regarding the market value of by-products derived from human excreta.

13.8 LOOKING BEYOND SOCIAL ACCEPTANCE: EXAMINING THE ROLE OF BUSINESS MODELS IN THE SUCCESS OF EXCRETA RESOURCE RECOVERY

Even in situations where there is high social acceptance and demand for resource recovery, businesses may still struggle to succeed. One of the key reasons for failure is the effectiveness of the business model. In order for sanitation reuse enterprises to scale up, they must be able to generate profit from the waste. Business models

TABLE 13.4
Market Value of Excreta-derived by-products

Byproduct	Area studied	Volume used in calculation	Ave. market value in $	Source
Fuel combustion	Uganda	16 tons per day	566,467	(Diener et al., 2014)
Protein	Senegal	40,435 tons per day	129,168	(Diener et al., 2014)
Biogas	Uganda	252,984 tons per day	158,743	(Diener et al., 2014)
Soil conditioner	Senegal	12,480 tons per day	81,120	(Diener et al., 2014)
Organic fertilizer	Haiti	3 tons per month	250	(Moya et al., 2019a)
Organic fertilizer	Kenya	48 tons per month	380	(Moya et al., 2019a)
Struvite–magnesium sulphate	India	250 L/day	37	(Etter et al., 2011)
Struvite–magnesite	Nepal	250 L/day	13	(Etter et al., 2011)
Struvite–bittern	India	250 L/day	24	(Etter et al., 2011)
Dry urine	Global	550 L/yr	86	(Senecal and Vinnerås, 2017)
Pit humus	Cameroon	50 kg	5.70	(Thuries et al., 2019)
Dewatered sludge	USA	7 million dry tons	250	
Compost	Romania	1 kg	0.2	(Petrescu-Mag et al., 2020)
Biochar	USA	1 mg	754.68	(Nematian et al., 2021)
Fly ash sludge	China	1 ha	131	(Hadas et al., 2021)
Irrigated water	China	1 m3	0.41	(Hadas et al., 2021)
Energy briquettes	Uganda	250 kg	1.13	(Atwijukye et al., 2018)
Black soldier fly larvae	Global	1kg	0.60	(Verbeke et al., 2015)

for agricultural reuse typically involve a reversed cash flow, where farmers pay for the treated product and this offsets the operational and maintenance costs at the beginning of the sanitation chain (Renting et al., 2013). A drawback of this type of business model is the seasonality in demand for FS. One prevalent business model in composting involves forming partnerships with various entities, including local governments, private businesses, and community-based organizations, to secure the initial investment capital. The operational and maintenance costs for composting are often low, and the farmers who receive the treated product can cover these expenses. To market their compost products, these businesses typically utilize an established distribution and marketing system.

However, such funding streams are not sustainable as they may end when the external funder withdraws. Other business models involve providing micro-finance, technical support for agricultural production as well as marketing support. This business model involves farmers not only producing agricultural goods, but also utilizing agro-sanitation assets as consumers (Ushijima et al., 2015). Effective business models for resource recovery in the sanitation sector should consider cost recovery throughout the supply chain through innovative strategies such as the production of a diversified portfolio of pelletized and enriched waste products. Savings in energy production can play a significant role in boosting cost recovery, as energy and transport costs can be a significant expense that can be minimized to reduce overall costs.

13.9 IMPLICATIONS

13.9.1 Implications for Policy and Practice

The motivation for this chapter is the inadequate management of human waste and sewage sludge, which leads to unsafe disposal and/or reuse of excreta, potentially causing serious health and environmental problems. However, this situation also presents an opportunity to recover valuable nutrients from excreta and convert them into profitable products, while minimizing disease transmission and environmental contamination. By investing in resource recovery infrastructure and promoting the commercialization of excreta-derived products, the sanitation sector can achieve financial, environmental, and public health benefits. The chapter emphasizes the need for a shift in perspective, from considering human waste as a disposal problem to recognizing it as a potential source of revenue through resource recovery.

Based on the literature reviewed, social acceptance is a crucial factor in the successful implementation and scalability of resource recovery enterprises that aim to derive products from human waste. Factors such as cultural beliefs, perceptions of disgust and shame, and lack of awareness and education can significantly hinder social acceptance.

Moreover, the market value of excreta-derived products, such as compost and fuel, indicates the potential for economic benefits for farmers, businesses, and the environment. However, cost-effectiveness and optimal business models are necessary to make these enterprises profitable and sustainable.

To address the challenges of social acceptance and market viability, policy interventions must prioritize education and awareness campaigns that target communities, government officials, and industry stakeholders. Moreover, policies should incentivize the development and implementation of innovative and cost-effective business models that ensure cost recovery and promote the sustainable commercialization of excreta-derived products.

More specifically, regulatory standards should be developed to guide industry actors on producing quality products and being accountable to the public. Ideally, these guidelines should include guidelines for labelling products, i.e., what information to provide, safety of products, pre- and post-production product testing and consumer education. In so doing, this can promote public trust in the product and incentivize their wide uptake.

Investing in resource recovery infrastructure alongside sanitation technology is also essential to provide financial, environmental, and public health benefits. Overall, there is a need for a paradigm shift toward viewing human excreta as a source of revenue generation through resource recovery, rather than a disposal problem, to achieve the goal of safe and sustainable sanitation for all.

13.9.2 Implications for Future Research

Based on the literature on social acceptance and markets for products derived from human waste, several implications for future research can be identified. First, further research is needed to better understand the factors that influence social acceptance of excreta-derived products. This includes investigating cultural, religious, and psychological factors that affect people's perceptions and attitudes toward these products. In addition, more research is needed to explore effective strategies for promoting social acceptance, such as targeted education campaigns and community engagement initiatives.

Second, there is a need for more research to identify and address barriers to the commercialization of excreta-derived products. This includes investigating market dynamics, identifying optimal business models, and developing effective marketing and distribution strategies. There is also a need to explore the potential for public-private partnerships and other innovative financing mechanisms to support the development of sustainable and profitable resource recovery enterprises.

Third, research is needed to better understand the environmental and public health impacts of different resource recovery technologies and practices. This includes investigating the effectiveness of different treatment processes for reducing pathogen and contaminant levels, as well as exploring the potential for different excreta-derived products to support sustainable agriculture and other uses.

Overall, future research should prioritize interdisciplinary and collaborative approaches that involve a range of stakeholders, including policymakers, researchers, practitioners, and communities, to identify and address the complex challenges associated with resource recovery from human waste.

13.10 CONCLUSION

This chapter provides insights into the use of human excreta-derived by-products, the factors influencing their acceptance, and the challenges faced in implementing them on a large scale. Adopting resource recovery of human excreta is an essential step toward creating a circular economy. Conclusions from the literature suggest that social acceptance is critical for the widespread application of these products. The study highlights a range of resource recovery technologies that exist or are under development, indicating that achieving a circular economy in the sanitation sector is feasible. The widespread adoption and commercialization of excreta-derived products faces significant challenges such as lack of public trust, unclear regulations and policies, perceived product risk, and inadequate product information. The chapter concludes that decision-makers must establish strong standards and regulations to guide the recovery of resources from human waste, increase education and awareness among users, and further explore factors that hinder the widespread application of these products. Policies should also encourage the development and implementation of innovative and cost-effective business models that promote cost recovery and sustainable commercialization of excreta-derived products. This research contributes to the knowledge base needed for transitioning to a circular economy and safe fecal sludge management.

REFERENCES

Ajzen, I., 1991. The theory of planned behavior. Organizational Behavior and Human Decision Processes, Theories of Cognitive Self-Regulation 50, 179–211. https://doi.org/10.1016/0749-5978(91)90020-T

Al Mamun, A., Hayat, N., Masud, M.M., Makhbul, Z.K.M., Jannat, T., Salleh, M.F.Md., 2022. Modelling the Significance of Value-Belief-Norm Theory in Predicting Solid Waste Management Intention and Behavior. *Frontiers in Environmental Science* 10.

Atwijukye, O., Kulabako, R., Niwagaba, C., & Sugden, S. (2018). Low cost faecal sludge dewatering and carbonisation for production of fuel briquettes. https://repository.lboro.ac.uk/articles/conference_contribution/Low_cost_faecal_sludge_dewatering_and_carbonisation_for_production_of_fuel_briquettes/9592649/1

Bahri, A., Drechsel, P., Raschid-Sally, L., & Redwood, M. (Eds.). (2009). *Wastewater Irrigation and Health: Assessing and Mitigating Risk in Low-income Countries*. London: Routledge. https://doi.org/10.4324/9781849774666

Bigliardi, B., Campisi, D., Ferraro, G., Filippelli, S., Galati, F., Petroni, A., 2020. The Intention to Purchase Recycled Products: Towards an Integrative Theoretical Framework. *Sustainability* 12, 9739. https://doi.org/10.3390/su12229739

Boyd Williams, N., Quilliam, R.S., Campbell, B., Ghatani, R., Dickie, J., 2022. Taboos, toilets and biogas: Socio-technical pathways to acceptance of a sustainable household technology. *Energy Research & Social Science* 86, 102448. https://doi.org/10.1016/j.erss.2021.102448

Buit, G., Jansen, K., 2016. Acceptance of Human Feces-based Fertilizers in Fecophobic Ghana. *Human Organization* 75, 97–107. https://doi.org/10.17730/0018-7259-75.1.97

Castine, S., McKinnon, A., Paul, N., Trott, L., & de Nys, R. (2013). Wastewater treatment for land-based aquaculture: Improvements and value-adding alternatives in model systems from Australia. *Aquaculture Environment Interactions*, 4(3), 285–300. https://doi.org/10.3354/aei00088

Chen, W., Liu, J., Zhu, B.-H., Shi, M.-Y., Zhao, S.-Q., He, M.-Z., Yan, P., Fang, F., Guo, J.-S., Li, W., Chen, Y.-P., 2022. The GHG mitigation opportunity of sludge management in China. *Environmental Research* 212, 113284. https://doi.org/10.1016/j.envres.2022.113284

Cucarella, V., & Renman, G. (2009). Phosphorus sorption capacity of filter materials used for on-site wastewater treatment determined in batch experiments-a comparative study. *Journal of Environmental Quality*, 38(2), 381–392. https://doi.org/10.2134/jeq2008.0192

Diener, S., Semiyaga, S., Niwagaba, C.B., Muspratt, A.M., Gning, J.B., Mbéguéré, M., Ennin, J.E., Zurbrugg, C., Strande, L., 2014. A value proposition: Resource recovery from faecal sludge—Can it be the driver for improved sanitation? *Resources, Conservation and Recycling* 88, 32–38. https://doi.org/10.1016/j.resconrec.2014.04.005

Donatello, S., & Cheeseman, C. R. (2013). Recycling and recovery routes for incinerated sewage sludge ash (ISSA): A review. *Waste Management*, 33(11), 2328–2340. https://doi.org/10.1016/j.wasman.2013.05.024

Drangert, J.-O., 1998a. Fighting the urine blindness to provide more sanitation options. *Water SA* 24, 157–164.

Drangert, J.-O., 1998b. Urine blindness and the use of nutrients from human excreta in urban agriculture. *GeoJournal* 45, 201–208. https://doi.org/10.1023/A:1006968110875

Dursun, I., kabadayı, E., Tuğer, A., 2017. Application of Value-Belief-Norm Theory to Responsible Post Consumption Behaviors: Recycling and Reuse. www.researchgate.net/publication/320087242_Application_of_Value-Belief-Norm_Theory_to_Responsible_Post_Consumption_Behaviors_Recycling_and_Reuse?enrichId=rgreq-7b732f231cc04c7bd2c6177f946a3c5c-XXX&enrichSource=Y292ZXJQYWdlOzMyMDA4NzI0MjtBUzo1ODg5MjA2MjI0MjgxNjVAMTUxNzQyMTAyNjI1Mw%3D%3D&el=1_x_2&_esc=publicationCoverPdf

Eskelinen, T., Sydd, O., Kajanus, M., Fernández Gutiérrez, D., Mitsou, M., Soriano Disla, J.M., Sevilla, M.V., Ib Hansen, J., 2022. Fortifying social acceptance when designing circular economy business models on biowaste related products. *Sustainability* 14, 14983. https://doi.org/10.3390/su142214983

Etter, B., Tilley, E., Khadka, R., & Udert, K. M. (2011). Low-cost struvite production using source-separated urine in Nepal. *Water Research*, 45(2), 852–862. https://doi.org/10.1016/j.watres.2010.10.007

Ferguson, C., Mallory, A., Anciano, F., Russell, K., Valladares, H. del R.L., Riungu, J., Parker, A., 2022. A qualitative study on resource barriers facing scaled container-based sanitation service chains. *Journal of Water, Sanitation and Hygiene for Development* 12, 318–328. https://doi.org/10.2166/washdev.2022.218

Ghazali, E.M., Nguyen, B., Mutum, D.S., Yap, S.-F., 2019. Pro-environmental behaviours and value-belief-norm theory: Assessing unobserved heterogeneity of two ethnic groups. *Sustainability* 11, 3237. https://doi.org/10.3390/su11123237

Gwara, S., Wale, E., Odindo, A., Buckley, C., 2021. Attitudes and perceptions on the agricultural use of human excreta and human excreta derived materials: A scoping review. *Agriculture* 11, 153. https://doi.org/10.3390/agriculture11020153

Hadas, E., Mingelgrin, U., & Fine, P. (2021). Economic cost–benefit analysis for the agricultural use of sewage sludge treated with lime and fly ash. *International Journal of Coal Science & Technology*, 8(5), 1099–1107. https://doi.org/10.1007/s40789-021-00439-z

Hein, N. (2022). Factors Influencing the Purchase Intention for Recycled Products: Integrating Perceived Risk into Value-Belief-Norm Theory. *Sustainability*, 14, 3877. https://doi.org/10.3390/su14073877

Huang, Y., Zhang, Z., Zhang, Y., Wang, Z., 2022. Perceptional differences in the factors of local acceptance of waste incineration plant. *Frontiers in Psychology* 13, 1067886. https://doi.org/10.3389/fpsyg.2022.1067886

Iacovidou, E., Hahladakis, J.N., Purnell, P., 2021. A systems thinking approach to understanding the challenges of achieving the circular economy. *Environmental Science and Pollution Research* 28, 24785–24806. https://doi.org/10.1007/s11356-020-11725-9

Kalmykova, Y., Sadagopan, M., Rosado, L., 2018. Circular economy–From review of theories and practices to development of implementation tools. *Resources, Conservation and Recycling, Sustainable Resource Management and the Circular Economy* 135, 190–201. https://doi.org/10.1016/j.resconrec.2017.10.034

Kambo, H. S., & Dutta, A. (2015). A comparative review of biochar and hydrochar in terms of production, physico-chemical properties and applications. *Renewable and Sustainable Energy Reviews*, 45, 359–378. https://doi.org/10.1016/j.rser.2015.01.050

Laborda, E., Del-Busto, F., Bartolomé, C., Fernández, V., 2023. Analysing the social acceptance of bio-based products made from recycled absorbent hygiene products in Europe. *Sustainability* 15, 3008. https://doi.org/10.3390/su15043008

Latawiec, A., Królczyk, J., Kuboń, M., Szwedziak, K., Drosik, A., Polańczyk, E., Grotkiewicz, K., Strassburg, B., 2017. Willingness to adopt biochar in agriculture: The Producer's perspective. *Sustainability* 9, 655. https://doi.org/10.3390/su9040655

López-González, J.A., Estrella-González, M.J., Lerma-Moliz, R., Jurado, M.M., Suárez-Estrella, F., López, M.J., 2021. Industrial composting of sewage sludge: Study of the bacteriome, sanitation, and antibiotic-resistant strains. *Front Microbiol* 12, 784071. https://doi.org/10.3389/fmicb.2021.784071

MacArthur, E., 2013. Towards the circular economy Vol. 2: opportunities for the consumer goods sector. https://ellenmacarthurfoundation.org/towards-the-circular-economy-vol-2-opportunities-for-the-consumer-goods

Mallory, A., Akrofi, D., Dizon, J., Mohanty, S., Parker, A., Rey Vicario, D., Prasad, S., Welivita, I., Brewer, T., Mekala, S., Bundhoo, D., Lynch, K., Mishra, P., Willcock, S., Hutchings, P., 2020. Evaluating the circular economy for sanitation: Findings from a multi-case approach. *Science of The Total Environment* 744, 140871. https://doi.org/10.1016/j.scitotenv.2020.140871

Manga, M., Camargo-Valero, M.A., Anthonj, C., Evans, B.E., 2021. Fate of faecal pathogen indicators during faecal sludge composting with different bulking agents in tropical climate. *International Journal of Hygiene and Environmental Health* 232, 113670. https://doi.org/10.1016/j.ijheh.2020.113670

Manga, M., Evans, B.E., Ngasala, T.M., Camargo-Valero, M.A., 2022a. Recycling of faecal sludge: Nitrogen, carbon and organic matter transformation during co-composting of faecal sludge with different bulking agents. *International Journal of Environmental Research and Public Health* 19, 10592. https://doi.org/10.3390/ijerph191710592

Manga, M., Kolsky, P., Rosenboom, J.W., Ramalingam, S., Sriramajayam, L., Bartram, J., Stewart, J., 2022b. Public health performance of sanitation technologies in Tamil Nadu, India: Initial perspectives based on E. coli release. *International Journal of Hygiene and Environmental Health* 243, 113987. https://doi.org/10.1016/j.ijheh.2022.113987

Manga, M., Muoghalu, C., Camargo-Valero, M.A., Evans, B.E., 2023. Effect of turning frequency on the survival of fecal indicator microorganisms during aerobic composting of fecal sludge with sawdust. *I International Journal of Environmental Research and Public Health* 20, 2668. https://doi.org/10.3390/ijerph20032668

Mantovi, P., Baldoni, G., & Toderi, G. (2005). Reuse of liquid, dewatered, and composted sewage sludge on agricultural land: Effects of long-term application on soil and crop. *Water Research*, 39(2), 289–296. https://doi.org/10.1016/j.watres.2004.10.003

McConville, J., Niwagaba, C., Nordin, A., Ahlström, M., Namboozo, V., Kiffe, M., 2020. Guide to Sanitation Resource-Recovery Products & Technologies: a supplement to the Compendium of Sanitation Systems and Technologies. Report / Department of Energy andechnologyy, SLU.

McConville, J.R., Metson, G.S., Persson, H., 2023. Acceptance of human excreta derived fertilizers in Swedish grocery stores. *City and Environment Interactions* 17, 100096. https://doi.org/10.1016/j.cacint.2022.100096

Mensah, J., 2020. Theory-anchored conceptual framework for managing environmental sanitation in developing countries: Literature review. *Social Sciences & Humanities Open* 2, 100028. https://doi.org/10.1016/j.ssaho.2020.100028

Moya, B., Parker, A., Sakrabani, R., 2019a. Challenges to the use of fertilisers derived from human excreta: The case of vegetable exports from Kenya to Europe and influence of certification systems. *Food Policy* 85, 72–78. https://doi.org/10.1016/j.foodpol.2019.05.001

Moya, B., Sakrabani, R., Parker, A., 2019b. Realizing the circular economy for sanitation: Assessing enabling conditions and barriers to the commercialization of human excreta derived fertilizer in Haiti and Kenya. *Sustainability* 11, 1–15.

Muoghalu, C., Owusu, A., Lebu, S., Nakagiri, A., Semiyaga, S., Iorhemen, O., Manga, M., 2023. Biochar as a novel technology for treatment of onsite domestic wastewater: A critical review. *Frontiers in Environmental Science* 11, 202. https://doi.org/10.3389/fenvs.2023.1095920

Nematian, M., Keske, C., & Ng'ombe, J. N. (2021). A techno-economic analysis of biochar production and the bioeconomy for orchard biomass. *Waste Management*, 135, 467–477. https://doi.org/10.1016/j.wasman.2021.09.014

Peal, A., Evans, B., Blackett, I., Hawkins, P., Heymans, C., 2014. Fecal sludge management (FSM): analytical tools for assessing FSM in cities. *Journal of Water, Sanitation and Hygiene for Development* 4, 371–383. https://doi.org/10.2166/washdev.2014.139

Petrescu-Mag, R. M., Petrescu, D. C., & Azadi, H. (2020). A social perspective on soil functions and quality improvement: Romanian farmers' perceptions. *Geoderma*, 380, 114573. https://doi.org/10.1016/j.geoderma.2020.114573

Polyportis, A., Mugge, R., Magnier, L., 2022. Consumer acceptance of products made from recycled materials: A scoping review. *Resources, Conservation and Recycling* 186, 106533. https://doi.org/10.1016/j.resconrec.2022.106533

Renting, H., Cabannes, Y., Kranjac-Berisavljevic, G., Cas, C., Barendse, J., Foundation, R., Shakya, P., Post, V., 2013. Sustainable Financing for WASH and Urban Agriculture. www.academia.edu/5456086/Sustainable_Financing_for_WASH_and_Urban_Agriculture

Rose, C., Parker, A., Jefferson, B., Cartmell, E., 2015. The characterization of feces and urine: A review of the literature to inform advanced treatment technology. *Critical Reviews in Environmental Science and Technology* 45, 1827–1879. https://doi.org/10.1080/10643389.2014.1000761

Segrè Cohen, A., Love, N.G., Nace, K.K., Árvai, J., 2020. Consumers' Acceptance of Agricultural Fertilizers Derived from Diverted and Recycled Human Urine. *Environmental Science & Technology* 54, 5297–5305. https://doi.org/10.1021/acs.est.0c00576

Senecal, J., Nordin, A., Simha, P., & Vinnerås, B. (2018). Hygiene aspect of treating human urine by alkaline dehydration. *Water Research*, 144, 474–481. https://doi.org/10.1016/j.watres.2018.07.030

Senecal, J., & Vinnerås, B. (2017). Urea stabilisation and concentration for urine-diverting dry toilets: Urine dehydration in ash. *Science of the Total Environment*, 586, 650–657. https://doi.org/10.1016/j.scitotenv.2017.02.038

Semiyaga, S., Okure, M.A.E., Niwagaba, C.B., Katukiza, A.Y., Kansiime, F., 2015. Decentralized options for faecal sludge management in urban slum areas of Sub-Saharan Africa: A review of technologies, practices and end-uses. *Resources, Conservation and Recycling* 104, 109–119. https://doi.org/10.1016/j.resconrec.2015.09.001

Simha, P., Lalander, C., Vinnerås, B., & Ganesapillai, M. (2017). Farmer attitudes and perceptions to the re-use of fertiliser products from resource-oriented sanitation systems—The case of Vellore, South India. *Science of the Total Environment*, 581–582, 885–896. https://doi.org/10.1016/j.scitotenv.2017.01.044

Stern, P., Dietz, T., Abel, T., Guagnano, G., Kalof, L., 1999. A value-belief-norm theory of support for social movements: The case of environmentalism. *Human Ecology Review* 6, 81–97.

Strande, L., Ronteltap, M., Brdjanovic, D., & Brdjanovic, D. (2014). *Faecal sludge management: Systems approach for implementation and operation*. IWA Publishing.

Strydom, W.F., 2018. Applying the theory of planned behavior to recycling behavior in South Africa. *Recycling* 3, 43. https://doi.org/10.3390/recycling3030043

Surendra, K. C., Tomberlin, J. K., van Huis, A., Cammack, J. A., Heckmann, L.-H. L., & Khanal, S. K. (2020). Rethinking organic wastes bioconversion: Evaluating the potential of the black soldier fly (Hermetia illucens (L.)) (Diptera: Stratiomyidae) (BSF). *Waste Management*, 117, 58–80. https://doi.org/10.1016/j.wasman.2020.07.050

Tarpeh, W.A., Clark, B.D., Nelson, K.L., Orner, K.D., 2023. Reimagining excreta as a resource: Recovering nitrogen from urine in Nairobi, Kenya, in: Madon, T., Gadgil, A.J., Anderson, R., Casaburi, L., Lee, K., Rezaee, A. (Eds.), *Introduction to Development Engineering: A Framework with Applications from the Field.* Springer International Publishing, pp. 429–462. https://doi.org/10.1007/978-3-030-86065-3_16

Thuries, L., Ganry, F., Joel, S., Oliver, R., Parrot, L., Simon, S., Montange, D., & Fernandes, P. (2019). Cash for trash: An agro-economic value assessment of urban organic materials used as fertilizers in Cameroon. *Agronomy for Sustainable Development*, 39. https://doi.org/10.1007/s13593-019-0598-7

Tilley, E., Ulrich, L., Luthi, C., Reymond, P., & Zurbrügg, C. (2014). *Compendium of Sanitation Systems and Technologies.* Swiss Federal Institute of Aquatic Science and technology (Eawag): Dübendorf, Switzerland, 2014.

Trimmer, J.T., Miller, D.C., Byrne, D.M., Lohman, H.A.C., Banadda, N., Baylis, K., Cook, S.M., Cusick, R.D., Jjuuko, F., Margenot, A.J., Zerai, A., Guest, J.S., 2020. Re-envisioning sanitation as a human-derived resource system. *Environmental Science & Technology* 54, 10446–10459. https://doi.org/10.1021/acs.est.0c03318

Ushijima, K., Funamizu, N., Nabeshima, T., Hijikata, N., Ito, R., Sou, M., Maïga, A.H., Sintawardani, N., 2015. The postmodern sanitation: Agro-sanitation business model as a new policy. *Water Policy* 17, 283. https://doi.org/10.2166/wp.2014.093

Verbeke, W., Spranghers, T., Clercq, P. D., Smet, S. D., Sas, B., & Eeckhout, M. (2015). Insects in animal feed: Acceptance and its determinants among farmers, agriculture sector stakeholders and citizens. *Animal Feed Science and Technology*, 204, 72.

Vögeli, Y., Lohri, C. R., Gallardo, A., Diener, S., & Zurbrügg, C. (2014). Anaerobic digestion of biowaste in developing countries. *Practical information and case studies.* www.semanticscholar.org/paper/Anaerobic-digestion-of-biowaste-in-developing-and-V%C3%B6geli-Lohri/7f60122bca72fd5dbdac8ab628a1223faa0924c8

Wiśniowska, E., 2019. 9–Sludge activation, conditioning, and engineering, in: Prasad, M.N.V., de Campos Favas, P.J., Vithanage, M., Mohan, S.V. (Eds.), *Industrial and Municipal Sludge.* Butterworth-Heinemann, pp. 181–199. https://doi.org/10.1016/B978-0-12-815907-1.00009-X

Wolsink, M., 2018. Social acceptance revisited: gaps, questionable trends, and an auspicious perspective. *Energy Research & Social Science* 46, 287–295. https://doi.org/10.1016/j.erss.2018.07.034

World Health Organization, 2021. Progress on household drinking water, sanitation and hygiene 2000–2020: five years into the SDGs.

14 Promotion of Circular Economy Through Waste Management Policies

Shriya Bhatia, Bakul Rao, and Ajay Deshpande

14.1 INTRODUCTION

Approximately 377 million people living in urban India generate 62 million tons of waste annually, of which more than 80% are disposed at dumpsites. There will be a daily requirement of 3.40 lakh cubic meters of landfill space if all the waste is simply dumped without treatment. By 2031, 165 million tons of waste is expected to be produced, which would require 66,000 hectares of landfill area, or 1240 hectares each year, assuming a 10-meter-high garbage pile. (Ministry of Housing & Urban Affairs, GoI, 2019). Numbers on the waste generation quantities are contested, and new numbers emerge; there are data issues on how and at which point of the waste cycle the quantities are measured. Whatever the actual numbers, a picture speaks a thousand words, and Figure 14.1 is a testament to the solid waste management (SWM) challenge we face in India. While we have come a long way in developing waste management policies, we still have a long way to go.

A few decades ago, solid waste challenges were attributed to an accelerated rate of urbanization, population growth, and industrialization and were seen as an urban phenomenon (Arlosoroff & Rushbrook, 1991). However, with increased mass production and consumption, waste is no longer an urban problem. Plastic waste has reached unconnected and far-flung villages, mountain peaks, and ocean depths. The inappropriate disposal of waste creates local pollution issues leading to health risks and greenhouse gas (GHG) emissions which accelerate climate change.

The global model of development is dependent on materialistic consumption practices. The more we produce, the more we buy; and the more we buy, the more we produce. The foundation of this model is on resource availability, which is under threat due to the take-make-throw practices of the linear economy. We have begun to realize that our resource stocks are diminishing, and their quality is deteriorating, which may threaten the global economic development trajectory. Global instabilities are disrupting supply chains, increasing costs, and impacting industrial growth. Therefore, nations have begun to focus on decoupling economic growth from environmental degradation. *Waste* is a cross-cutting environmental concern linked to various developmental sectors, including livelihood, human rights, health,

DOI: 10.1201/9781003364467-14

FIGURE 14.1 Solid waste disposal into the streams in Mumbai.

developmental finance, and environmental protection. The waste management (WM) sector creates jobs and stimulates the green and circular economy through the recovery and recycling sectors. Globally, waste management policies have graduated upwards on the waste management hierarchy (WMH). From earlier days of focusing on scientific disposal, countries are graduating to methods of waste avoidance. Thus, shifting from a linear approach to a circular approach provides co-benefits on climate change mitigation, resource security, energy security, and environmental protection. The waste management policy canvas is expanding along with the development of allied policies on resource efficiency and circular economy. The chapter's objective is to understand the trajectory of WMP and reflect on how they can contribute to a circular economy.

14.2 DRIVERS OF WASTE MANAGEMENT

Before moving to concepts of CE, it is important to understand the advancements on waste management which have similar development drivers and objectives. Public health concerns in the nineteenth century, followed by environmental protection post-1970s, brought focus on collection systems and controlled scientific disposal. While developing countries are still struggling with it, technology development and financial investments in developed countries are driven by the need for climate change mitigation through GHG capture and energy recovery measures. The resource value of waste, recovered by the informal sector, has historically been an essential driver in developing countries and continues to gain importance globally due to resource security issues. The latest trends in industrialized nations is toward 'Closing the loop' or 'circularity,' and resource management which moves away from the concept of 'end-of-pipe.' Clean development mechanism (CDM) and private sector participation as a financial mechanism is encouraged by international financial institutions.

It has been a technical driver for waste management to promote environmental protection, and climate change mitigation. Other institutional drivers come from public awareness, private sector interests, and international financial institutions (Wilson, 2007). Currently, it is the Sustainable Development Goals (SDGs) and the international climate change negotiations that are driving WM and CE policy action.

14.2.1 Integrated Sustainable Waste Management (ISWM)

The above drivers led to a comprehensive and nuanced framework on Integrated Sustainable Waste Management (ISWM). The initial conceptualization by the Collaborative Working Group on waste management in middle and low-income countries in 1996 recognized three 'dimensions' of a waste management system:

(i) physical components of collection, treatment, and disposal;
(ii) sustainability aspects, including the environmental, social, health, legal, political, institutional, economic, technical, and financial;
(iii) various stakeholders involved, i.e., governments, users, NGOs, private players, informal sector, and external support agencies.

This was simplified by the UN-Habitat and further streamlined and used in the Global Waste Management Outlook, 2015. The ISWM framework have two overlapping 'triangles': (i) physical hardware: collection, 3R's, and safe disposal, and (ii) 'Software': the three governance aspects, namely inclusivity (all stakeholders to contribute and benefit); financial sustainability; and sound institutions and policies (Wilson et al., 2012; Wilson et al., 2015; Wilson, United Nations Environment Programme et al., 2015).

14.2.2 Waste Management Hierarchy (WMH)

It is vital to see ISWM alongside the concept of the waste management hierarchy presented in Figure 14.2, which dates back to 1977 when it was introduced in the European Union's (EU) Second Environmental Action Program of 1977–1981. As per the WMH, waste avoidance is given first priority; then waste must be minimized. Next comes recycling and reuse. Following that, one should focus on treatment with the goals of energy recovery and waste reduction for eventual disposal. The residue alone should then be landfilled (Wilson, 1996). The 10th meeting of the Conference of the Parties to the Basel Convention on the "Control of Transboundary Movements of Hazardous Wastes and their Disposal" at Cartagena, in 2011, formulated the "Strategic framework for the implementation of the Basel Convention for 2012–2021." It spelled out the guiding principles that recognized the WMH; prioritizing prevention, minimization, reuse, recycling, other recovery including energy recovery, treatment and scientific. It offers a prioritization to the technological choices and waste management approaches, taking into account how to advance and maximize benefit. It encourages strategies that deliver excellent environmental outcomes, considering life cycle thinking. This is now the basis for formulating many WMPs.

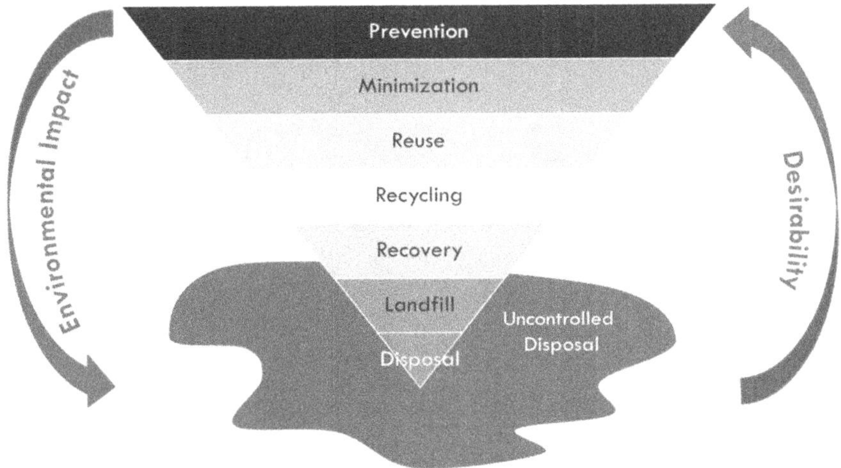

FIGURE 14.2 Waste management hierarchy.

Source: Iyamu et al., 2020; Wilson, 1996.

14.3 EMERGENCE OF CE CONCEPT AND EXPANDING ON R'S

The introduction of the CE can be traced to Pearce and Turner in 1989 (Ghisellini et al., 2016; Su et al., 2013). It examined the interlinkages between environment and its economic functions. Nature as a resource base, a sink to absorb the externalities of economic activities, and offering ecological services, that maintain life on earth. This understanding can be attributed to the anthropocentric approach used in neoclassical environmental and economic analysis, which attributes the utility of the environment for humans measured through the economic function (Andersen, 2007).

CE has been predominantly driven by practitioners and focused on private businesses as the key enablers (Kirchherr et al., 2017). The Ellen MacArthur Foundation's work on Circular Economy has significantly influenced global CE discourse (Geissdoerfer et al., 2017). It defines 'CE as an industrial system that is restorative or regenerative by intention and design. It replaces the 'end-of-life' concept with restoration, shifts towards renewable energy, eliminates the use of toxic chemicals, which impair reuse, and aims to eliminate waste through the superior design of materials, products, systems, and business models (Ellen MacArthur Foundation, 2014).

A total of 114 definitions were analyzed based on which Kirchherr et al., in 2017, came up with a broad, all-encompassing definition as follows:

> An economic system based on business models which replace the 'end-of-life' concept with reducing, alternatively reusing, recycling and recovering materials in production/distribution and consumption processes, thus operating at the micro level (products, companies, consumers), meso level (eco-industrial parks) and macro level (city, region, nation and beyond), with the aim

to accomplish sustainable development, which implies creating environmental quality, economic prosperity, and social equity, to the benefit of current and future generations.

Beyond CE, discussions on R's have been around for a while, including the 3 R's (Reduce, Reuse, Recycle) concerning waste management. The 3 R's have been commonly used, interpreted in different ways, and expanded further into 4 R's, 5 R's, 6 R's, 7 R's, 9 R's, and 12 R's. The 12 R's include: Refuse, Rethink, Reduce, Re-design, Reuse, Repair, Refurbish, Renovate, Remanufacture, Return, Recover, and Recycle (Modak, 2021). Another version re-conceptualized by Ellen MacArthur Foundation is the ReSOLVE Framework, i.e. Regenerate, Share, Optimize, Loop, Virtualize, and Exchange (EMF, 2015). All these are essentially targeted toward minimizing waste production and dependency on virgin raw materials. While prioritizing circularity it is crucial to understand which R gains attention and which R has a higher impact. Figure 14.3 below explains the same in a comprehensive manner. The worldwide application of CE focuses on recycling rather than reuse, achieving improved recycling rates in developed countries (Ghisellini et al., 2016). While conceptualizing CE policies for India, one needs to be conscious of the bias toward recycling, carefully articulate priorities, and incorporate the existing practices of the higher order R's, which are primarily supported by the informal sector.

14.4 POLICY HISTORY ON WASTE MANAGEMENT

History provides a memory function as a storage of past successes and failures, of productive and unproductive ideas. It offers methodological principles and implications incorporating time, context, and complex processes while studying societies and efforts to bring change. History provides a reference point to understand how institutions have come about through the political process of contestations of ideas between distinctive groups. It can help develop an alternate perspective on policy problems by providing a vantage point to frame a long-term solution (Woolcock et al., 2011).

The study of policy history helps us understand the significant factors that shaped the policies at different periods. The below sections explain the significant changes in the development of WMP which are also relevant in the context of the global interest in CE. They are discussed through the three broad phases; prior to 2000, 2000 to 2014, and post-2014, primarily because a significant shift is observed in intent, interest, types of instruments, and institutional models. Table 14.2 explains the waste-management related policies in India post-2000 concerning critical interventions, actors, policy instruments, actors, focus on WMH, and consequences.

14.4.1 Prior to MSW Rules 2000

Mismanagement of waste has had far-reaching implications for environmental and human health. The relationship between the quality of the environment and human health gained significance due to the Bombay Plague of 1896, attributed to overcrowded living conditions and accumulated garbage. It led to an institutional

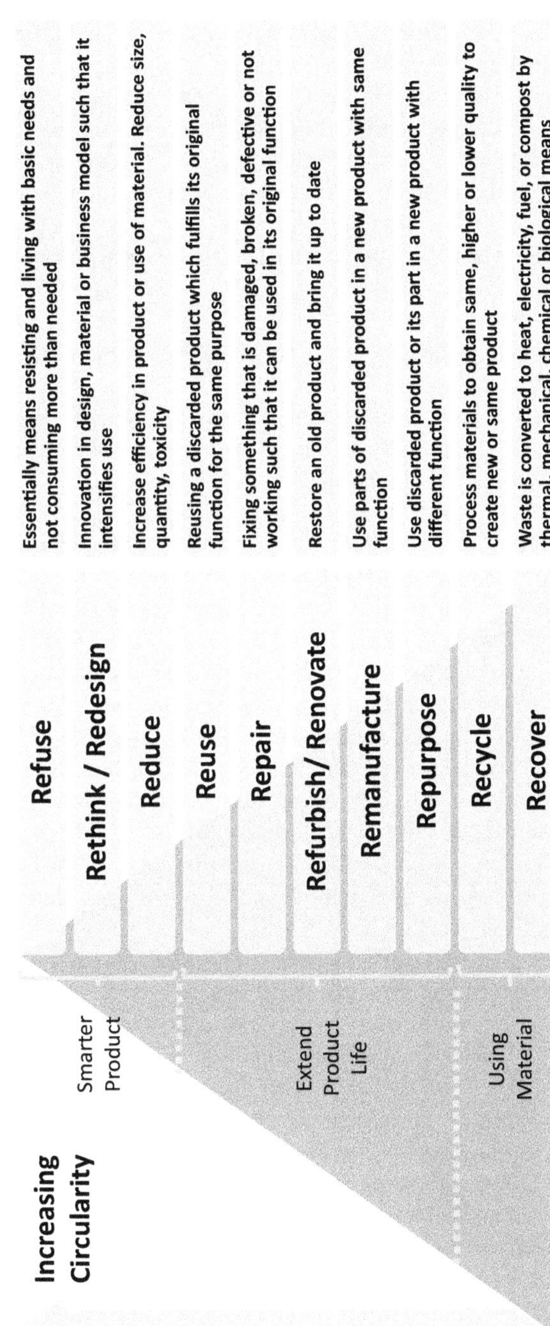

FIGURE 14.3 Prioritizing circular economy strategies.

Source: Modak, 2021; Potting et al., 2017.

change by creating the Bombay City Improvement Trust in 1898 and passing the Epidemic Disease Act of 1897. The Bombay Municipal corporation sanitized all areas and proposed removing refuse beyond city limits by rail. Low-lying areas were identified, and Bombay Improvement Trust reclaimed land using waste in a planned and controlled manner (Mirza, 2019). However, it is essential to note here that the Bombay Act No. III of 1888 defined the roles and responsibilities of the municipal corporation and the occupier concerning refuse, rubbish, or waste. An institutional link between waste and health remains today, as some municipalities have waste management under the Chief/City Health or Medical Officer.

While this happened in Bombay two centuries ago, we saw history repeat itself with the pneumonic plague outbreak in Surat in 1994. The primary cause of the outbreak was flooding in low-lying areas due to high rainfall, faulty drainage systems, and inadequate waste management systems. In addition to being a human tragedy, it was a severe setback to the nation's economy. It caused fear and tarnished India's reputation abroad. Local authorities across the country immediately began cleaning up with business as usual the year after. The plague brought policy attention to the waste management sector. The J.L. Bajaj Committee, established in 1995, made numerous recommendations, including encouraging private sector participation, waste segregation at source, primary collection, user fees, use of upgraded equipment and vehicles, composting, and landfilling (Furedy, 1995).

The Bombay Plague led to a series of Public Interest Litigations (PIL); the most noticeable one was the Almitra Patel vs Union of India (1996). The Supreme Court recognized the fundamental right to life with an expanded interpretation to include the right to a clean and healthy environment, thus ensuring proper waste disposal. Multiple high-powered committees were constituted at different periods making different recommendations on institutional, financial, health, operational, and legal aspects of waste management, but none were implemented. Finally, the Supreme Court's interventions to the PILs led to the enactment of the Municipal Solid Waste (Management and Handling) Rules, 2000. The courts continue to be significant with their supplementary judicial interventions in the follow-up litigations (Chintan Environmental Research and Action Group, 2018).

The MSW Rules need to be seen alongside the 74th Constitutional Amendment Act (CAA), 1992, and the Municipal Acts, which determined SWM as the primary responsibility of the Urban Local Bodies (ULBs). Pre-2000, the government was seen as a service provider for waste management, which saw a gradual shift post-2000 to being a facilitator.

Other than the MSW Rules, hazardous and biomedical waste was becoming a significant environmental and health concern. The Hazardous Waste (Management and Handling) Rules, 1989, were implemented through the State Pollution Control Boards. It provided for control of generation, collection, storage, transport, treatment, and disposal. The Guidelines for the Management and Handling of HWs issued by the MoEF in 1991 provided protocols for transport, storage, and disposal facilities. It specified the manifest system providing protocols for the movement of HW. Apart from the rules, the government provided fiscal incentives for industries to comply with environmental provisions, thus creating a business environment for waste management (Kumar et al., 2007).

14.4.2 Between MSW Rules 2000 to Swachh Bharat Mission 2014

The MSW Rules allocated the primary institutional responsibility of waste management to the ULBs, monitored by the State and central institutions responsible for urban development and environmental protection. It provided timelines, quality standards, guidelines, and operational procedures for composting, incineration, and scientific landfills. Subsequent issues on lack of enforcement of rules brought to light the need for institutional capacity building of the ULBs, action planning at the State and ULB level, technology support, and strengthening the financial capacities of the ULB. Fund allocations were made to support these rules through the 12th Finance Commission in July 2005 and subsequently with additional funding through the Jawaharlal Nehru National Urban Renewal Mission (JNNURM) and Urban Infrastructure Development Schemes for Small and Medium Towns (UIDSSMT) which were launched in December 2005. Several large-scale SWM projects were funded under the JNNURM scheme. However, the sustainability of capital-intensive centralized projects and waste-to-energy plants, particularly, were questioned due to shutdowns and local conflicts (Chintan Environmental Research and Action Group, 2018).

Multilateral agencies have occupied an influential role in the formulation of solid waste management policy in India, determining its design, structure, and implementation. They were brought in to support the technical, financial, and managerial deficiencies and limitations at the ULBs level. The policy support came in the form of knowledge series, policy papers, reports, toolkits for public-private partnerships, technical manuals on waste-to-energy power plants and integrated solid waste management facilities, and manuals for SWM DPR preparation that provided technical and capacity-building support. Various expert committees were set up, such as the Technology Advisory Group on MSWM and the Inter-Ministerial Task Force on 'Integrated Plant Nutrient Management from city compost', Task Force on 'Waste-to-Energy (WTE)', to support the Public-Private Partnerships (PPP) (ICRA Management Consulting Services Limited, 2010).

The design of the policy focused on the lack of collection, segregation, transportation, and scientific disposal of waste through technocratic and capital-intensive solutions. The 12th and 13th Finance Commission and the Jawaharlal Nehru National Urban Renewal Mission (JNNURM) provided technical and financial support for infrastructure development. It emphasized urban governance reforms and infrastructure modernization. It advised the ULBs to explore PPP in SWM projects (Chintan Environmental Research and Action Group, 2018).

In the wake of India's transformation to a neoliberal economy, there was increasing pressure for ULBs to become accountable, efficient, and financially self-sustainable. Historically, waste was the liability of the municipalities. In the past, waste was essentially an 'urban commons' used gratis by the informal sector to earn a living. With the popularization of the 'waste to wealth' concept, urban waste became a viable prospect for revenue. It led to the capitalization of waste through privatization by new ventures (Chaturvedi & Gidwani, 2011). The JNNURM supported capital-intensive public-private partnerships in waste management. The central belief is that technology and private capital will ensure system efficiency and, therefore, are seen as the panacea for managing the waste crisis and practicing a circular economy (Chaturvedi et al., 2019).

Neoliberalism has been the dominant ideology infusing the public policies of governments through international agencies. It considers economic growth as means of achieving human development. It propagates a truncated role for the State, with a rollback of the redistribution policies, relinquishing societal control to the invisible hand. The ideological neutrality of the market principle is dangerous as it assumes no social responsibility and accountability toward growing disparity or inequality. It allows for a government of the market, by the market, for the market, where social equity conflicts with profit-making goals (Chatterjee, 2011).

The Municipal Waste Management Rules, 2000 encouraged the privatization of waste management services due to the Municipalities' low capacity. This favored contractors and correspondingly denied rag pickers access to waste. It is also important to note that there have been counter currents through NGOs advocating for the waste picker rights and organizing them through waste picker collectives/associations. They have been instrumental in skilling them, upgrading their lives, providing access to finance, and identification cards, providing alternate job opportunities, improving their working conditions, and fighting for their right to access waste. It brought to attention the need to look at decentralized SWM by recognizing the waste pickers' role. New institutional models came up through formalizing waste picker cooperatives, NGO engagements, and RWAs involvement. ULBs like Delhi Municipal Corporation, Municipal Corporation of Greater Mumbai, and Pune Municipal Corporation were pioneers in developing partnerships with waste picker cooperatives/associations and NGOs. They also developed institutional mechanisms to mobilize communities in segregating and composting activities through the Advanced Locality Management (ALM) and Mohalla Committees.

The complexity of handling various waste streams and the environmental and health risks posed by them led to the enactment of Batteries (Management and Handling) Rules, 2001, Bio-medical Waste Management Rules, 1998 (updated in 2009, replaced in 2016 and amended till 2019), Hazardous Waste (Management, Handling, and Transboundary Movement) Rules, 2008 (first one in 1989, updated in 2008, replaced in 2016), E-Waste (Management and Handling) Rules, 2011 (replaced in 2016), Plastic Waste (Management & Handling) Rules, 2011 (replaced in 2016), and Construction and Demolition Waste Management Rules, 2016. These rules carefully defined procedural and institutional directives for safe transportation, management, and disposal. These distinctively differentiated them from the Municipal Solid Waste Rules. It generated a new market for private enterprises to provide waste collection, transportation, treatment, and disposal services.

14.4.3 Post Swachh Bharat Mission 2014

The multiple waste management rules and the latest Swachh Bharat Mission (SBM) are critical drivers for developing the Public-Private Partnership (PPP) model in delivering waste management services. Private sector involvement has led to the de-bundling services into a door-to-door collection, street sweeping, transportation, storage, treatment, and disposal for ease of management.

The Extended Producers Responsibility (EPR) introduced in Plastic Waste and E-Waste Rules is gaining acceptance and making manufacturers accountable for collection and safe disposal. There is an expansion of policy instruments beyond regulations to include service-level benchmarking, tax rebates, service/user charges, ICT for awareness, viability gap funding, subsidies and grants for capital expenditure, performance-based grants, land use planning, provision of government land for SWM projects, awards, and monitoring mechanisms. These positively impact segregation efficiency, collection efficiency, processing, treatment, and safe disposal. There is a surge in SWM projects due to SBM's mission mode approach, primarily driven by performance targets. The sudden rush on projects has led to the concern about the authenticity, compost quality, long-term commitment, viability, and sustainability of these efforts. The current policy has also expanded the roles of WM to stakeholders beyond the ULB, leading to collaborations with large businesses and entrepreneurs supported by venture capitalists, NGOs, CBOs, and SHGs toward pooling financial resources, human resources, and technical expertise. The wealth and resource opportunity in the sector is expected to drive growth in private investments, creating a competitive landscape that may threaten small informal players. There is a growing realization of the relationship between land requirement and good practices in SWM; hence cities like Mumbai have provided space at decentralized and centralized levels, incorporating solid waste management facilities in the Development Plan 2014–34 published in 2018. Table 14.1 below summarizes the waste management-related policies that have been driving CE.

From integrated solid waste management concepts, we are now gearing up to concepts like resource efficiency and circular economy which are also anchored on similar principles of 3R's to 12R's. The formulation of the Indian Resource Panel (IRP) in 2015, the Draft National Resource Efficiency Policy, 2019, and the Vehicle Scrappage Policy, 2021 is a step in that direction. The European Union is driving the resource efficiency agenda in India through their European Union Resource Efficiency India (EU-REI), supported by Deutsche Gesellschaft für Internationale Zusammenarbeit (GIZ). The agenda setting for the Draft Resource Efficiency Policy, 2019, can be attributed to international advocacy. Strategy papers were prepared jointly by NITI Aayog and EU-Delegation (MoEFCC, 2019). There were other agencies, like Ellen MacArthur Foundation, FICCI, CII, Accenture, IL&FS, and The Energy and Resources Institute (TERI), that supported through advisory, technical reports, conferences, training workshops, manuals, and toolkits.

Table 14.2 is a detailed mapping of WM-related policies. It is observed that there is inherent incrementalism in Indian policymaking. The changes have happened either by expanding policy on the various waste streams, allocating roles to different stakeholders, recognizing technological solutions, or expanding on policy instruments over time. Reflecting on the current and past narratives driving waste management in mainstream media or policymaking, "health," "hygiene," and "cleanliness" has been primary. However, the narratives on climate change mitigation, circular economy, resource security, and energy security have gained momentum.

TABLE 14.1
Waste Management and Related Policies Driving Circular Economy

Pre-2000	2000 to 2014	2014 onwards
a) Bombay Act No. III, 1888 (MMC Act) b) Bombay Plague of 1896 c) Epidemic Disease Act, 1897 d) Bombay City Improvement Trust in 1898 e) Removal of Waste by Rail beyond city limits. Waste used to reclaim land f) Hazardous Waste Rules, 1989 g) Surat Plague, 1994 h) J.L. Bajaj Committee, 1995 i) Public Interest Litigation (Almitra Patel vs. Union of India, 1996) j) Bio-medical Waste Management Rules, 1998 k) 74th Constitutional Amendment Act, 1992	a) MSW (M&H) Rules, 2000 b) Batteries (M&H) Rules, 2001 c) 12th and 13th Finance Commission, 2002/07 d) JNNURM & UIDSSMT, 2005 e) UNFCC's Clean Development Mechanism, Kyoto Protocol, 2005 f) Maharashtra Non-Biodegradable Garbage (Control) Act, 2006 g) Maharashtra Plastic Carry Bags (Manufacture & Usage) Rules, 2006 h) Greater Mumbai (GM) Cleanliness and Sanitation Byelaws 2006 i) GM Construction, Demolition & Desilting Waste (Management & Disposal) Rules, 2006 j) NUSP, National Award Scheme, 2008 k) E-waste (M&H) Rules, 2011 l) Plastic Waste (M&H) Rules, 2011 m) Task Force on WTE, 2013	a) Swachh Bharat Mission, 2014 b) Swachh Bharat Kosh, 2014 & CSR c) Solid Waste Management Rules, 2016 d) The National Action Plan for Municipal Solid Waste Management, 2016 e) C&D Rules, 2016 f) Update of E-waste, Plastic waste, Biomedical Waste, Batteries, and Hazardous Waste Rules in 2016 g) Maharashtra Plastic and Thermocol Products, Notification, 2018 h) Draft National Resource Efficiency Policy, 2019 i) Motor Vehicles (Registration & Functions of Vehicle Scrapping Facility) Rules, 2021

14.5 POLICY INSTRUMENTS

Over the years, there has been significant progress in using multiple kinds of policy instruments with a two-pronged approach—stick and carrot. Firstly, it has mainly been driven by command-and-control mechanisms such as penalties imposed under the Environmental Protection Act, 1986. Secondly, alternate financing mechanisms, such as grants and subsidies, have been used to encourage businesses. Thirdly, there are instruments to provide technical support and drive change through monitoring, benchmarking, and collaboration.

Table 14.2 points out the policy instruments corresponding to various government actions in the three phases of policy development on waste management. Table 14.3 is a summary of the policy instruments indicative of the policy progress.

TABLE 14.2
Mapping of Waste Management-related Policies in India

Year	Policy Interventions	Key Actors	Consequences	Instruments	WMH
2016–2020 **2020**	**Swachh Survekshan (SS), 2016[1] onwards** **Swachh Survekshan League–2020**	Minister of Housing and Urban Affairs (MoHUA), ULBs, NGOs/ CBOs, Citizens, Informal sector	The monthly MIS, independent on-field assessment, evidence verification, citizen feedback, garbage-free star rating, and city ranking improved accountability and created competition among cities. It has helped reduce data discrepancies and bring consistency and parity to SWM data. IEC has led to awareness and resulted in behaviour change. Indicators are linked to the mandates of SWM Rules, 2016 and has helped track implementation. Although 3R is one of the focus, it's weak on waste reduction measures and continues to focus on cleanliness rather than circularity.	Monitoring, data management, service level benchmarking, ICT, and awards	Reduce Recycling Recovery Safe Disposal
2019	**Draft National Resource Efficiency Policy, 2019[2]**	Ministry of Environment Forest and Climate Change (MoEFCC), NREA (proposed)	Yet to be approved. It proposes the creation of a National Resource Efficiency Authority (NREA). Suggestions on green public procurement and ecolabeling. Focused on automotive, plastic packaging, building construction, electronic equipment, solar photo voltaic, steel, and aluminium sectors.	Guidelines and Standards	Reduce Reuse Recycling Recovery Safe Disposal

(continued)

TABLE 14.2 (Continued)
Mapping of Waste Management-related Policies in India

Year	Policy Interventions	Key Actors	Consequences	Instruments	WMH
2018	**Maharashtra Plastic and Thermocol Products (Manufacture, Usage, Sale, Transport, Handling, and Storage) Notification, 2018**[3]	GoM, MCGM, Bombay High Court	Ban on single-use plastic and thermocol. 'Extended Producers and Sellers/Traders Responsibility' "Buy Back Depository Mechanisms. It created a market for eco-friendly alternatives to plastic. It has created environmental awareness and behaviour change, and shift to environmentally conscious lifestyles. Plastic disposals were replaced by biodegradable options. It has created jobs for making paper and cloth bags.	Command and control regulations, Bans, fines to citizens, penalty for non-compliance	Reduce Recycling
2018/ 2020	**Revised guidelines for its waste-to-energy (WTE) program**[4]	Ministry of New and Renewable Energy (MNRE)	Private companies are increasingly interested in investing in WTE, but the beginning had been slow due to past conflicts. Nonetheless, companies are interested in bio-methanation based WTE plants.	Financial Assistance (Grants)	Recovery Safe Disposal
1989 superseded in 2008, 2016	**Hazardous and Other Wastes (Management and Transboundary Movement) Rules, 2016**[5] **HW (M, H & T M) Rules, 2008, HW (M & H) Rules, 1989**	MOEFCC, CPCB	It provides SOPs for safe collection, handling, packaging, storage, transportation, processing, treatment, reuse, recycling, reprocessing, conversion, destruction, and disposal of HW.	Institutional Directives, Procedural directives Manifest Systems	Reduce Reuse Recycle Recover

2011 replaced in 2016	**E-Waste Management Rules, 2016**[6]	MoEFCC, CPCB, MPCB, Manufacturers, producers, E-waste recycling companies	It led to diversion of e-waste from municipal waste. It created a market for formal e-waste collection, dismantling, refurbishing, and recycling companies. Producers started setting up joint mechanisms for collection. It also focused on the reducing hazardous materials in the manufacturing of electronics. 2016 rules acknowledged refurbishing and introduced the Producer responsibility organization and also the need for an EPR plan by the producer	Institutional Directives, Procedural directives Command and control, Monitoring	Reduce Reuse Recycling Recovery Safe Disposal
2003 replaced in 2016	**Plastic Waste (Management and Handling) Rules, 2016**[7]	MoEFCC, PCB, Generators, manufacturers, Recyclers	Registration of plastic waste recyclers. It led to the Maharashtra Plastic and Thermocol Products Notification, 2018, and enforcement of the plastic ban in Mumbai	Institutional Directives, Procedural directives Penalties, Monitoring	Reduce Recycling
1998 replaced in 2016	**Bio-medical Waste Management Rules, 2016**[8]	MoEFCC, CPCB, MPCB, Private companies as facility operators and transporters	It led to diversion of biomedical waste from municipal waste. Biomedical waste was segregated into further categories. Incinerators were set up by hospitals & municipalities. Authorization of transporters and facility operators. City-level and institutional-level management systems in place.	Emission Standards, Institutional and Procedural Directives, Penalties, Monitoring & Reporting. Manifests	Reduce Recycling Safe Disposal

(*continued*)

TABLE 14.2 (Continued)
Mapping of Waste Management-related Policies in India

Year	Policy Interventions	Key Actors	Consequences	Instruments	WMH
2016	**Solid Waste Management Rules, 2016**[9] Rules were now applicable to rural areas. It mandated DoUD in the States to prepare a policy and SWM strategy focusing on waste reduction, reuse, recycling, recovery, and optimum utilization of waste. ULBs were expected to prepare SWM Plan, identifying facilities and funds for segregation, storage, collection, treatment, disposal of wastes, and scientific closure of dumpsites.	MoEFCC, MoUD CPHEEO, CPCB, SPCB, ULBs, Waste Generators, Rag Pickers, Facility Operators	It encouraged decentralized collection and treatment systems with participation of NGOs, SHGs, and RWAs. It acknowledged the role of the waste pickers and their cooperatives. Space for waste in Development plans, land allocation for material recovery facilities and landfill sites, and buffer around landfill sites. It provided guidelines for locating landfills, New authorized MSW facilities are coming up. MPCB issued directions to ULBs to create provision for 25% in the annual budget for the management of municipal solid waste and management, and treatment of sewage. Standing committees of Municipal Corporations have approved the same. Action on waste reduction is missing. Monitoring aspects yet unclear on particular aspects. Penalty clause open for interpretation by ULBs The Development Plan took cognizance of the land requirement for solid waste management facilities beyond landfills. 46 DWSC and 4 more proposed in the DP[10]. MCGM is undertaking scientific closure of the Mulund dumpsite and GoM has allocated additional land for the landfill.	Institutional Directives Procedural directives Tipping Fee, User Fee, Command & Control regulations	Recycling Recovery Safe Disposal

2016	**The National Action Plan for Municipal Solid Waste Management**[11] Indicative formats for DPRs and SW Action Plans	CPCB, ULBs, MoUD,	Acted as a guidance document to prepare SWM Action Plans. This became a requirement under MSW Rules 2016		Recycling Recovery Safe Disposal
2016	**Construction and Demolition Waste Management Rules, 2016**[12] Siting, facility development, and management criteria are mentioned. Allocation of responsibility to ULB, CPCB, Service provider, and facility operator mentioned	MoEFCC, ULBs, SPCB, BIS, IRC, infrastructure development/ construction companies	Approvals and setting up of C&D waste processing plants have been slow due to the lack of capacity of the municipality to provide market assistance, fewer private players, land constraints, lack of financial feasibility, technology availability and competitive market pricing of products, and lack of enforcement. The service provider's duties are all-encompassing, thus reducing the responsibility and accountability of the ULB.	Institutional Directives Procedural directives Command & Control regulations	Reduce Recycling Safe Disposal
2016	**Policy on Promotion of City Compost**[13] Currently withdrawn	Ministry of Chemicals and Fertilizers, Fertilizer companies, Farmers, waste facility operators	MDA subsidy has been meagre, and companies have yet to receive payment. Conflict of interest for the firm producing chemical fertilizers explains lack of interest in compost. Only 5% of organic waste being composted. Subsidy design and implementation have been a failure. Poor Quality of waste due to presence of domestic hazardous chemicals and lack of monitoring of standards.	Financial Subsidy	Reduce Recovery

(continued)

TABLE 14.2 (Continued)
Mapping of Waste Management-related Policies in India

Year	Policy Interventions	Key Actors	Consequences	Instruments	WMH
2014	**Swachh Bharat Kosh, 2014**[14]	Companies, Foundations, NGOs, ULBs and village panachayats	SBK was setup to facilitate channelization of CSR toward SBM. The Companies Act 2013, Schedule VII was modified to include SBK. Funds have been allocated mainly for toilet construction in schools.	Financial Support CSR	Potential for Processing
2014	**Swachh Bharat Mission (Urban)**[15]	MoHUA, MoNRE, ULBs	It has increased private sector interest and involvement in waste management projects and services. It has also initiated CSR funding in decentralized waste management through CSOs. It has also seen the integration and formalization of waste picker groups, while also being marginalized in contexts where waste access is being controlled. ULBs have started the empanelment of waste management service providers. MCGM has set up 46 dry waste segregation centres in partnership with NGOs/CBOs/Foundations run by Waste pickers or employing waste pickers. Segregation through 719 ALMs[16]	Fund allocation Technical & financial guidelines. Technology assessments. Performance linked grants, CSR, ICT Awareness & Capacity Building	Recycling Recovery Safe Disposal

2014	**Task Force on Waste to Energy (WTE) under Dr. Kasturirangan by Planning Commission**[17] Prioritization of waste treatment technologies based on waste composition. Suggestions on composting and/or biogas generation supported by segregation of C&D waste and recyclables. RDF for non-recyclables with a high calorific value	Planning Commission, MoUD, MOEF	WTE power plants were set up with little consideration and control of the waste compositions that fuelled the plants. The techno-economic feasibility of plants was poor and dependent on viability gap funding, capital subsidy rather than performance-based subsidy. Conflicts from local communities due to the environmental and health risks posed by the plant further derailed such projects. Thus, leading to failure.	Technology support, Grants, Viability Gap funding, Tipping fee, Sale of products/power, Service/ Management Contracts PPP/ DBFOT model	Recovery Safe Disposal
2008	**National Award Scheme for Sanitation**	MoUD, ULBs	It was linked to National Urban Sanitation Policy, 2008. A survey was used to procure data on cities and encouraged them to do better	Data Management	Safe Disposal
2006	**Maharashtra Plastic Carry Bags (Manufacture and Usage) Rules, 2006**[18]	GoM, ULBs	Ban on plastic bags of less than 50 microns. Mumbai enforced the ban through special squads conducting periodic raids and actions against manufacturers and traders.	Command & Control regulations, Penalties	Reduction Recycling

(continued)

TABLE 14.2 (Continued)
Mapping of Waste Management-related Policies in India

Year	Policy Interventions	Key Actors	Consequences	Instruments	WMH
2005	**UNFCC's Clean Development Mechanism, Kyoto Protocol**[19]	MoEF, UNFCC, MCGM International Private companies	An emission trading scheme was implemented through emission-reduction projects in developing countries by earning from saleable certified emission reduction (CER) credits paid by developed countries. 22 waste management projects in India were developed under the CDM[20].	Financial support	Recovery, Safe Disposal
2005	**Jawaharlal Nehru National Urban Renewal Mission (JNNURM)**	MoUD GoI, Government of Maharashtra, MCGM	Many capital-intensive SWM projects got a kick start due to technical and financial support through JNNURM. Due to high capital and operation expenses, there was a focus on creating bankable and financially sustainable projects.	Financial support, technical support, Urban reforms, Institutional Capacity building	Recycling Recovery Safe Disposal
2000	**Municipal Solid Wastes (Management and Handling) Rules, 2000**[21]	MoEF, SPCB, ULBs, DoUD, Private facility operators	Private sector participation in waste management, initiation of centralized waste management processing and treatment facilities, sanitary landfill sites, and WTE plants. Projects on scientific closure and capping of dumpsites. It also led to conflicts between the community and ULBs due to WTE impacts, and issues related to access to waste led to conflicts between waste pickers and private contractors. SPCB issued ULBs notices of non-compliance.	Quality standards, Institutional directives, Standard Operating procedures for establishing and operating WM facility. Command and control Land provisioning	Recycling Recovery Safe Disposal

Case of Mumbai: (1) Integrated Solid Waste Management Project through Privatization and standardization
(2) Allocation of Land and PPP model for Scientific Landfill & Treatment and processing of waste through DBOT
(3)Implementation of decentralized Schemes such as Advanced Locality Management & Slum Adoption[22]
(4) Enactment of Greater Mumbai Cleanliness and Sanitation Byelaws 2006[23]–Penalties through Clean up Marshals schemes since 2007
(5) Notified Construction & Demolition & Desilting Waste (Management & Disposal) Rules 2006

TABLE 14.3
Progress on Policy Instruments

Pre-2000	2000 to 2014	2014 onwards
• Municipal Budgets • Constitutional Law as a remedy • Penalties • Trade Refuse Charges • Tipping Fees • Sub-contracting for collection and disposal	• Command and control (Penalties) • Institutional Directives • Procedural Directives (Manifest systems) • Financial Support • Environmental Quality Standards • Instructions for Hazardous Material Concentrations • Technical Support through Manuals, Guidelines and Toolkits • CDM–Carbon credits	• Command and control (Penalties) • Institutional directives, • Procedural directives • Technical support • User charges • Incentives, tax rebates • Subsidies and grants • Land at subsidized cost • EPR Targets • Monitoring tools • Benchmarking • Awards • ICT, awareness, capacity building • Development plan reservations • Citizen participation, • Corporate Social Responsibility

Figure 14.4 explains how the various policy actions have focused and progressed on the WMH over the years.

14.6 POLICY ACTORS AND INSTITUTIONAL MODELS

The policies regarding waste management, resource efficiency, and circular economy mainly emerge from India's environmental management policy sub-system. However, some policies emanate from the urban, infrastructure, and industrial development sub-system, that impact waste and resource management. Figure 14.5 explains the various policy sub-systems and policy actors. Various waste management rules and the Draft National Resource Efficiency Policy (NREP), 2019 have been prepared by the Ministry of Environment Forests and Climate Change (MoEFCC), and the Swachh Bharat Mission (Urban), 2014 instituted by the Ministry of Housing & Urban Affairs (MoHUA), while the Vehicle Scrappage Policy, 2021 was prepared by the Ministry of Road Transport and Highways (MoRTH). Other national state actors like the Ministry of New and Renewable Energy (MNRE), Ministry of Chemicals and Fertilizers, Ministry of Electronics and Information Technology, Ministry of Steel, and Ministry of Mines are expected to provide a supportive role for these policies. Waste management functions are the primary responsibility of the ULBs at the local level, but their policy actions are now being governed by the central government, which is influenced by advocacy efforts from international agencies. There are international actors like the World Bank (WB), Asian Development Bank (ADB),

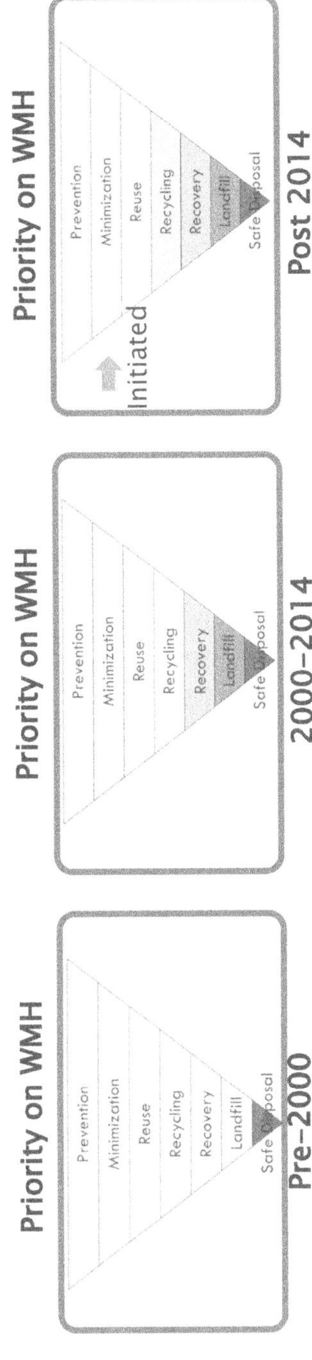

FIGURE 14.4 Progress of waste management policies on the waste management hierarchy.

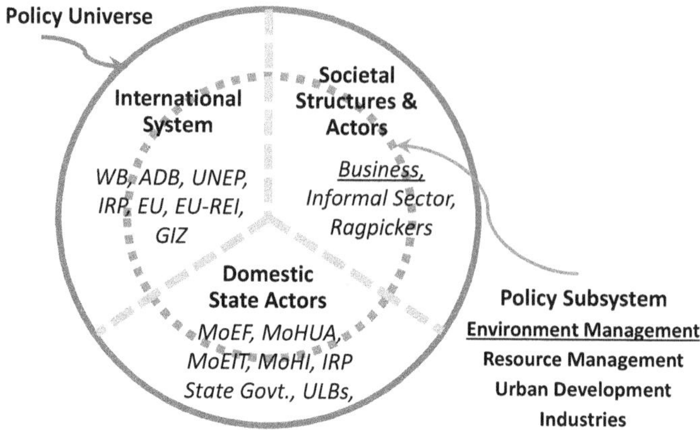

FIGURE 14.5 Waste Management and circular economy policy sub-system and policy actors.

United National Environmental Program (UNEP), European Union (EU), GIZ, and others who are advocating for a shift from the public waste management paradigm to a private sector driven circular economy. Societal actors supporting advocacy and implementing waste management practices include private businesses, the informal sector, industry associations like CII, and NGOs like TERI.

The MSW Rules, 2000, JNNURM, and other waste-specific rules encouraged the privatization of waste management services. The policy focused on private capital and technology as the primary savior of the waste management crisis. Prior to 2000, ULBs sub-contracted collection and disposal services along with use of tipping fees. PPP was the most sought after institutional model post 2000 to engage private sector to design, build, and operate SWM facilities while ULBs provided the land and access to waste (Refer to Section 5.3 for more details). Privatization displaces informal sector workers that either provided the service directly or depended on the waste materials from the spaces and processes that are being contracted to private firms. The Swachh Bharat Mission (SBM), Atal Mission for Rejuvenation and Urban Transformation (AMRUT), and Smart Cities, continue to be critical drivers of the Public Private Partnerships model that supports capital-intensive, centralized, technological solutions to waste management. While new SWM rules and SBM have acknowledged the role of the informal sector in facilitating waste recovery, however the privatization model's profits are dependent on large-scale material and energy recovery operations, thus generating conflicting claims for the same good. This threatens the integrated, inclusive, sustainable approach needed for successful waste management.

Initially, contracts for waste management services were limited to collection and transportation from municipal bins to disposal sites. ULBs have outsourced door-to-door collection as end-to-end waste management services, including treatment and disposal to a single or consortium of firms. Cities have seen more failures than success of the privatization experiments. Many firms have underbid and then failed to render services of adequate quality. There have been irregularities and inconsistencies with contracts. The financial sustainability of capital-intensive centralized projects and WTE, particularly, were questioned with most plants mired in controversy. WM plants in Nagpur, Hyderabad, Bangalore, Delhi, and Gurgaon, have complaints of poor maintenance, poor quality of compost, shutdowns, and local protests due to health impacts of dioxins, nuisance, smell, and mismanagement. One of the reasons for the failure of WTE is attributed to the low calorific value of waste due to the recovery of high calorific value waste by the informal sector. The cost of further segregation and pollution control makes the plant unviable. The residents protested against these plants due to their close proximity and release of dioxins and furans in the air, which put the resident population at health risk. Other wet processing plants also face viability issues since there is a lack of compost demand because of market competition from government-subsidized chemical fertilizers and poor quality of compost as a result of mixed waste streams (Narain & Sambyal, 2016).

While centralized models were under threat, new decentralized waste picker-led models were being acknowledged. These are explained in Section 5.2, which discusses the phase between 2000 to 2014 when the initiation of the model was undertaken. The Pune Municipal Corporation (PMC) and SWaCH Cooperative Partnership for waste management is the most globally recognized example. In 2008, SWaCH entered an MoU with PMC authorizing them to undertake waste collection at doorstep along with other waste management services, including segregation and material recovery, for five years. The contracts kept extending, and their services expanded. However, the model had been under threat since the contract ended in December 2020. The municipality hadn't given them the promised safety equipment, life insurance, medical assistance, and allowances. It had been renewing the contract on an ad hoc basis by only providing one or two-month extensions (Banerjee, 2021). The contract had finally been renewed in 2021 due to societal pressures. Despite being celebrated as a successful decentralized institutional model, it hasn't been scaled up or replicated and continue to face challenges threatening its existence.

Another interesting example, contrary to the centralized model, is the Dry Waste Collection Centres (DWCC) that are set up through a partnership between the ULB, NGO, and Corporates. ULBs provide the land and electricity. At the same time, the NGOs operate the DWCC and Corporates support it through CSR funds. One such case is the ITC's Wealth Out of Waste (WOW) Program in Bengaluru, which operated DWCCs in 29 wards. They realized that the door-to-door collection of recyclables was not economically viable and tried to rope in contractors and waste pickers who were used to extracting valuable recyclables. It generated new tensions between the waste picker and the Corporates. Nonetheless, there were some successful cases of MoUs with waste picker cooperatives and ITC (Chintan Environmental Research and Action Group, 2018).

With SBM, CSR, and EPR, a new institutional model is emerging which involves multiple stakeholders including, the ULB, the private sector, and the NGOs/ CBOs.

Figure 14.6 illustrates the significant shifts in institutional models adopted over the last few decades.

FIGURE 14.6 Major shifts in typical institutional models for waste management.

14.7 CONCLUSION

The study of policy history helps understand the significant factors that shape the policies at different periods. The economic, social, political, regulatory, and environmental contexts have influenced policies in waste management. Be it the plague, the public interest litigations, the economic liberalization post 1991, or the climate change agreements; all of which have shaped current policies. It helps understand the role of specialized institutions and the linkages between National, State, and Local institutions that determine resource allocations, thus influencing policy implementation.

Law is central to the policymaking process in India. The domains of law and policy interfaces include rights, the legislature's role, administrative law, regulatory agency functions, and planning laws. Court judgments show that 'public interest' is built into the legalism of our constitution. Hence, we see PILs leading to laws such as the MSW Rules, which have made their way into public policy actions at the local level. Globalization has led to transformations in regulatory functions and planning laws. It has advanced its agenda through policy changes, and social movements are left with legal recourse as the only means to counter it. This agenda is rooted in the Washington Consensus and has led to progressive disinvestment of state responsibility (Mathew et al., 2019). The new models of urban infrastructure development have led to privatization in the delivery of public goods.

The discourse on waste management has matured and is now slowly being taken over by the circular economy agenda. In both cases, the role of the informal sector in the Global South is well acknowledged. The informal economy employs over 50% of the global working population and the majority of the labour force in developing nations, and we continue to see its expansion in new forms. The precariousness of this informal workforce and its contributions though well understood, remain unrecognized and is extensively stigmatized (Chen & Carré, 2020). Our policies are still grappling with socio-economic and political complexities and are struggling to create inclusive solutions to the social equity concerns. It would be useful to examine how some scholars have developed methods of including them within the larger waste management framework. The analytical framework and tool for integrating the informal recycling sector (IRS) in waste and resource management in developing countries were developed by Velis et al., 2012. The tool was applied to 10 cities around the world. It demonstrated the interdependencies between the informal sector and its link to the SWM system, the materials and value chain, society as a whole, and aspects of organization and empowerment. It enlisted actions which included the following: the legal right to access waste, adjusting zoning and land use planning to enable sorting, local agreements, memorandum of association or formal contracts with waste pickers and informal recycling sector, providing space for sorting and storage, providing access to low-cost capital, institutionalizing policies regarding IRS for partnerships and cooperation, documenting the role and benefits of the informal sector, providing public health support, creating cooperatives, training and capacity building.

The new interests in resource efficiency are being supported by multilateral agencies that advocate the shift from cradle-to-grave into a cradle-to-cradle approach. However, its main agenda is to capture material through recycling to ensure resource security for big manufacturing companies, thus driving economic growth. Instead of being residue-focused, circularity principles need us to embrace waste minimization which is also the top goal of the waste management hierarchy. Waste minimization can be achieved through sharing, reuse, repair, and refurbishment, which extends the life of products and thus reduces consumption.

Waste management policies must recognize and deconstruct the complex economic, social, and political ecosystems within which multiple stakeholders contest access. These contestations manifest in space across different geographies. Our policies must strengthen the multi-stakeholder partnership model and build a decentralized network of actors. As an important stakeholder at the local level, the informal sector needs to be supported to scale up its circular practices. NGOs/CBOs can help create behavioral change by encouraging sustainable lifestyles. Policy instruments need to build on current economic incentives and policies on providing land for centralized and decentralized circular economy activities.

NOTES

1 Further details on *Swachh Survekshan Survey Toolkit* (2020). Ministry of Housing and Urban Affairs, Government of India. www.swachhsurvekshan2020.org

2 Further Details on Ministry of Environment Forest and Climate Change, Government of India (2019). *Draft National Resource Efficiency Policy 2019*. MoEFCC, Government of India. https://moef.gov.in/wp-content/uploads/2019/07/Draft-National-Resourc.pdf

3 Further Details on 'Department of Environment, GoM. (2018). *Maharashtra Plastic and Thermocol Products (Manufacture, Usage, Sale, Transport, Handling and Storage) Notification, 2018*. Government of Maharashtra. www.mpcb.gov.in/sites/default/files/plastic-waste/rules/plasticwasteenglish119102020.pdf

4 Further details on Ministry of New and Renewable Energy, GoI (2020). *Revised Guidelines of Waste to Energy Programme on Energy from Urban/Industrial/Agricultural and Municipal Solid Waste*. Government of India. https://mnre.gov.in/waste-to-energy/schemes

5 Further Details on Ministry of Environment Forest and Climate Change, GoI (2016b). *Hazardous and Other Wastes (Management and Transboundary Movement) Rules, 2016*. Government of India. https://cpcb.nic.in/displaypdf.php?id=aHdtZC9IV01fUnVsZXNfMjAxNi5wZGY

6 Further Details on Ministry of Environment, Forest and Climate Change, GoI (2016b). *E-Waste (Management) Rules, 2016*. Government of India. https://cpcb.nic.in/displaypdf.php?id=RS1XYXN0ZS9FLVdhc3RlTV9SdWxlc18yMDE2LnBkZg==

7 Further Details on Ministry of Environment, Forest and Climate Change, GoI (2016a). *Plastic Waste (Management and Handling)Rules, 2016*. Government of India. www.mppcb.nic.in/proc/Plastic%20Waste%20Management%20Rules,%202016%20English.pdf
8 Ministry of Environment Forest and Climate Change, GoI (2016a). *Bio-Medical Waste Management Rules, 2016*. Government of India. https://dhr.gov.in/sites/default/files/Bio-medical_Waste_Management_Rules_2016.pdf
9 Further Details on Ministry of Environment, Forest and Climate Change, GoI (2016d). *Solid Waste Management Rules, 2016*. Government of India. https://cpcb.nic.in/uploads/MSW/SWM_2016.pdf
10 Further details on the Municipal Corporation of Greater Mumbai web portal https://dpremarks.mcgm.gov.in/dp2034/
11 Further details on 'Central Pollution Control Board (2016). *The National Action Plan for Municipal Solid Waste Management: In compliance with Hon'ble National Green Tribunal Order Dated 5th February, 2015 in the Matter of OA No. 199 of 2014, Almitra H. Patel &Anr. Vs Union of India &Ors.* Central Pollution Control Board. https://cpcb.nic.in/uploads/MSW/Action_plan.pdf
12 Further details on Ministry of Environment, Forest and Climate Change, GoI (2016c). *Construction and Demolition Waste Management Rules, 2016*. Government of India. https://cpcb.nic.in/displaypdf.php?id=d2FzdGUvQyZEX3J1bGVzXzIwMTYucGRm
13 Further details on Agarwal, R. (2019, February 15). *India's city compost policy needs overhauling*. Down to Earth. www.downtoearth.org.in/blog/agriculture/india-s-city-compost-policy-needs-overhauling-63248 and https://pib.gov.in/PressReleaseIframePage.aspx?PRID=1654529#:~:text=The%20Government%20has%20introduced%20a,made%20out%20from%20city%20waste.
14 Further details on Ministry of Corporate Affairs, GoI (2014). *Notification on CSR contribution towards Swachh Bharat Kosh*. Government of India. https://jalshakti-ddws.gov.in/sites/default/files/Swachh_Bharat_Kosh.pdf
15 Further details on Ministry of Housing & Urban Affairs, GoI (2017). *Guidelines for Swachh Bharat Mission—Urban*. Ministry of Housing & Urban Affairs, Government of India. http://swachhbharaturban.gov.in/writereaddata/SBM_GUIDELINE.pdf
16 Further details on the Municipal Corporation of Greater Mumbai web portal, https://portal.mcgm.gov.in/irj/go/km/docs/documents/MCGM%20Department%20List/Solid%20Waste%20Management/Docs/DWSC%20-%20List%20of%20Centres%20PDF.pdf
17 Further details on Planning Commission (2014). *Report of the Task Force on Waste to Energy*. Planning Commission. http://swachhbharaturban.gov.in/writereaddata/Task_force_report_on_WTE.pdf
18 Further Details on Department of Environment, GoM (2006). *Maharashtra Plastic Carry Bags (Manufacture and Usage) Rules, 2006*. Government of Maharashtra. http://mahenvis.nic.in/Pdf/Laws/Cbag_nonbio.pdf'
19 Further details on UNFCC's Clean Development Mechanism can be found on https://unfccc.int/process-and-meetings/the-kyoto-protocol/mechanisms-under-the-kyoto-protocol/the-clean-development-mechanism
20 Potdar, A., Singh, A., Unnnikrishnan, S., & Naik, N. (2016). Innovation in solid waste management through Clean Development Mechanism in India and other countries. *Process Safety and Environmental Protection*, 160–169.
21 Further details on Ministry of Environment and Forests, GoI (2000). *Municipal Solid Wastes (Management and Handling) Rules, 2000*. Government of India. www.mpcb.gov.in/sites/default/files/solid-waste/MSWrules200002032020.pdf
22 Further details on the Advanced Locality Management: www.mumbaidp24seven.in/reference/New_Practices_of_Waste_Management_Case_of_Mumbai.pdf
23 Further details on the Municipal Corporation of Greater Mumbai web portal, https://portal.mcgm.gov.in/irj/go/km/docs/documents/MCGM%20Department%20List/Solid%20Waste%20Management/Docs/Bye%20laws/02%20Greater%20Mumbai%20Cleanliness%20Byelaws%20-%202006.pdf

REFERENCES

Andersen, M. S. (2007). An introductory note on the environmental economics of the circular economy. *Sustainability Science*, *2*(1), 133–140. https://doi.org/10.1007/s11625-006-0013-6

Arlosoroff, S., & Rushbrook, P. (1991). Developing countries struggle with waste management policies. *Waste Management & Research,* 9(1), 491–494.

Banerjee, S. (2021, June 29). Denied benefits for pandemic work, Pune's waste collectors protest civic body's apathy. *The Hindu.* www.thehindu.com/news/states/denied-benefits-for-pandemic-work-punes-waste-collectors-protest-civic-bodys-apathy/article35046237.ece

Chatterjee, I. (2011). From red tape to red carpet? Violent narratives of neoliberalizing Ahmedabad. In W. Ahmed, A. Kundu, & R. Peet (Eds.), *India's New Economic Policy: A Critical Analysis* (pp. 154–178). Routledge.

Chaturvedi, A., Gaurav, J. K., & Gupta, P. (2019). The many circuits of a circular economy. In P. Schröder, M. Anantharaman, K. Anggraeni, & T. J. Foxon (Eds.), *The Circular Economy and the Global South* (1st ed., pp. 25–42). Routledge. https://doi.org/10.4324/9780429434006-2

Chaturvedi, B., & Gidwani, V. (2011). The right to waste: Informal sector recyclers and struggles for social justice in post-reform urban India. In A. Waquar, A. Kundu, & R. Peet (Eds.), *India's New Economic Policy: A Critical Analysis* (pp. 125–153). Routledge.

Chen, M. A., & Carré, F. J. (Eds.). (2020). *The informal economy revisited: Examining the past, envisioning the future.* Routledge.

Chintan Environmental Research and Action Group. (2018). *State of Waste in India: Eighteen Years After the First National Rules.* Chintan Environmental Research and Action Group.

Ellen MacArthur Foundation. (2014). *Towards the Circular Economy Volume 1.* EMF. https://ellenmacarthurfoundation.org/towards-the-circular-economy-vol-1-an-economic-and-business-rationale-for-an

EMF. (2015). *Delivering the circular economy: A toolkit for policymakers.* Ellen MacArthur Foundation. https://ellenmacarthurfoundation.org/a-toolkit-for-policymakers

Furedy, C. (1995). Plague and garbage: Implications of the Surat Outbreak (1994) for urban environment management in India. *Development, Environment, Productivity and Policy Implications for South Asia.* Learned Societies Conference, Montreal.

Geissdoerfer, M., Savaget, P., Bocken, N. M. P., & Hultink, E. J. (2017). The Circular Economy–A new sustainability paradigm? *Journal of Cleaner Production, 143*, 757–768. https://doi.org/10.1016/j.jclepro.2016.12.048

Ghisellini, P., Cialani, C., & Ulgiati, S. (2016). A review on circular economy: The expected transition to a balanced interplay of environmental and economic systems. *Journal of Cleaner Production, 114*, 11–32. https://doi.org/10.1016/j.jclepro.2015.09.007

ICRA Management Consulting Services Limited. (2010). *Toolkit for Public Private Partnership frameworks in Municipal Solid Waste Management* (Knowledge Series on Public-Private Partnership). Ministry of Urban Development, Government of India. https://smartnet.niua.org/sites/default/files/resources/India_SolidWasteMgmt_PPP_Tookit-Volume-I_EN.pdf

Iyamu, H. O., Anda, M., & Ho, G. (2020). A review of municipal solid waste management in the BRIC and high-income countries: A thematic framework for low-income countries. *Habitat International, 95*, 102097. https://doi.org/10.1016/j.habitatint.2019.102097

Kirchherr, J., Reike, D., & Hekkert, M. (2017). Conceptualizing the circular economy: An analysis of 114 definitions. *Resources, Conservation and Recycling, 127*, 221–232. https://doi.org/10.1016/j.resconrec.2017.09.005

Kumar, S., Mukherjee, S., Chakrabarti, T., & Devotta, S. (2007). Hazardous Waste Management System in India: An Overview. *Critical Reviews in Environmental Science and Technology, 38*(1), 43–71. https://doi.org/10.1080/10643380701590356

Mathew, B., Pellissery, S., & Narrain, A. (2019). Why Is Law Central to Public Policy Process in Global South? *SSRN Electronic Journal.* https://doi.org/10.2139/ssrn.3458919

Ministry of Housing & Urban Affairs, GoI. (2019). *25th Report of the Standing Comittee of Urban Development on Solid Waste Management including Hazardous Waste, Medical Waste and E-Waste pertaining to the Ministry of Housing and Urban Affairs*. Loksabha Secretariat. https://eparlib.nic.in/handle/123456789/800929?view_type=browse

Mirza, S. (2019). Becoming waste: Three Moments in the Life of Landfills in Mumbai City. *Economic & Political Weekly*, 54(47), 36–41.

Modak, P. (2021). *Practicing circular economy* (First edition). CRC Press.

MoEFCC, M. of E., Forest and Climate Change. (2019). *National Resource Efficiency Policy 2019 (Draft)*. Government of India. http://moef.gov.in/wp-content/uploads/2019/07/Draft-National-Resourc.pdf

Narain, S., & Sambyal, S. S. (2016). *Not in my backyard: Solid waste management in Indian cities* (S. Banerjee & A. A. Parrey, Eds.). Centre for Science and Environment.

Potting, J., Hekkert, M., Worrell, E., & Hanemaaijer, A. (2017). *Circular Economy: Measuring Innovation in the Product Chain*. PBL Netherlands Environmental Assessment Agency. www.pbl.nl/sites/default/files/downloads/pbl-2016-circular-economy-measuring-innovation-in-product-chains-2544.pdf

Rodić, L., & Wilson, D. (2017). Resolving governance issues to achieve priority sustainable development goals related to solid waste management in developing countries. *Sustainability*, *9*(3), 404. https://doi.org/10.3390/su9030404

Su, B., Heshmati, A., Geng, Y., & Yu, X. (2013). A review of the circular economy in China: Moving from rhetoric to implementation. *Journal of Cleaner Production, 42*, 215–227. https://doi.org/10.1016/j.jclepro.2012.11.020

Velis, C. A., Wilson, D. C., Rocca, O., Smith, S. R., Mavropoulos, A., & Cheeseman, C. R. (2012). An analytical framework and tool ('*InteRa*') for integrating the informal recycling sector in waste and resource management systems in developing countries. *Waste Management & Research: The Journal for a Sustainable Circular Economy*, *30*(9_suppl), 43–66. https://doi.org/10.1177/0734242X12454934

Wilson, D. C. (1996). Stick or carrot?: The use of policy measures to move waste management up the hierarchy. *Waste Management & Research, 14* (4), 385–398.

Wilson, D. C. (2007). Development drivers for waste management. *Waste Management & Research*, *25*(3), 198–207. https://doi.org/10.1177/0734242X07079149

Wilson, D. C., Rodic, L., Cowing, M. J., Velis, C. A., Whiteman, A. D., Scheinberg, A., Vilches, R., Masterson, D., Stretz, J., & Oelz, B. (2015). 'Wasteaware' benchmark indicators for integrated sustainable waste management in cities. *Waste Management, 35*, 329–342. https://doi.org/10.1016/j.wasman.2014.10.006

Wilson, D. C., Rodic, L., Scheinberg, A., Velis, C. A., & Alabaster, G. (2012). Comparative analysis of solid waste management in 20 cities. *Waste Management & Research: The Journal for a Sustainable Circular Economy*, *30*(3), 237–254. https://doi.org/10.1177/0734242X12437569

Wilson, D. C., Rodic, L., Modak, P., Soos, R., Carpintero, A., Velis, K., ... & Simonett, O.. (2015). *Global waste management outlook*. Report United Nations Environmental Programme, UNEP

Woolcock, M., Szreter, S., & Rao, V. (2011). How and why does history matter for development policy? *Journal of Development Studies*, *47*(1), 70–96. https://doi.org/10.1080/00220388.2010.506913

15 Integrated Waste Recycling Parks

Bringing Circularity into Waste Management

Vijai Singhal

15.1 INTRODUCTION

Enhancing resource efficiency (RE) and promoting the use of secondary raw materials (SRM) is very essential to ensure resource security. It requires an integrated, concentrated, and collaborative approach to address the challenge of increasing demands to fulfil the needs of a vast and growing population with finite supply of abiotic resources.

Circular economy (CE) is the new sustainable policy by which we can reduce waste generation by reusing and recycling the products to form a loop. The new resources are utilized less in this policy and the waste generation is minimized (Waikar and Sadgir 2022). In such a case, waste becomes a source of value for other stakeholders in the supply chain, or for those operating in similar or different supply chains (Ünal et al. 2019). Waste recycling is an important pillar of circular economy and is paramount for material conservation, environmental protection, and for addressing the issues related to climate change. The waste streams for which recycling facilities are normally installed include end of the life vehicle (ELV), lead acid battery waste, used oil, e-waste, plastic waste, etc.

Presently, recyclers of various waste streams such as plastic waste, e-waste, hazardous waste, etc. are located in different places and carrying out recycling activities. In the recycling process, many waste streams are also generated which the recycler may not be able to recycle. For example, an e-waste recycler will also generate plastic waste, metal waste, used oil, foam, etc. for which it may not have adequate recycling infrastructure. Likewise, an ELV recycler will generate used lead acid batteries, e-waste, plastic waste, aluminum, iron, etc. for which the recycler may not be having requisite recycling facilities in-house.

In such situations, the option with the recycler is either to set up such recycling facilities by itself or look to another recycler who can recycle such waste streams. However, setting up of recycling facility for all types of waste by one recycler may not be economically viable. Further, looking for another recycler for recycling other waste streams involves transportation and other logistic issues and may increase recycling costs. Apart from the economic cost of transportation and other logistic,

DOI: 10.1201/9781003364467-15

such recycling is less efficient and exert more pressure on common environmental resources, thus exerting high environmental cost.

15.2 CONCEPT OF IWRP

Industrial symbiosis (IS) refers to a novel approach of exchange of under-utilized resources (material, energy, etc.) across companies and industries in order to yield environmental and economic benefit (Chertow 2000). The benefits result from higher resource utilization due to reduction of waste (downstream) and primary inputs (upstream). Lombardi and Laybourn (2012) further broadened this definition to include also the sharing of knowledge and human resources through the network, emphasizing mutual learning and sustainable innovation. In IS, separate industries engage themselves in a collective approach involving physical exchange of materials, energy, water, and/or by-products. The keys to IS are collaboration and the synergistic possibilities offered by geographic proximity. Application of IS can promote transition from linear to circular production systems, and promote sustainable production and consumption (Kerdlap 2020). Starodubova et al. (2022) proposed a method for evaluating the relationship between eco-industrial parks with circular economy technologies by introducing a specialization index which considered four types of technologies, i.e. waste treatment, industrial symbiosis, water resource efficiency, and renewable energy.

The concept of IS is paramount in IWRP where recyclers of different waste streams are co-located in a well-defined industrial area to take advantage of synergy in their operations in such a way that the waste generated from one recycler becomes raw material for another member recycler of the park. It is a system approach in which the individual recyclers are not viewed in isolation but in concert to optimize the total materials cycle. Such a park will strive to optimize resources, energy, and capital.

The waste streams may include plastic waste, e-waste, hazardous waste, waste from PV solar panels, metal scrap, ELV etc. Apart from the recyclers, such park will have facilities for training and capacity building for the workforce engaged in recycling. An awareness center may also be established in the park for creating awareness among children and other stakeholders on issues related to various types of waste and how these wastes are recycled keeping all the environmental safeguards in mind. The park may also include a material testing laboratory, R&D laboratory, repair shops, warehouses for materials, raw material suppliers, etc. to enhance its value and making such facilities cost effective.

The IWRP seeks stakeholders' involvement in which administration, industries, and citizens are in unity to aim at creating sound recycling eco-system and in harmony with environment. In the park, a community of recyclers and service providers seek to enhance environmental and economic performances through collaboration in managing environmental resource and materials and converting them into useful product by recycling. Over and above, the park should be designed keeping in mind the principles of sustainability incorporating green building parameters. The park should strive to become water positive and should maximize use of renewal energy or carbon neutral fuels so as to reduce greenhouse gas emissions.

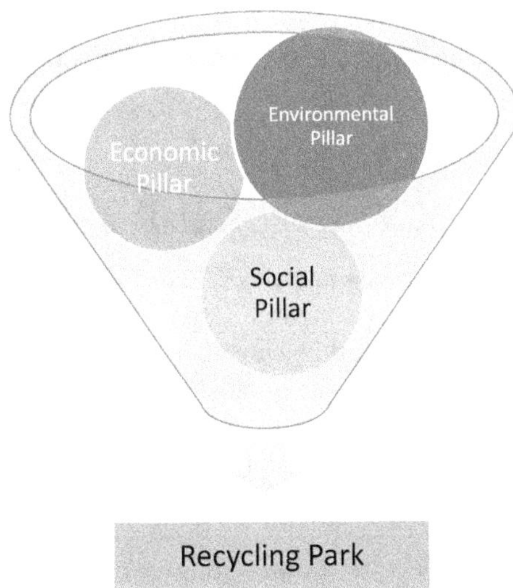

FIGURE 15.1 Three sustainability pillars for successful operation of IWRP.

Source: GIZ India 2022.

In a nutshell the park should be established keeping three sustainability pillars of economic, environment, and social in mind to run the park successfully for the benefit of the society as depicted in Figure 15.1.

15.3 HOW IWRP ARE DIFFERENT FROM ECO-INDUSTRIAL PARKS (EIP)

Industrial parks are known by different names and cover industrial areas: industrial zones, industrial investment regions, special economic zones, and industrial corridors, among others. These parks can cause both negative and positive impacts on the environment. Thus, sensitive planning and management are needed to mitigate negative outcomes and to optimize economic, social, and environmental gains.

Broadly, an EIP is defined as a dedicated area for industrial use at a suitable site that supports sustainability through the integration of social, economic, and environmental quality aspects into its siting, planning, management and operations (World Bank 2021). Key drivers for EIPs are reducing environmental footprints; promoting efficiency gains and cost effectiveness; enabling community cohesion; resilience to various types of risks, providing better access to finance and technical support and enhancing competitiveness. United Nation Industrial Development Organization (UNIDO) has compared 33 examples of EIPs in 12 developing and emerging economies, including their policy context (UNIDO 2016).

IWRP and EIP look similar in approach, however, in an IWRP, central focal point is recycling of different waste streams within a well-defined industrial zone. It integrates sustainability management aspects of the eco-Industrial Park concept however, it primarily specializes on recycling activities, while also integrating on-site common facilities that are able to process waste streams into marketable products. In case of EIP, the industries may not necessarily be recycling industries and may manufacture diverse products. However, exchange of waste streams generated from these industries is still an important aspect of EIP.

15.4 ENVIRONMENTAL AND ECONOMIC BENEFITS OF IWRP

Setting of IWRP has many environmental and economic benefits which may not be available when the recyclers are located in different locations without the common infrastructure and services. By working together, businesses strive for a collective benefit greater than the sum of individual benefits that could be achieved by acting alone. Co-locating the recyclers and by effective networking among the member companies may lead to significant economies in environmental management related to infrastructure, information flows, and regulatory enforcement, as well as to decrease conflict over land use.

The park will help the recyclers by matching wastes from one company to the resource needs of another, thus increasing the recycling efficiency and better economic gains. The concentration of companies can foster innovation, technological learning, and company growth. Economies of scale of the supply of services and facilities will reduce the costs for companies, thus successful IWRP will contribute to high growth in regions and national economic development. The park will also present a unique investment opportunity for recyclers of various waste streams in the region and is likely to attract large investments and create job opportunities for the people of the region.

The park may give boost to recycling in the region as more waste will be recycled and generate new jobs, thus contributing to overall development of the region. Since the park will have common facilities like material testing laboratory, storage yards for raw materials, environmental testing laboratory, etc., it will result in reduced installation and service cost to the member companies.

The park will have common facilities for training, skill development, and public awareness which can be very beneficial for the member companies as it will get trained and skilled manpower at the reasonable cost. The awareness facilities developed at the park may be used by students, citizens, and other stakeholders to develop a better understanding of waste and its recycling. It will also help in casting aside the public distrust, anxiety, and discomfort for waste and waste recycling.

The park can also encourage research and development activities related to waste management and recycling for which an R&D center can be established. In this regard, governments can encourage cooperative efforts for smaller firms with joint research and development on environmental issues.

15.5 SALIENT DESIGN CONSIDERATIONS FOR IWRP

Various design considerations such as selection of suitable site, master planning and zoning, basic common and specific facilities to be established at the park, governance schemes for the park, integration of informal sector, etc. are very important for the success of IWRP. In this section, these key parameters have been examined with respect to conceptualization, designing, and implementation of the IWRP.

15.5.1 Identification and Selection of Waste Recyclers

Successful exchange of waste streams among the recyclers is the corner stone for the eventual success of the IWRP. Accordingly, while planning for such a park, identification of waste streams generated by each recycler, their characterization and quantification and potential recyclers among the members of the park who will be able to recycle these waste streams in techno-economic viable fashion needs to be examined in detail. We also need to look for recyclers which can play the anchor role for the park by become major suppliers of various waste streams to the other member recyclers.

Typically, in such a park, recyclers which could be included are ELV, lead acid battery waste, used oil, e-waste, plastic waste, iron scrap, aluminum scrap, etc. Figure 15.2 depicts how waste streams generated by each of these recyclers could become possible raw material feed for the other member recyclers in the park, thus closing the material loop. Ideally, the park should be able to recycle all the waste generated from the member recyclers by carefully selecting the recyclers in a way that no waste is required to be sent outside the park. A major recycler such as ELV or e-waste could become the anchor around which the other recyclers could be selected.

Beside the recyclers, the park may also in-house suppliers of various raw materials and units which are engaged in quality upgradation of raw material (waste) as per the requirement of member units. For example, plastic waste recycler may need a specific kind of plastic which it proposes to recycler. Such suppliers and units engaged in quality upgradation of raw material may provide it with the desired quality and quantity of raw material, thus enhancing the recycling efficiency and ultimately the profitability of the recycler.

15.5.2 Important Considerations for Selection of Suitable Site for IWRP

Selection of suitable site is the first step in setting up of the waste recycling park. Site selection considers availability and suitability of land which includes indicators such as topography, geology, hydrology, and natural disaster. Availability of waste for member recyclers of the park and markets for the end products manufactured by these recyclers will be very important while deciding the location of the park. Connectivity of the site and transport linkage is critical as the cost and efficiency of operation depends on the same. Infrastructure which includes system for supply of water, heat, power, gas, communication cables, etc. are also critical considerations as it will impact the viability and profitability of the park. The relationship between the park and the city also needs to be explored in detail as the city's infrastructure can be used for park development. Availability of ample labor and other local services in the vicinity can be very beneficial in the initial phase while the units are being set up.

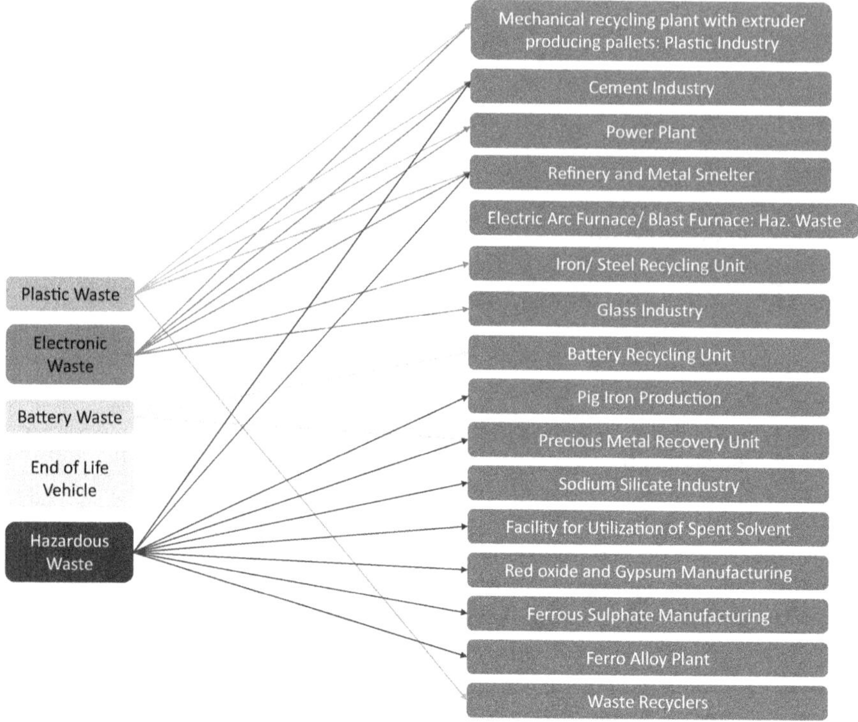

FIGURE 15.2 Mapping of Recyclers and Waste Streams Generated/Exchange.

Source: GIZ India 2022.

15.5.3 Master Planning and Zoning of IWRP

Environmental planning of the waste recycling park is paramount for its ultimate success. Recyclers in the park need to be located in a way that the waste generated by one recycler can be taken for recycling by another recycler with minimum transportation with maximum synergy in their operation. Further, different areas such as industrial zone, service zone, administration area, parking, etc. are also needs to be demarcated keeping the functionality of different areas.

Accordingly, following points needs to be kept in mind while carrying out zoning of the park:

- Type and quantity of waste recycled and technology employed for recycling by each recycler needs to be studied in detail so as to decide optimum requirement of land for their operation, storage, and other utilities.
- Potential for expansion for each member unit (e.g., additional production plant, warehouse, increased use of energy, water, steam etc.), probability of expansion, and whether such a change will occur in the short term, or in the long term to decide how much land is to set aside, and whether additional land can be made available near the tenant's current location.

- Need to conduct a detailed mass, energy, and water balance evaluation to ensure that the proposed symbiotic linkages indeed exist without any one member of the network being put at risk from being led to rely on a temporally depleting byproduct source.
- Detailed characterization and quantity of waste streams generated by each recycler should be documented so as to explore the synergies in the waste streams for recycling purposes to maximize beneficial symbiotic relationships among the recyclers.
- Details of utilities such as water, power, steam etc. to be used by each member of the park should be known so that sizing and location of the utilities could be optimized with proximity of shared infrastructure,
- Requirement of disposal of residual waste, treatment of industrial wastewater and sewage by each member so that sizing of the common facilities required can be carried out.
- Whether the industries have any potential for waste heat recovery/utilization by other member unit so that such units may be located in close proximity.
- Local zoning regulations followed at the state level or at the development agency level needs to be adhered to.
- Should consider landscape design strategies (e.g., buffer zones) that mitigate industrial noise, air pollution, and prevent degradation of wildlife and native plant species in the area.
- Location of each recycler will also be governed by specific operational requirements of each member units. For example, certain tenants may require immediate access to the road for emergency personnel access, higher frequency of incoming and outgoing truckloads, or other transportation related requirements. If this is the case, the tenants in question may be located closest to the roads within the park. If another tenant executes operations that produce a large amount of noise, then such member unit may be located farther from the boundaries of the park to preserve the noise level of the park neighborhood.
- The buildings within the recycling park should comply with the green building rating systems such as GRIHA, IGBC, LEED; other certifications such as ISO certification for quality of recycled products etc. The use of local materials/ recycled products in building construction should be encouraged. The occupational safety and health (OSHA) guidelines should be met to ensure a safe workspace for the employees.
- Rain water harvesting should be done by channelizing water from top of the built structures and through pits along the road. Pervious surfaces in open areas may also be considered for rainwater harvesting and to recharge the ground water.
- The building should be oriented such that the longer side aligns with the east-west direction, and windows wherever applicable, should preferably be south and west facing to minimize heat gain.
- Prevailing wind direction should be considered for positioning of industries which should lie in the downwind side to prevent air pollution. Commercial facilities and social infrastructure like housing units, etc. should be placed in the upwind side.

Ultimately, the optimal layout for the park may be determined based on waste exchange and sensitivity analysis of key variables so as to maximize economic and sociological benefits while minimizing adverse environmental impacts. Further, the park needs to be designed on sustainability principles aiming for zero waste outside the park, maximum use of renewal or carbon neutral energy and minimum use of ground water aiming for the park as a whole to become water positive.

15.5.4 Basic Common Facilities to be Established at the IWRP

Common facilities established in the park will strive to reduce total environmental burden, increase cost-effectiveness, and result in synergetic waste management within the park. Some basic facilities like water, power, roads, firefighting, weigh-bridge, bank, dispensary, post office, parking area, sewage treatment plant, etc. are provided in any industrial area or industrial cluster by the project development agency. Accordingly, these facilities will also be part of IWRP as these are required for efficient operation of any industrial area.

15.5.5 Specific Common Facilities to Be Established at the IWRP

Besides the basic common facilities which any industrial area requires, there are some common facilities which will be specific to the IWRP looking to the special needs of the waste recyclers and overall philosophy and guiding principles of the park. Following are the important specific facilities which may be required for IWRP:

1. **Material Testing Laboratory:** Quality of the waste which is the basic raw material for recycling is very important as it impacts the recycling efficiency and even recyclability to a large extent. Establishing a common material testing lab will be cost effective and it can be designed keeping the specific requirements of the member units in the park. The lab could be run on built and operate principal, though space for housing the lab can be provided by project development agency and the same can be decided while carrying out detailed planning for the park.
2. **Training, Awareness and Skill Development Center**: The basic philosophy for such a park is not just to have recycling industries, however, it aims to create better business environment to these units by creating all other required facilities to them. Training and skill development is an important requirement for any industry, however, more so in recycling business as there is acute shortage for trained manpower in this sector. This center may cater to the needs of all the member industries of park and strive to provide well-trained manpower and also provide skill upgradation from time to time. The facilities of training and skill development can also be used by other industries located outside the park which are engaged in similar businesses.

Awareness about different wastes, how it is recycled and its environmental implications are not well understood by public and other stakeholders like NGOs, students etc. Therefore, besides the training and skill development activities, the park can also carry out public awareness activities on waste and recycling of waste. It will immensely benefit students, visitors to the park, people employed in the park, NGOs and general public and enhance their understanding and knowledge on these critical issues. It will also enhance reputation of the park and increase its interaction with outside world. It will also help in gaining the trust and support from local communities and private stakeholders.

All these activities can be out sourced to some expert agency which has adequate experience in running such centers. However, infrastructure required for taking up the activities may be created by the park development agency so that overall service cost is less and the services are affordable to all the stakeholders.

3. **R&D Laboratory:** The park, at a later stage, may also set up a R&D laboratory to develop new recycling technologies, new recycled products or carry out research on any other issues relating to waste recycling. Such a lab may provide a boost to recycling of locally available wastes by exploring technologies for their better recycling. Preferably, such a lab should be operated by an expert institution having adequate experience in the field of waste recycling and waste management.
4. **Environmental Monitoring and Analysis Laboratory:** The member units of the park and the park as a whole will need to follow all the environmental regulations applicable in that region and will also continuously monitor quality of ambient air, ground water, surface water, soil and other components of ambient environment. For that purpose, the park may have its own environmental laboratory or it may outsource the work to some reputed recognized laboratory. In initial stage of the park development, services from an outside laboratory may be preferred and later on, park administration may decide to set up its on laboratory after examining its financial viability. However, tentative location and adequate space may be set aside at initial planning stage itself.
5. **Renewable Power Generation:** The park will be designed on environmentally sustainable principle for which use of renewable energy is very vital. Therefore, the park may establish solar power generation facilities in open areas or on building roof tops. Alternatively, it may buy renewal power to avoid fossil fuel based grid power. Ultimately, the park should aim at offsetting both electrical & thermal energy through renewable energy to maximum extent.
6. **Rainwater Harvesting:** The park may have rainwater harvesting system for roof and non-roof areas to offset its dependence on ground and surface water. It may use existing structures in low-lying areas for harvesting rainwater which can later be used by the park to meet its raw water needs.
7. **Raw material Storage Yard:** Recyclers of waste need to store large quantities of waste, however, there may be seasonal or other variation in the quantities required to be stored. Further, small recyclers and service providers may not need raw material storage area throughout the year. In such a situation, the park may develop common storage yards which may be made available to the members on rental basis. This will save a lot of space in individual

industries as well as save capital cost for the members which need storage only intermittently.

8. **Repairing and Servicing shops:** First principle of waste management is to avoid generation of waste. The park may provide space to repair shop of electrical/electronic items, shops selling refurbished goods, etc. to encourage repair and use of pre-owned goods to increase their potential life and reduce generation of waste.

15.5.6 Possible Governance Schemes for Waste Recycling Park

Effective and good governance structure of the park is the key for its successful establishment, operation, and to achieve its main objectives. The park operation must be fully compliant with local and national legislation, and both national and international standards in environmental protection and social development. The park management will serve as a centralized management authority which will not only ensure the park's smooth operation but will also connect the tenant companies with relevant authorities like pollution control regulation authorities, municipalities, industrial development agencies, etc. and provide them relevant services and engage with all the stakeholders.

A number of models for management of waste recycling parks can be adopted as shown in Figure 15.3 depending on the local regulations and opinion of various stakeholders of the park. These models may include the following.

- **Associative Management:** The park's member companies themselves form an association or a Special Purpose Vehicle (SPV) to manage the park with minimal intervention from the government.

Government Associative Management: In this model of governance the initial development of the park is carried out by the government agencies and once the park has sufficient member units and acquire enough capacity to manage the park by themselves, the management of the park may be transferred to the association or the SPV formed by the member units. The government may nominate one or two government officials as members in the SPV to keep an eye on the overall operation of the park and guide the park management on critical and contentious issues.

- **Government Management**: The park is entirely managed by government authorities which may include different government agencies related to

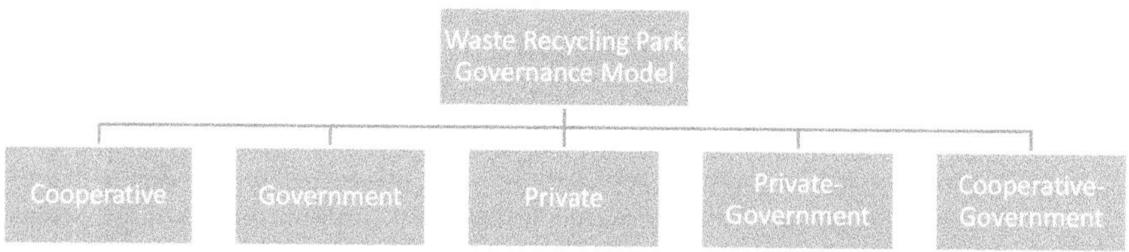

FIGURE 15.3 Various Governance Models for Waste Recycling Park.

industrial development, pollution control, municipal administration, etc. or it may be primarily managed as a single agency, say, agency related to industrial development and other agencies only support the primary agency.

- **Public-Private Management:** It is a cooperative approach in which a government managed park also receives financial and technical support from private companies, non-governmental organizations, associations, or foundations to service provision, shared between a city and the private sector. The arrangement can be permanent in which a government representative is permanently employed with the park or it may be temporary in that case the private sector is part of a capacity-building process until the government can perform all park management functions.
- **Private Management**: In this model, a private legal entity establishes or manages the park. The agency will be responsible for making all the investment and for creating all the facilities in the park. Revenue generation forms their core objective, which is typically achieved by renting out plots of land or charging fees for infrastructure or services.

15.5.7 Informal Sector Integration in Activities of IWRP

Waste collectors or waste pickers are an integral part of the waste management system and plays important role in the collection of valuable waste fractions from dumpsites, commercial areas, markets, etc. These informal sector actors play an important role in waste management economy, especially in developing countries where a large part of waste collection and segregation is still carried out by the informal sector. However, the informal sector is involved in very different and sometimes quite conflictual ways compared to the formal sector, which is why a strategy that encourages participation and integration of the informal sector can be helpful to achieve waste management targets set out by IWRP.

Waste pickers and waste collectors have a very deep understanding about waste and can play an important source of raw material to the recycling units located in the park. With their network, long-standing personal relationships and knowledge about local waste flows, the informal sector players can take part in waste management activities such as collection, separation, dismantling, and transportation. A potential integration of informal sector is a win-win situation for the informal sector also as it will result into good service conditions for the workers, better health facilities and social status. Additionally, there is also an inherent environmental incentive for municipalities as the waste streams, that were earlier treated unscientifically within the informal system will now be available for the formal treatment systems where treatment will happen with all the environmental safeguards and with due compliance of prevailing environmental regulations.

15.6 DRIVERS FOR SUCCESSFUL IMPLEMENTATION OF IWRP

It is true that co-location of recyclers of various waste streams can foster innovation, technological learning, and company growth. Economies of scale of the supply of services and facilities can reduce the costs for companies, thus successfully contributing to their high growth and growth of economies of the regions and national economy. However, successful establishment and operation of IWRP will depend on many enabling drivers which need to be carefully considered. We will discuss some of these factors in this section.

1. **Regulatory Instruments:** Regulatory policies create an enabling environment for greener technology transfer, development of technical guidance and national guideline for recycling parks, among others. It can provide set targets for the development, issue new standards and encourage cooperation programs. For example, national guideline for development of such park can provide a broad framework and help the development agencies in better planning and execution of such facilities. Further, green procurement policy and standards for recycled products will help in better marketing and value realization for the recycled products, thus making recycling business economically viable. Government may also promote energy efficiency and low carbon development by setting respective policies and implementing a complementary regulatory framework.
2. **Economic Instruments**: Waste recycling business is not very well established, especially in developing countries and availability of good quality raw materials and skilled labour, marketability of recycled products, etc. are still a big challenge. Therefore, government and other development agencies can play a supportive role by brining economic instruments which help the recycling business to overcome these obstacles. Making land available at subsidized cost, seed funds to support infrastructure, regulating fees and charges of common services, providing subsides for park development are some of the instruments which may provide initial support to the park and ultimately prove critical for successful operation of the park. Such instruments can also become an important factor in attracting reputed companies to the park.
3. **Sound and Effective Industrial Symbiosis:** The main premises of waste recycling park is the exchange of waste streams among the member units so as to encourage waste recycling and increase economic gains for all the members. Therefore, a detailed pre-feasibility and material flow study of waste streams and potential for their practical exchange needs to be worked out very carefully to make sure that companies can exchange waste in close industrial loops. There should be sufficient flow of materials to make industrial symbiosis worthwhile. The park management will be required to continually help the member units in overcoming hurdles, if any, in inter-company issues and resolve them quickly and effectively. In case the availability and quality of the waste required by any member unit becomes problematic then it needs to be addressed and necessary solutions to be worked out after consulting all stakeholders. It is also quite helpful if the park management have various domain experts on waste management so that comprehension and subsequent resolution of the problems is timely and effective.
4. **Engagement with various Stakeholders and Information Dissemination:** The companies are often reluctant to establish synergistic relationships with other member companies not only due to a lack of knowledge of the industrial

symbiosis mechanism (Mauthoor 2017) but also due to a lack of knowledge of other companies with the potential to receive or provide waste. Therefore, the park management should maintain a transparent and continual flow of information sharing and consultation with all the stakeholders for successful operation of the park. The various modes of communication may include newsletters, television programs, websites (and portals). Other modes of communication can be awareness raising programs on recycling parks and education and training activities. Meetings and open discussions with park members, various service providers, raw material suppliers, nearby community leaders etc. may be organized on periodic basis to understand the issues and reach to a solution through consultative process. A structured and time bound mechanism for grievance redressal can also go a long way in quick disposal of the problems faced by the park members.

5. **Support from International Organizations:** Setting up of IWRP is a novel concept and not experimented at many places. However, similar concept of EIPs has been implemented in many countries successfully. International agencies like UNIDO, GIZ and World Bank have extensively work in this field which may provide lot of learnings for implementation of IWRP. Countries like Japan, UK, Switzerland has good experience in the area which may be utilized for these facilities.
6. **Strong Organizational Structure for the Park:** A strong leadership with clear vision is required for the park to be successful. The park management should have experts from administration, finance and technical domains to implement the policies of the park and to take necessary decisions. However, management and policy making process should essentially be collaborative and consultative. With dedicated responsibilities and functions of the park management entity, the overall strategy for resource efficiency, circular economy application, economic, social and environmental performance can be driven effectively in the park.

15.7 CASE STUDIES FOR IWRP

IWRP is a relatively new concept in waste management which explores mutual cooperation among the members of the park to harness industrial symbiosis by applying the principles of industrial ecology. Accordingly, not many industrial zones have been developed, designed, and implemented on the principles of IWRP as detailed in this chapter. In this section a few case studies are being presented to gain clear understanding of the concepts of IWRP.

15.7.1 Eco Industrial Recycling Park at Kalundborg, Denmark

The model of industrial symbiosis was first fully implemented and realized in the eco-industrial park at Kalundborg, Denmark (Chertow 2000). This network arose spontaneously from a self-organizing initiative between companies to address water scarcity (Zhang and Chai 2019). The park initially had collaborative agreements between a few industrial concerns to share key resources such as water and take advantage of waste resources (heat, steam, and gas). However, in the long term, the collaborations have deepened with firms exploiting new opportunities for initiating closed loop collaborative practices (Valentine 2016). The Kalundborg Park which has been in existence for more than 40 years can provide valuable learnings into the process that encourage environmental networks to transition from one-off cooperative ventures to sustainable collaborative relationships.

15.7.1.1 Member Units and Material Exchange in the Park

The eco-system at Kalundborg consists of the following six members:

1. Asnæs power station – coal-fired plant producing electricity.
2. Statoil – an oil refinery.
3. Novo Nordisk – producer of insulin and industrial enzymes.
4. Gyproc – producer of plasterboard for the building industry.
5. The town of Kalundborg, which receives excess heat from Asnaes for its residential district heating system.
6. Bioteknisk Jordrens – a soil remediation company.

As shown in Figure 15.4, these firms exchange various materials/wastes like treated wastewater, steam, fly ash, residual heat, biomass, yeast slurry, sulfur, gypsum, etc. Statoil refinery supplies its treated wastewater as well as its used cooling water to Asnæs power station, thus reducing consumption of fresh water by about one million cubic meters per year. In turn, Asnæs power station supplies steam to Statoil and Novo Nordisk for heating of their processes. The Statoil refinery treats its excess gas to recover sulfur which is sold to sulfuric acid manufactures. The cleaned gas is

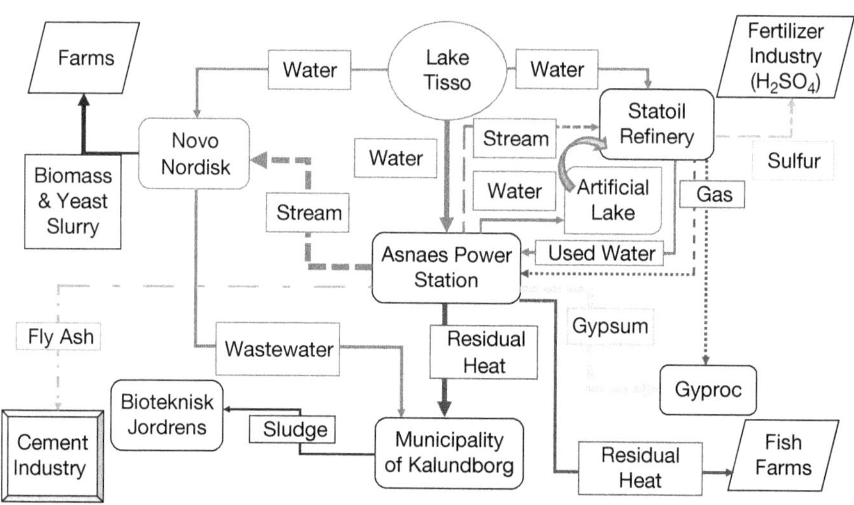

FIGURE 15.4 Material Flow in Kalundbarg Eco Industrial Recycling Par.

Source: Christensen 1999.

TABLE 15.1
Annual Saving of Resources at Kalundbarg Eco Industrial Recycling Park through Interchange of Wastes

Location	Resource	Savings
Statoil	Water	1.2 million cubic meters (from Asnaes)
Asnaes	Coal	30, 000 tons (by using statoil fuel gas)
Novo Nordisk	Oil	19,0000 tons (by using statoil fuel gas)
General	Fertilizer	Equivalent to 800 tons Nitrogen and 400 tons phosphorous
General	Sulfur	2,800 tons
General	Gypsum	80,000 tons

Source: Wikipedia.

then supplied to Asnæs power station and to Gyproc as an energy source. In the desulfurization process installed by Asnæs power station, calcium sulphate (gypsum) is recovered as waste which is given to the nearby plaster board manufacturing industry. The plastic board manufacturing unit, by doing this, was able to replace natural gypsum by synthetic "waste" gypsum from Asnæs power station. Novo Nordisk generates a large quantity of used bio-mass coming from its synthetic processes and it is used by farmers as a fertilizer since it contains nitrogen, phosphorus, and potassium. Asnæs power station provides the residual heat for the district heating system of the town.

It is estimated that US$160 million has been saved so far (Christensen 1999). The payback time of a project is less than 5 years on average. Table 15.1 shows annual saving of resources at Kalundbarg Eco Industrial Recycling Park through interchange of wastes.

15.7.1.2 Management Structure of the Park

Kalundborg Park is managed by a general assembly, a board, and a secretary employed by the board. General assembly is the highest body for management of the park (Kalundborg Symbiosis website 2023). Each member unit appoints a senior employee who represents it on the board. The Board determines the budget and the membership fee according to the rules of procedure. The budget and the quota are to be submitted to the general assembly for approval. The association is managed by a secretary who has overall responsibility for the day-to-day management of the association.

15.7.2 Remondis Lippe Eco Industrial Recycling Park, Lünen, Germany

Remondis Lippe Plant covers an area of 230 hectares and is home to a wide range of innovative recycling systems which makes it Europe's largest industrial recycling center (Remondis Lippe Plant website 2023). The area on which the park is developed used to be an aluminum production facility, however, the existing machines, plants, and buildings of the previous plant were not demolished but deliberately converted and transformed into recycling facilities.

There are over 14 different waste management facilities distributed over the park. Among the different facilities, synergetic uses of resources help to increase the number of material life cycles that can be closed. Around 1400 people work at the facility and it recycles and recovers a whole range of material from organic material, to metal slag, all the way through to plastics so that it can be returned to production cycles. The experts at the Lippe Plant develop patented processes so that residual materials can be transformed into new top-quality products. Around 1.4 million tons of residual materials are delivered to the Lippe Plant each year for processing.

15.7.2.1 Material Exchange at the Park

The site of the Lippe Plant includes a certified laboratory, facilities for recycling slaughterhouse waste, waste wood processing/biomass power plant production of gypsum-bearing building products from industrial gypsum, and a chemical plant for the production of special aluminum salts (Walkowiak 2020). Furthermore, plastic granulate is obtained from plastic waste and high-quality metal granulate from slags and furnace waste materials from stainless steel and non-ferrous metal production. There is also a building rubble recycling area, a composting plant, and a processing plant for industrial waste to produce substitute fuels for power generation in power plants. Since 2006, biodiesel has also been produced from animal fats and vegetable raw materials. Further, electronic and electrical waste is recycled and a biomass power plant is operated. The plant produces approx. 500,000 tons of raw materials for industrial customers. According to a study drawn up by Prognos AG and the INFA Institute, the activities at the Lippe Plant reduce carbon emissions by 488,000 tons every year (website, Remondis Lippe Plant).

15.7.2.2 Common Facilities at the Park

All facilities utilize shared infrastructure that include roadways, weighbridges, safety and fire safety department, as well as the administration building. Processing steam as well as energy is directly supplied by the fluidized-bed power station to the member units. Additionally, electricity requirements may also be supplied by the battery storage unit or the biomass power plant. There is a wastewater treatment plant for all facilities as well as common water supply system. The treated waste water is reused as park internal water supply.

Furthermore, the members in the park can also use the state-of-the-art environmental laboratory to analyze input and output materials, if necessary. The Umwelt Control Labor (UCL) performs a very important task at the Lippe Plant as all of the material delivered to the site is examined by UCL experts to find out what is the best treatment option for a particular waste stream. By carrying out in-depth analysis, they can explore the recycling options in detail and, if necessary, develop new options. UCL also examines all the hazardous materials which are delivered to the fluidized-bed power station for incineration.

The implemented synergies also help to close the material cycles of the different waste treatment facilities. Residual material which is left over at the end of a recycling process may be processed further in a different facility. In some instances, processes and business areas also depend directly on each other and mutually benefit each other. For example, the sodium aluminate produced from waste containing aluminum serves as the basis for the production of a particularly high-quality white mineral.

15.7.2.3 Governance Model of the Park

The park is operated and managed under a private governance, with Remondis being the private operator and real estate agent of the park. The facilities themselves are either managed by Remondis or by subsidiary companies of Remondis.

The marketing of the products is carried out by the individual subsidiary companies in the park and the units in the park are free to market their waste products themselves. The environmental laboratory is a major source of revenue as it is utilized to analyze processing samples as well as carry out regulatory compliance monitoring for environmental authorities and offer training of laboratory staff from within and outside Germany. The laboratory makes use of IT tools to deliver results in real time. Plant also generates revenue by selling recycled products and from commissions on waste treatment.

15.7.2.4 Selection Process of the Recyclers

An iterative approach based on utilizing the existing facilities of the park is followed for the selection of new recyclers and waste management facilities. Depending on the circumstances, specific waste management facilities can be chosen as a matter of coincidence. An example of this was the Bovine Spongiform Encephalopathy (BSE), crisis, also known as "mad cow disease," during which animal carcass incineration became a very relevant matter. The Lippe Plant formulated a plan to address these issues and constructed the processing plant for high-risk animal material, which utilized the existing incineration facilities.

One of the learnings from the selection criteria of waste management facilities of the Lippe Plant is that mono-waste streams are typically more suited for industrial processing. This is specifically relevant for the Waste Electrical and Electronic Equipment (WEEE) treatment facility, which has recently been transformed to a fridge processing site. This makes the processing of waste fractions a lot easier, as they become more homogenous and less diverse, allowing for more effective recovery of raw materials.

15.7.2.5 Awareness Creation, Training and Education Activities at the Park

Most people have heard of recycling and yet the whole subject often remains a little abstract. Very few people understand what actually happens to their old plastic, organic waste, and other materials. The Lippe Plant tries to provide a firsthand experience of the recycling process to the visitors.

The plant offers guided tours for visitors of its facilities. Visitors that live in nearby areas are able to visit free of charge. Tours can be adjusted depending on group needs. On their website, the plant also has a Web Application which allows for a virtual study tour through the Lippe Plant. The facilities have different viewing spots, which enable visitors close up views of processing steps. Visitors can also gain insights into the recycled products made at the plant, which are displayed at the pillars in the entrance hall of the old head office building.

Since the Lippe Plant also acts as headquarter to the REMONDIS group, it also provides onsite programs for apprenticeships and trainee positions. The Environmental Testing Laboratory also plays an important role in providing training and continuing educational opportunities. The plant also frequently collaborate with local universities and other educational institutions.

15.7.3 Integrated Waste Recycling Park at Jaipur, Rajasthan, India

State Government of Rajasthan is establishing an Integrated Waste Recycling Park at Jaipur, Rajasthan in technical collaboration with GIZ India. Consultation meetings were held with various recyclers and waste management companies in the country on need of such a waste recycling park and most of the participating entities overwhelmingly supported setting up such a park looking at the environmental and economic benefits it provides. The park also has financial support from central government for setting up of the park.

The proposed park will have waste recyclers of different waste streams such as ELV, e-waste, plastic waste, used oil, etc. A suitable site measuring 44 hectares at village Tholai has been selected for the park. This is a greenfield site chosen for development. A pre-feasibility for setting up of the park has been completed by GIZ India and it has recommended setting up of such a Park at Jaipur. The work of zoning/master planning of the park is in process.

Besides the recyclers of different waste streams, the park also proposes to have common facilities like material testing laboratory, training and skill development center, public awareness activities, sewage treatment plant, rainwater harvesting system, etc. Besides the above, the park may have all the common facilities such as entry gates, road network, landscaping and green area development, water supply infrastructure, electric supply infrastructure, fire station, weighbridge, administrative building, bank, post office buildings, parking area, etc. which are normally part of any well-developed industrial area. Table 15.2 also presents a few case studies for the best practices in IWRPs in the world.

15.8 CONCLUSION AND WAY FORWARD

As per the analysis carried out by Beers et al. (2020) of 50 EIPs established in 8 countries across the world, higher average current performance against the International Framework can be found in Colombia (68%), Indonesia (67%), and Vietnam (63%). Ukraine and South Africa have the highest improvement potential (27% and 25%, respectively). A low compliance with specific pre-requisites and performance indicators under park management, economic, environmental, and social performance indicate a need by the industrial park for technical assistance.

Though the concept of EIP has been demonstrated in many countries, however, the concept of IWRP needs to be applied in different economies to draw on the learnings and increase our understanding and refine the concept. Further, different variety and experimentation will shed light on what works and what doesn't, what the significant

TABLE 15.2
A Few Case Studies for the Best Practices in Integrated Waste Recycling Parks

Sr. No.	Name of the Park	Location	Description of the Facility	Recycling Facilities installed	Common Facilities Installed	Source
1	**Remondis Lippe Eco Industrial Recycling Park**	Lünen, Germany	Over 14 different waste management facilities are spread over an area of 230 hectares and is home to a wide range of innovative recycling systems which makes it Europe's largest industrial recycling center	• Metal Slag Processing Unit • Processing Plant for high-risk animal material • Organic material Treatment Facility • Biodiesel Plant • FGD gypsum Refinery • Alkali refinery • Fluidized-Bed power station • Transfer Centre and disposal facility for medical waste • Biomass-fired power plant • Plastic waste treatment facility • WEEE Dismantling Facility	Roadways, weighbridges, safety and fire safety, administration building. processing steam as well as electricity, waste water treatment plant, common water supply system, environmental laboratory	Website (www.remondis-sustainability.com/en/acting/lippe-plant)
2	**Eco Industrial Recycling Park at Kalundborg**	Kalundborg Denmark	The park came into existence in 1959 with the startup of the Power Station. Originally not planned for industrial symbiosis and its current state of waste heat and materials sharing developed over a period of 20 years.	• Coal-fired plant • Oil refinery. • Insulin and industrial enzymes production facility • Plasterboard production • Excess heat from power plant for residential heating system. • Soil remediation facility	Kalundborg Utility is common services provider in Kalundborg. It supplies approximately 6,150 households with drinking water and distributes district heating to approximately 5,050 households in Kalundborg city. It also treats the entire wastewater from the Kalundborg municipality.	Website (www.symbiosis.dk/en)

(continued)

TABLE 15.2 (Continued)
A Few Case Studies for the Best Practices in Integrated Waste Recycling Parks

Sr. No.	Name of the Park	Location	Description of the Facility	Recycling Facilities installed	Common Facilities Installed	Source
					In addition, the utility supplies surface water to some of the large companies in Kalundborg.	
3	**Integrated Waste Recycling Park (proposed)**	Village Tholai Jaipur (Rajasthan) India	Its unique park proposed mainly for recyclers of various waste streams and will be spread on 44 ha land. The park is piloted by government of Rajasthan with funding from Government of India under Prime Minster Gati-Shakti Program. GIZ India is providing technical support in setting up the park.	• End of the Life Vehicle Recycling Plant • Plastic Waste Recycling Plant • e-Waste Recycling Plant • Battery Recycling Plant • Hazardous waste Recycling plant	Fire station, weighbridge, sewage treatment plant solar power generation, rainwater harvesting, training, awareness and skill development centre, material testing lab, raw material storage sheds.	GIZ India 2022

4	**Eco-Park Hong Kong**	Lung Mun Road, Tuen Mun	Eco-Park is first recycling-business park of Hong Kong. The facility has been developed by the Environmental Protection Department, especially for recycling industry waste with a site area of 200,000 m^2. The Park has been developed in two phases and provides a rentable area of 140,000 m^2 for the recycling industry. The park is in operation since 2007. It promotes the conversion of waste into resources by turning recyclable materials into products.	Currently, Eco-Park has nine tenants who recycle waste cooking oil, waste metals, waste wood, waste WEEE, waste plastics, waste batteries, construction and demolition waste, waste glass and waste rubber tyres. Total waste processed in year 2022 was 213784 tons and recycled products manufactured were 232210 tons. 14378 tons of waste was disposed in year 2022 (Source: Annual Environmental Monitoring & Audit Report 2022, published by Eco-Park Hong Kong)	the administrative building includes visitor centre, plastic recycling and education centre, meeting room, seminar room and conference room (for tenants to hire), office for the operator, EPD berthing spaces for the loading and unloading of marine cargo, weighing waste intake, eco-garden and eco-carpark. Urban Property Management Limited (UPML) manages, maintains and markets the Eco-Park on behalf of EPD.	Website (www.ecopark.com.hk/en/index.aspx)

(*continued*)

TABLE 15.2 (Continued)
A Few Case Studies for the Best Practices in Integrated Waste Recycling Parks

Sr. No.	Name of the Park	Location	Description of the Facility	Recycling Facilities installed	Common Facilities Installed	Source
5	**Kitakyushu Eco-Town Project**	Kitakyushu, Japan	Kitakyushu eco-town project is a business that aims to build a resource-recycling society with the aim of utilizing all kinds of waste as raw materials for other industrial fields and ultimately reducing waste to zero emissions. The Project has been considered as a unique regional policy example by integrating an “Environmental Conservation Policy” and an “Industry Promotion Policy”. Kitakyushu Eco-Town Project is acclaimed as an international model for environmental improvement.	• PET Bottles Recycling • Discarded Office Equipment Recycling • Automobile Recycling • Discharged Home Appliances Recycling • Used Fluorescent Tubes Recycling • Used Medical Equipment Recycling • Mixed Wastes Recycling • Recycling of parts from Home Appliances and Automobiles • Used cellular phones, Home Electronics, and Circuit Boards Recycling • Waste Wood and Plastic Recycling • Beverage Containers Recycling • Automobile Recycling • Discarded cooking oil Recycling • Used Organic Solvent Recycling	**Eco-Town Centre**–It is an environmental learning centre to disseminate knowledge about projects at Eco-Town. It also supports location for enterprises at Eco-Town. **Kitakyushu Recycling Port**–To handle recycling resources safely and efficiently. **Joint Storage and Logistics Centre**–For storing used parts. **Kitakyushu Science and Research Park**– Environmental policy concept establishment and basic research, training in human resources. It also acts as a base for industry-academia cooperation.	• Website (www.kitaq-ecotown.com) • Global Environment Centre Foundation, 2005, Eco-Towns in Japan-Implications and Lessons for Developing Countries and Cities, available online (https://wedocs.unep.org/bitstream/handle/20.500.11822/8481/Eco_Towns_in_Japan.pdf)

6	**Envi Grow Park**	Forssa, Finland	Established in the 1990s, the Envi Grow eco-industrial park recycles materials, energy, expertise and information within the waste management, food and construction industries in a closed-loop system. Total investment for "Envi Grow Park" was EUR 285 600, of which the EU's European Regional Development Fund contributed EUR 114 240 from the "Southern Finland ERDF Programme for the 2007 to 2013 programming period. Thematic objective of the park is to support the shift toward a low-carbon economy	• Hosts more than 20 companies, employing 200 experts that specialize in recycling and waste management or operate with recycled materials and renewable energy. • A new biomethane filling station for cars and trucks is the latest addition to the park. • Production of local organic food in large modern greenhouses. • Production of green covers (sod turf) and bio-fertilizers. • Biogas, bioethanol and synthetic diesel for use as transport fuel; the use of carbon dioxide separated from biogas. • Industrial aquaculture using bioenergy. • Innovative recovery and reuse of by-products from the food industry.	Regional Development Company (Forssan Yrityskehitys Ltd.) has been established to increase collaboration and business relationships among the actors in the city as well as contribute to the development of the local environmental strategy. It actively attends and organizes various events to promote IS. The regional businesses in general meet regularly with the IS companies to discuss their needs.	• Website (www.fyk.fi) • European Commission–Envi Grow Park Project (https://ec.europa.eu/regional_policy/en/projects/finland/envi-grow-park-a-virtuous-circle-of-recycling-and-waste-management-in-southern-finland) • Uusikartano et al. 2022

risks are, how regulatory hurdles can be overcome, what can be financed, and what is most environmentally beneficial.

There is a huge opportunity for the waste recyclers to showcase IS as a logical extension of resource productivity. Governments can focus on IWRP as another way to redevelop brownfields. The roles of various public and private actors will sort themselves out over time. There is a strong need for technical assistance including a stronger focus on knowledge dissemination, experience sharing, and peer-to-peer learning among existing IWRPs and regulating authorities.

BIBLIOGRAPHY

Beers S V, Tyrkko K, Flammini A, Barahona C and Susan C (2020): Results and Lessons Learned from Assessing 50 Industrial Parks in Eight Countries against the International Framework for Eco-Industrial Parks. *Sustainability*, 12, 611; doi:10.3390/su122410611. www.mdpi.com/journal/sustainability

Chertow M R (2000): Industrial Symbiosis: Literature and Taxonomy. *Annu. Rev. Energy Environ.*, 25 , 313–37.

Christensen J (1999): Proceedings of the Industry & Environment, Workshop, held at the Indian Institute of Management, Ahmedabad, India, 1999.

GIZ India (2022): A (Pre) Feasibility Study for Setting up India's first Resource Recovery Park (RRP) / Waste Recycling Park (WRP) in Tholai, Rajasthan, India, Prepared for Rajasthan State Pollution Control Board (RSPCB) as part of the EU-REI project.

Kalundborg Symbiosis website (2023): www.symbiosis.dk/en/governance, accessed on 18.12.2022 at 7.15 pm

Kerdlap P, Low JSC, Tan DZL, Yeo Z and Ramakrishna S (2020): A Methodology for Multi-level Life Cycle Environmental Performance Evaluation of Industrial Symbiosis Networks. *Resour. Conserv. Recycl.*, 161: 104963. doi:10. 1016/j.resconrec.2020.104963

Lombardi D R and Laybourn P (2012): Redefining Industrial Symbiosis. *J. Ind. Ecol.* 16: 28e37. https://doi.org/10.1111/j.1530-9290.2011.00444.x

Mauthoor S. (2017): Uncovering Industrial Symbiosis Potentials in a Small Island Developing State: The Case Study of Mauritius. *J. Clean. Prod.*, 147: 506–513.

Remondis Lippe Plant, Website (2023): www.remondis-lippe-plant.com/en/an-overview-of-the-site/#:~:text=REMONDIS'%20Lippe%20Plant%20has, accessed on 19.12.2022 at 10.30.am

Starodubova A, Dinara Iskhakova D, and Chulpan Misbakhova C (2022): The Role of Eco-Industrial Parks in Promoting Circular Economy Technologies in the Regions. *BIO Web of Conferences*, 48: 01024. https://doi.org/10.1051/bioconf/20224801024

Ünal E, Urbinati A, Chiaroni D and Manzini, R (2019): Value Creation in Circular Business Models: the case of a US small medium enterprise in the building sector. *Resour. Conserv Recycl.*, 146: 291e307.

UNIDO (2016): Global Assessment of Eco-industrial Parks in Developing and Emerging Countries.

Uusikartano J, Saha P and Aarikka-Stenroos L (2022): The Industrial Symbiosis Process as an Interplay of Public and Private Agency: Comparing Two Cases. *J. Cleaner Prod.*, 344: 130996, https://doi.org/10.1016/j.jclepro.2022.130996

Valentine S V (2016): Kalundborg Symbiosis: Fostering Progressive Innovation in Environmental Networks. *J. Cleaner Prod.*, 118 (2016): 65–77.

Waikar, M and Sadgir, P (2022): Sustainable Development in Circular Economy: A Review. Part of the Lecture Notes in Civil Engineering book series (LNCE,volume 260) https://doi.org/10.1007/978-981-19-2145-2_48

Walkowiak B B (2020): Eco-innovation in Germany, Country Profile 2018–2019: Germany, Wuppertal Institute for Climate, Environment and Energy.

Wikipedia (2023): https://en.wikipedia.org/wiki/Kalundborg_Eco-industrial_Park, accessed on 12.02.2023..

World Bank (2021): International Framework for Eco-Industrial Parks v.2. Washington, DC: World Bank.

Zhang X and Chai L (2019): Structural Features and Evolutionary Mechanisms of Industrial Symbiosis Networks: Comparable Analyses of Two Different Cases. *J. Clean. Prod.*, 213: 528–539.

Index

For Product Safety Concerns and Information please contact our EU representative GPSR@taylorandfrancis.com
Taylor & Francis Verlag GmbH, Kaufingerstraße 24, 80331 München, Germany

www.ingramcontent.com/pod-product-compliance
Lightning Source LLC
LaVergne TN
LVHW020601110826
845149LV00002B/341
9781032428215